AF333082

Ngoc Thanh Nguyen and Radoslaw Katarzyniak (Eds.)

New Challenges in Applied Intelligence Technologies

Studies in Computational Intelligence, Volume 134

Editor-in-Chief

Prof. Janusz Kacprzyk
Systems Research Institute
Polish Academy of Sciences
ul. Newelska 6
01-447 Warsaw
Poland
E-mail: kacprzyk@ibspan.waw.pl

Ngoc Thanh Nguyen
Radoslaw Katarzyniak
(Eds.)

New Challenges in Applied Intelligence Technologies

 Springer

Ngoc Thanh Nguyen
Institute of Information Science and Engineering
Wroclaw University of Technology
Str. Janiszewskiego 11/17
50-370 Wroclaw
Poland
E-mail: thanh@pwr.wroc.pl

Radoslaw Katarzyniak
Institute of Information Science and Engineering
Wroclaw University of Technology
Str. Janiszewskiego 11/17
50-370 Wroclaw
Poland
E-mail: Radoslaw.Katarzyniak@pwr.wroc.pl

ISBN 978-3-540-79354-0   e-ISBN 978-3-540-79355-7

DOI 10.1007/978-3-540-79355-7

Studies in Computational Intelligence   ISSN 1860949X

Library of Congress Control Number: 2008925307

Typeset & Cover Design: Scientific Publishing Services Pvt. Ltd., Chennai, India.

Printed in acid-free paper

9 8 7 6 5 4 3 2 1

springer.com

Preface

To built intelligent systems that can cope with real world problems we need to develop computational mechanisms able to deal with very large amounts of data, generate complex plans, schedules, and resource allocation strategies, re-plan their actions in real time, provide user friendly communication for human-device interactions, and perform complex optimization problems. In each of these tasks *intelligence technologies* play an important role, providing designers and creators with effective and adequate computational models.

The field of intelligence technologies covers a variety of computational approaches that are often suggested and inspired by biological systems, exhibiting functional richness and flexibility of their natural behavior. This class of technologies consists of such important approaches as data mining algorithms, neural networks, genetic algorithms, fuzzy and multi-valued logics, rough sets, agent-oriented computation, often integrated into complex hybrid solutions. Intelligence technologies are used to built machines that can act and think like living systems, solve problems in an autonomous way, develop rich private knowledge bases and produce results not foreseen and programmed in a direct way by designers and creators.

This book consists of 37 chapters authored by participants to Special Session on New Challenges in Applied Intelligence Technologies, co-located with 21st International Conference on Industrial, Engineering & Other Applications of Applied Intelligent Systems (IEA-AIE 2008). The chapters discuss examples of applications of intelligence technologies to five general fields, reflecting main streams of practical and scientific interest of computer related community: agent and multi-agent systems; personal assistants and recommender systems; knowledge modeling and processing; optimization and combinatorial problems; computer and telecommunication systems.

The book is divided into 5 parts. Part 1 consists of six chapters in which applications of intelligence technologies to agent and multiagent systems are presented and discussed by authors. Agent and multiagent systems are the class of computer based systems that in a natural way call for advanced intelligence, strongly inspired by natural agents. Part 2, consisting of five chapters, covers few cases of intelligence technologies applied to implementations of personal assistants and recommender systems. Part 3 consists of ten chapters in which intelligence technologies are applied to multiple tasks of knowledge processing, all defined on various levels of knowledge conceptualization. In this part of the book a few examples of processing both symbolic and sub-symbolic knowledge representations are proposed, including interesting cases of inconsistent and incomplete knowledge processing. Part 4 covers the case of optimization and combinatorial problems which are a promising field for direct applications of intelligence technologies. This class of solutions is strongly inspired by multiple natural systems, in which advanced intelligence and intelligent behavioral models of processing and problem solving need to be applied to cope with high computational complexity of real time optimization. This part of the book consists of ten chapters. Part 5 consists of 5 chapters, all devoted to applications of intelligence technologies to knowledge processing tasks related to computer and telecommunication systems.

The material of each part of this book is self-contained. We hope that the book can be useful for graduate and Ph.D. students in Computer Science, in particular participants of courses on Soft Computing, Multi-Agent Systems, Optimization, Computer and Telecommunication Systems. It can also be useful for researchers and readers working on advanced knowledge management and processing. It is the hope of editors that readers of this volume can find many inspiring ideas and define new challenges in creation and application of intelligence technologies, in order to design new classes of effective intelligent systems. Many such challenges are suggested by particular approaches and models presented in all chapters of this book.

We wish to express our great gratitude to Professor Janusz Kacprzyk, the editor of this series, for his interest and encouragement, and to all authors who contributed to the content of this volume. Thanks are also due to Wojciech Lorkiewicz, our PhD student from the Institute of Information Science and Engineering of Wroclaw University of Technology, for his excellent work during organization of Special Session on New Challenges in Applied Intelligence Technologies.

February 2008

Ngoc Thanh Nguyen
Radosław Piotr Katarzyniak

Contents

Part III: Knowledge Modelling and Processing

Part IV: Intelliengence Technologies in Optimization and Combinatorial Problems

---

Part V: Intelliengence Technologies in Computer and Telecommunication Systems

---

Part 1
Agent and Multiagent Systems

A Comparative Study between Human-Human Interaction and Human-Robot Interaction

KangWoo Lee[1], Jung-Hoon Hwang[2], and Dong-Soo Kwon[3]

[1] School of Media, Soongsil University
 1-1, Sangdo-5 Dong, Dongjak-Gu, Seoul, 156-743, Korea
 kangwooster@gmail.com
[2] Korea Electronics Technology Institute
 68 Yatap-Dong, Bundang-Gu, Seongnam, 463-816, Korea
[3] Dept. of ME, Korea Advanced Institute of Science and Technology
 Guseong, Yuseong, Daejon, 305-701, Korea

Abstract. In this comparative study the concept of common ground is introduced for understanding the process of knowledge sharing by human and robot, and applied to "questioning and answering" task during object identification experiments. Various cases of human-human and human-robot interactions were performed and the interaction patterns among these groups were investigated. In comparison to Human-human interaction, the performance of robot was determined by both its expertise level as well as its human partner's expertise level. This suggests that the robot's intelligence alone is not sufficient to communicate with human users.

Keywords: Human-robot interaction, common ground, comparative, state transition, interaction effort, perspective taking.

1 Introduction

With the multifaceted nature of Human-Robot Interaction (HRI), research efforts in the area have been made from different disciplines such as psychology, robotics, artificial intelligence, etc. This could be the reflecting evidence that there is a great interest on how to enable a robotic system to interact with human users. On the other hand, this also implies that no solid account has been made from any of these disciplines.

Recently, researchers in HRI realized the importance of common ground that has been a critical issue in social psychology or linguistics since the underlying mechanism of common grounds may provide the linkage between human and robot [2, 6, 9]. Many researchers have pointed out that common ground enhances communicative and collaborative activities of interlocutors and avoids misunderstanding and additional efforts. That is, all collective actions such as playing a duet, shaking hands, playing chess and so on are built on common grounds and its accumulation. In this sense, the establishment of common grounds between human and robot is expected to improve not only the robot's performance but also the convenience of human users.

Unfortunately, most studies on common grounds have been done with human subjects only. So, it is not surprising that direct applications of these studies to HRI

N.T. Nguyen, R. Katarzyniak (Eds.): New Chall. in Appl. Intel. Tech., SCI 134, pp. 3–12, 2008.
springerlink.com

domains would cause some unexpected problems. Moreover, few studies related to HRI are limited to either emphasizing its importance or announcing its role in the development of a robotic system, and neglected the connections between findings in human studies and its application to HRI domains.

In this paper, we present a comparative study between Human-Human Interaction (HHI) and HRI in which paired participants (human-human or human-robot) with different common grounds performed a 'questioning and answering' task. A major goal of this comparative study is to reveal the commonalities and discrepancies in the two interactions – HHI and HRI and to investigate the role of common ground in the communicative task. This may provide interesting insights on the issues of how findings in HHI can be applied to designing a robotic system.

2   Defining Interaction States

We first attempt to construct the model of a 3-way relationship between human, robot, and their surrounding environment using information theoretic terms. Conventionally, Artificial Intelligence (AI) has focused on the linkage between an intelligent system and its environment, whereas Human-Computer Interaction (HCI) or Human-Machine Interaction (HMI) has focused on the linkage between human user and artificial system. These linkages have been independently studied in those areas. However, as a robot moves in our daily environment and frequently interacts with ordinary people, a robotic system is demanded not only to be intelligent but also to be communicative. In this sense, the linkage between human, robot and environment are interwoven in HRI.

In Fig. 1(a) the interrelationship in the grounding process is presented. H denotes the knowledge possessed by a human user; R denotes the knowledge possessed by a robot; and E indicates external or environmental objects or events. Each area represents the ground that is associated with knowledge or the physical world. It is also closely related to the patterns of interaction, since the interaction pattern is different according to which ground the agents belong. The area also can be mapped into an agent's state that is evolved through the interaction. In this sense, interaction process can be considered as state transitions decided by utterance transaction. Fig. 1 (b) shows the sequential process

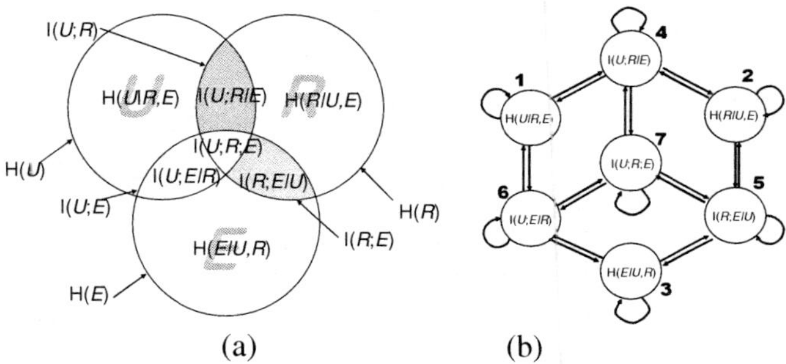

Fig. 1. (a) Three-way relationship between U, R and E (b) State Transition Diagram. The 3 way interaction can be decomposed into 7 different interactive states.

model of possible human-robot interaction. In Fig. 1 (a), possible states of agents are decomposed into different areas. The interactive states of communicators would be described using information theoretic terms as follows:

1) $H(E\,|\,U,R)$, s_3, represents the state of uncertainty, which is shared by neither a human nor a robot, is an unknown portion of the environment for the robot as well as for the human user.

2) $I(R;E\,|\,U)$, s_5, represents the state in which the environmental knowledge possessed by an agent R is not known to the other agent U. In human-robot interaction this means how little idea a user has of what the robot knows about its environment. During interaction, a user may learn something about the robot related to the environment.

3) $I(U;E\,|\,R)$, s_6, represents the state in which the environmental knowledge possessed by an agent U is not known to the other agent R. In human-robot interaction, this could be the major source of uncertainty for a robot system when it interacts with a human user.

4) $I(U;R;E)$, s_7, represents the state in which both agents share common knowledge about the environmental object. This state means that the interaction between two agents occurs on the shared ground that does not require additional exchange of information. The interaction between agents in this state is efficient and successful. In human-robot interaction, as interaction between human and robot occurs more frequently, dynamic changes in ground formation are expected to accompany.

It should be noted that the variables H and R are simply replaced by H_1 and H_2 in HHI.

3   Measuring Interaction

To investigate a communicative interaction between two agents, we have developed measures based on the extension of conceptualization of interaction described above. The measures have been designed to exploit 3 different aspects of communicative interactivity: state transition, efficiency, and efforts of interaction.

3.1   Measuring State Transition

As noted earlier, the change of interactive states is accompanied by the interaction between agents, and either leads them into common grounds or drive them out of common grounds. If they share little knowledge about external objects, the change may fluctuate over the course of interaction. In contrast, if they share more knowledge grounds about external objects, the change can be easily converged into a certain stable state. This implies that the patterns of interactivity can be estimated by how easily the interactive states are converged into a stable state. The state transition matrix can be obtained as we apply Markovian process to observe the frequency of each state.

4   Handybot

A virtual robot and a portable mediate interface, so-called 'Handybot,' were used in our experiment. A mediate interface is a medium between a human and a robot that have been recently studied [4, 10]. The portable mediated interface 'Handybot' spatially expands a robot's functions and remotely controls it.

4.1   Symbolic Language Interface

Symbolic language is designed to facilitate communication between a human user and a robot. The symbolic language system uses several parts of speech to describe an object: nouns, verbs and adjective. For instance, the noun symbols correspond to external objects such as a cup or chair, the adjective symbols correspond to objects' features such as color or shape, and the verb symbols represent the types of actions the robot can take. A sentence can be constructed via a combination of these symbols using a statistical method.

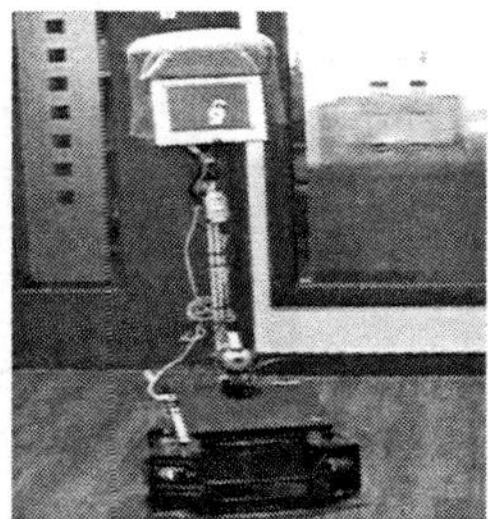

Fig. 2. Handybot (left) and main robot (right)

4.2   Bayesian Reasoning and Interaction Strategies for Handybot

The robot's knowledge of the environment is composed of objects that have features such as shape, color, or function. There are many formal ways to represent object knowledge or the relationship between an object and its features. In this study, Bayesian network was used, in which an object is linked with its features using conditional probabilities. The description of an object is based on the object's features such as its shape, color or function. The robot can reason from the features described by a human user through 'Handybot,' and make a decision based on the input descriptions. In addition, the robot has interaction strategies that govern the number of additional requests it can make to the user, the timing of the request, and the timing of decision.

First, as stated above, the robot's representation and reasoning system was built using a Bayesian network. A Bayesian network is a probabilistic model of variables that are represented using nodes and connections [8]. The objects in a Bayesian network are uniquely represented by the connective patterns between the nodes at each layer. Weights were assigned to the connections between objects and features, so that the conditional probability of an object can be determined by the joint probabilities of its feature variables.

The robot has two reasoning modes: expert and novice. The difference between them is how the weights were used to describe the object's features, functions and names [7]. In expert mode, weights were assigned more accurately for connecting the object's features to the words implemented in the 'Handybot.' Since the object may have many abstract descriptions, the adjustment of the correlation between the describing words and the object is used to determine the reasoning mode.

In the network, the probability of the object is represented by the conditional probability of the object over jointly distributed independent feature variables.

$$P(Obj_i \mid F_n) = \prod_{k=1}^{n} P(Obj_i \mid f_k) \tag{1}$$

where the features are $F_n = f_1, f_2 \ldots, f_n$, Obj_i is an object, and n is the number of features. The conditional probability is sequentially updated through the interactions between a human user and a robot agent. Therefore, Eq. (3) can be rewritten as follows:

$$P(Obj_i \mid F_n) = P(Obj_i \mid F_{n-1}, f_n) \tag{2}$$

Based on Eq. (2), the robot determines the object that has the maximal conditional probability for given features, and decides the object as its belief of what the user has indicated. If the belief is not what the user meant, the robot agent sets the conditional probability of the object to zero and replaces the belief with the object that has the second-highest maximal probability, or requests additional information.

Second, the robot has several interaction rules that govern the occasions and frequency of those occasions in which it either seeks additional information from the user, or shows what it has guessed.

1)   The robot requests additional information of the object until

$$\max P(Obj_i \mid F_n) \geq k[\sum_{j}^{m} P(Obj_j \mid F_n)/m] \tag{3}$$

where k is a constant, and m is the number of given features. Thus, when the robot's belief concerning the object reaches the threshold that is k times more than the average over its beliefs on other objects, it can present what it has guessed.

2) The robot is allowed to request information for up to six additional features. If the robot fails to guess correctly after the maximum allowable information, it initializes all previously collected information and requests information about the same object again from the user. 3) If there are multiple objects with greater than the threshold, the robot presents the three most probable objects in a descent order.

5   Experimental Methods

5.1 Participants

For a comparative study between HHI and HRI, 63 subjects were assigned to two different sets of experiments. 30 subjects participated in the HHI experiment and the other 33 subjects participated in the HRI experiment. Approximately, one-half of the subjects majored in mechanical engineering while the rest majored in various academic fields such as psychology, art, etc.

5.2 Stimulus

The stimuli used in the experiment were 33 mechanical parts such as a gear box, bolt, nut, etc. These materials are familiar to students majoring in engineering, but may not be familiar to the students of social science or art.

5.3 Experimental Design

Two independent sets of experiment were carried out. The stimuli used in those two experiments were identical.

- ***HHI experiment***

In the HHI experiment, subjects were divided into two groups, expert and novice, based on their familiarity with the set of stimuli, and paired into 3 possible conditions. Students majoring in engineering were classified as the expert group, whereas the students majoring in social science or art who were not familiar to the stimulus items were classified as the novice group. Subjects in the groups were assigned to one of 3 pairing conditions: expert–expert (E^H-E^H), novice–novice (N^H-N^H) and expert–novice (E^H-N^H) . The superscript simply indicates the human subjects.

In the experiment, two subjects were located in a chamber. A desk was placed between them. The experimental materials were displayed on the desk so both subjects could see them. Each pair performed a 'questioning and answering' task in which one of the two subjects explains an object, and the other answered what it might be. At each trial, an object's picture and its name are randomly presented at the monitor that could be seen only by the questioner. The questioner was asked to describe the object in terms of its features such as color, shape, function, etc., and the answerer had to decide what it might be among many others on the basis of the descriptions.

The experiment consisted of two sessions. In the first session, a subject from each pair took the role of questioner and switched the role in the second session (i.e., a questioner would then become an answerer). Each session consisted of 16 trials. A trial was completed if the answerer correctly responded and the questioner pressed the confirmation button. The next trial would start and continue until all items were correctly answered. To exclude all expressions other than the verbal expression, each subject was asked to wear a mask and no gestures were allowed. Fig. 3 shows the experimental setup where two subjects are separated across the table with mechanical parts for identification. The whole experimental process was recorded with a camcorder and the item selection process was determined by a computer program which randomly chose an object from the previous established database of the objects on the table

- ***HRI experiment***

In the HRI experiment, human subjects were classified into the two groups based on their familiarity with the stimulus. Each subject was paired with our 'Handybot' with one of two reasoning modes: expert or novice. Therefore, human subjects can be situated in one of 4 possible cases: human expert-robot expert (E^H-E^R), human novice-robot expert (N^H-E^R), human expert-robot novice (E^H-N^R), and human novice-robot novice (N^H-N^R).

In the experiment subjects were asked to carry out the 'questioning and answering' task. Basically, the procedure used in the HRI experiment was identical to that of the HHI experiment, but the human subject was paired with our 'Handybot' and took the role of a questioner then an answerer.

5.4 Evaluation Process of Interactive States

After the HHI experiment, the video records were analyzed and the states of each agent were evaluated and classified as defined by the operational definition below. Since the internal state of an agent could not be observed directly, it had to be inferred from observations. The operational definitions for the states are given as the following:

First, no shared ground on a particular object is assumed between the questioner and answerer before an interaction occurs. The asker is assumed to be in state s6 in which he has the knowledge of the object, while the responder is assumed to be in state s3. Second, the questioner is assumed to be in state s7 about the object with the answerer when he/she receives the evidence of acceptance such as "OK" or "I know." Third, if the answerer chooses the correct object on the basis of the given information such as the name or function, we assume that he/she already was in state s5. Fourth, if the answerer fails to choose the correct object, he/she is assumed to be in state s6. Fifth, the state of both participants is changed from state s5 or s6 to state s7 if the answerer correctly chooses the object.

For HRI, this evaluation process is not necessary because all responses during the interaction were tagged with time stamps and recorded into a log file by a computer program.

Fig. 3. The scenes form HHI (left) and HRI (right)

6 Results

6.1 State Transition Probability between HHI and HRI

The state transition probability matrices obtained from the HHI and HRI experiments were summarized in Table 1. The columns indicate the very previous state whereas the rows indicate the next state. The summed frequencies (SF) of states are shown at the left and bottom of tables.

At a glance, results obtained from HHI are quite similar to those from HRI. First of all, the E-E groups in both HHI and HRI complete the tasks with a smaller number of frequencies than other groups (χ^2=118.07, p < 0.0001 for HHI and χ^2=337.44, p < 0.0001 for HRI). This result can be compatible with the psychological findings in which the efficiency of communication task was dependent on the expertise of partners [3]. This implies that the more of the relevant common ground they have, the more efficient the interaction between them are. Second, the state transition in E-E groups in both interactions were easily moved to the common ground state, s_7, in

Table 1. State transition probabilities in HHI and HRI

$$E^H\text{-}E^H \text{ vs. } E^H\text{-}E^R$$

EE	S_5	S_6	S_7	SF 356	S_5	S_6	S_7	SF 521
S_3	0.714	0.286	0	5.9%	0.531	0.469	0	6.1%
S_5	0.629	0	0.371	42.3%	0.611	0	0.389	34.5%
S_6	0	0.556	0.444	17.7%	0	0.504	0.496	25.5%
S_7	0.091	0.074	0.835	34.1%	0.055	0.153	0.801	33.9%
SF	34%	14%	52%	100	26%	21%	53%	100

$$N^H\text{-}N^H \text{ vs. } N^H\text{-}N^R$$

NN	S_5	S_6	S_7	SF 695	S_5	S_6	S_7	SF 1301
S_3	0.467	0.533	0	6.5%	0.576	0.424	0	5.1%
S_5	0.812	0	0.188	41.4%	0.721	0	0.279	35.0%
S_6	0	0.811	0.189	41.9%	0	0.816	0.184	39.5%
S_7	0.141	0.070	0.789	10.2%	0.208	0.192	0.600	20.4%
SF	38%	38%	24%	100%	32%	38%	30%	100%

$$E^H\text{-}N^H$$

EN	S_5	S_6	S_7	FS 643
S_3	0.500	0.500	0	7.5%
S_5	0.811	0	0.189	49.4%
S_6	0	0.713	0.287	31.4%
S_7	0.160	0.080	0.760	11.7%
	46%	27%	27%	100

$$E^H\text{-}N^R \text{ vs. } N^H\text{-}E^R$$

5.4%	S_5	S_6	S_7	FS 1065	S_5	S_6	S_7	FS 1115
S_3	.386	.614	0	5.4%	0.483	0.517	0	5.2%
S_5	0	.768	.232	55.7%	0	0.698	0.302	30.3%
S_6	.587	0	.413	16.8%	0.761	0	0.239	41.0%
S_7	.145	.209	.647	22.1%	0.263	0.107	0.630	23.5%
	15%	51%	34%	100%	40%	26%	34%	100%

comparison to the other groups (χ^2 =129.31, p < 0.0001). But no significant difference between HHI and HRI in the state transition to the common ground state was found (χ^2=1.82, p > 0.15).

However, we found interesting discrepancies when we looked into the conditions in details. First, little difference in the state transition probability from s_7 to s_7 or to other states was found in E-E groups during HHI and HRI whereas significant difference in the state transition probability was found in N-N and E-N (N-E) groups during HHI and HRI (Fisher's exact test, p < 0.05). This result implies that the interaction in HRI moved toward and away from the common ground state during communication, and reached the common ground state by trial-and-error in some sense. In contrast, subjects seemed to take a 'step-by-step' strategy that narrowed down the possibilities of correct answers in HHI.

7 Conclusion

AI has been implemented in robots to afford them with the capacity 'to think.' Nevertheless, robots with artificial intelligence are still not capable of carrying out real world tasks with satisfactory performance. They are limited in perceptual, cognitive, executive, and emotional abilities, and frequently require human involvement during the fulfillment of their tasks. One solution to this problem is the development of an interaction system that allows human intervention if the robot's capabilities are not adequate for a given task. In this way, HRI can be characterized as an interaction between agents with different intelligent capabilities. However, it does not simply mean that the discrepancy is the quantitative difference caused by the poor functioning of a robot system. Rather, it can be the qualitative difference between two representation systems. Hence, in order to communicate and coordinate with an agent with different functional capabilities and representation systems, HRI is required to establish common ground mutually shared by interacting agents. To prepare for the era of human-robot symbiosis, it is necessary to find a way to merge the gap in intelligence and functional capabilities between human and robot by establishing common ground mutually shared by these interacting agents.

Acknowledgments. This work was partially supported by the Brain Korea Program supported by the Ministry of Education and the Intelligent Robotics Dev. Program of the 21C Frontier R&D Programs funded by the Ministry of Commerce, Industry and Energy of Korea.

References

1. Chi, M., Feltovich, P., Glaser, R.: Categorization and representation of physics problems by experts and novices. Cognitive Science 5, 121–152 (1981)
2. Clark, H.H., Brennan, S.E.: Grounding in Communication. In: Resnick, L.B., Levine, J.M., Teasley, S.D. (eds.) Perspectives on Socially Shared Cognition. American Psychological Association, Washington (1991)

3. Clark, H.H., Wilkes-Gibbs, D.: Referring as a Collaborative Process Cognition 22, 1–39 (1986)
4. Fong, T., Thorpe, C.: Robot As Partner: Vehicle Teleoperation With Collaborative Control. In: Schultz, A.C., Parker, L.E. (eds.) Multi-Robot Systems: From Swarms to Intelligent Automata, Kluwer Academic Publishers, Dordrecht (2002)
5. Fussell, S., Krauss, R.M.: The effects of intended audience on message production and comprehension: Reference in a common ground framework. Journal of Experimental Social Psychology 25, 203–219 (1989)
6. Kiesler, S.: Fostering common ground in human-robot interaction ROMAN 2005, pp. 729–734 (2005)
7. LaFrance, M. The Quality of Expertise: Implications of Expert-Novice Difference for Knowledge Acquisition. SIGART Newsletter, Number 108, Knowledge Acquisition Special Issue 6-14 (1989)
8. Russell, S.J., Norvig, P.: Probabilistic Reasoning Systems. In: Artificial Intelligence: A Modern Approach, 1st edn., pp. 436–470. Prentice Hall, Englewood Cliffs (1995)
9. Stubbs, K., Hinds, P.J., Wettergreen, D.: Autonomy and Common Ground in Human-Robot Interaction. A Field Study. IEEE Intelligent Systems 22(2), 42–50 (2007)
10. Hwang, J.-H., Kwon, D.-S.: A Portable Mediate Interface 'Handybot' for the Rich Human-Robot Interaction. Journal of Control, Automation and Systems Engineering 13(8), 735–742 (2007)

Dialogue and Argumentation in Multi-agent Diagnosis

Asma Moubaiddin[1] and Nadim Obeid[2]

[1] Department of Linguistics, Faculty of Arts,
  The University of Jordan
[2] Department of Computer Information Systems,
  King Abdullah II School for Information Technology,
  The University of Jordan
  `obein@ju.edu.jo`

Abstract. In this paper, we make a first step towards a formal model of dialogue and argumentation for a multi-agent (model-based) diagnostic system. We shall discuss some of the issues in multi-agent cooperative fault diagnosis, the theories of communicating agents and their reasoning capabilities. We propose a Partial Information State (PIS)-based framework for dialogue and argumentation. We shall employ a three-valued based nonmonotonic logic for representing and reasoning about partial information. We show via an example that the system can easily be customized to handle distributed problem-solving tasks.

Keywords: Multi-Agent Systems, Intelligent Interfaces, Model-based Diagnosis, Three-Valued Logic, Nonmonotonic Reasoning, Dialogue, Argumentation.

1 Introduction

Many Modern large systems are sophisticated and remotely located. They have a large number of components and sub-systems that interact in complex ways. In such systems, early diagnosis of an imminent fault and determining the root cause failures are essential. Such deep knowledge based reasoning goes beyond the capacities or knowledge of a single agent. Hence, there is a need for distributed monitoring and diagnosis. The employment of specialized agents, together with the realization that the knowledge of some agents is incomplete, suggests that the agents have to communicate in order to reach an agreement on the causes of observed anomalies.

The aim of this paper is to make a first step towards a formal model of communication for a multi-agent (model-based) diagnostic system. Section 2 will be concerned with issues in multi-agent cooperative fault diagnosis. In section 3, we discuss the theories of communicating agents and their reasoning capabilities. In section 4, we make a step towards a formal model of communication. Section 5 is concerned with single agent and multi-agent cooperative diagnosis, where an example is presented.

2 Some Issues in Multi-agent Cooperative Diagnosis

Let G and G1 be two agents involved in a diagnostic task. Assume that the faulty device D is not available to the specialist agent, say G1. Then G1 will have to be dependent on G that has access to D. G1 is also dependent on G cooperating by

N.T. Nguyen, R. Katarzyniak (Eds.): New Chall. in Appl. Intel. Tech., SCI 134, pp. 13–22, 2008.
springerlink.com

describing the state of D, carrying out some diagnostic actions on D, if needed, and reporting the results. However, G may not know the significance of the observable features in terms of a technical specification where a diagnosis can be identified. Furthermore, what constitutes an adequate problem formulation for G1 is not something that G can describe appropriately. G may not have the technical knowledge needed to appropriately articulate a description of the problem.

It is clear that both G1 and G may need to be involved in a process of reformulation and questioning until they reach a point with adequate specific and technical understanding so that G1 can make a correct diagnosis of the fault, plan the appropriate test(s), propose a repair plan and help G to execute the test(s), and repair plan. To maintain the involvement of both agents, there is a need to deviate from one complete diagnostic solution and go about developing diagnostic solutions in small steps refinements. For more details cf. [9].

3 Theories, Reasoning and Communication

In addition to the relevant diagnostic knowledge and reasoning capability of an agent, the model should include objects of dialogue such as the available alternatives to an agent and the criteria for evaluating these alternatives. Some agents may need to have the ability to reason about time, change and events. These agents will need to employ an appropriate logic such as TFONL presented below [12]. TFONL allows us to take advantage of both the model-based and the heuristic approaches to diagnosis. It also allows us to represent and reason with different types of knowledge such as structural, behavioral, causal and heuristic. Furthermore, it is suitable for argumentation and dialogue frameworks [10].

3.1 Temporal First Order Nonmonotonic Logic (TFONL)

The time structure employed in this paper is an interval-based theory of time. The agent's knowledge and reasoning capability are expressed in a Temporal First Order Nonmonotonic Logic (TFONL) [12]. The system is based on the quantified version of the non-temporal system T3 (cf. [11].

The language, L_{T3}, of T3 is that of Kleene's three-valued logic extended with the modal operators M (Epistemic Possibility) and P (Plausibility). In T3, L is the dual of M and N be the dual of P, i.e., $LA \equiv {\sim}M{\sim}A$ and $NA \equiv {\sim}P{\sim}A$. A truth-functional implication $\supset$ that behaves exactly like the material implication of classical logic is defined in [11] as follows: $A \supset B = M({\sim}A \& B) \lor {\sim}A \lor B$. Non-monotonic reasoning is represented via the *epistemic possibility operator* M and the *plausibility operator* P. Informally, MA states that A is not established as false. Using M, we may define the operators $UA \equiv MA\&M{\sim}A$ (*undefined*), $DA \equiv {\sim}UA$ (*defined*) and $\neg A \equiv DA \& {\sim}A$ (*classical negation*).

TFONL employs an explicit temporal representation with a theory of time that takes both points and interval as primitives. It employs an explicit representation of events. We may embody default logic into TFONL. It takes advantage of both the model-based and the heuristic approaches to diagnosis. Furthermore, it is suitable for argumentation and dialogue frameworks (cf. [10, 12] for more details).

3.2 Agent Communication

Agents collaboration requires a sophisticated Agent Communication Language (ACL). The main objective of an ACL is to model a suitable framework that allows heterogeneous agents to interact and to communicate with meaningful statements that convey information about their environment or knowledge [6]. There are two main ACLs: KQML [3] and FIPA [4,8]. FIPA protocols can successfully be used in simple applications. However, Both KQML and FIPA are not suitable for use by autonomous agents engaged in dialogue required to solve complex problems because an agent must follow the whole protocol in order to communicate. Therefore, it is preferable to use a set of small basic dialogue protocols that can be put together to construct complex protocols. It seems that the dialogue classification [16], presented in Table 3.1, provide such a needed flexibility.

Table 1. Dialogue Types

Dialogue Type	**Initial Situation**	**Feature**	**Goal**
Persuasion	Conflict of opinion	Proofs are important	Reaching consensus
Inquiry	Collective ignorance	- Cooperative - Proofs are essential	More reliable knowledge
Information-Seeking	Individual ignorance	Restricted	Spreading information
Negotiation	Conflict on an issue	different constraints & goals	Consensus on the issue

The distinction between the types of dialogue is based on collective goals, individual goals and reasons for starting the dialogue. It is important to note that in the course of communication, there often occurs a shift from one type of dialogue to another. Dialogue embedding takes place when the embedded dialogue is functionally related to the first one. For instance, a persuasion dialogue may require an information-seeking sub-dialogue.

3.3 Modeling Communication through Argumentation

Agents, in an MAS, are expected to have the ability to be involved in coherent conversations. Several approaches to modeling communication have been proposed (cf. [2,1].

We shall adopt the argument-based approach [1] in which the agents' reasoning capabilities are associated with the strength of their arguments. Arguments allow an agent to critically question the validity of information presented by another participant, explore multiple perspectives and/or get involved in belief revision processes. Arguments, especially in nonmontonic systems, are logical proofs where some of their steps can be defeated. It is possible in a nonmonotonic system to provide an argument for both a proposition and its negation. Hence, in a dialogue between two agents, a defeasible argument is a structured piece of information that might be defeated by other (defeasible) arguments. Unlike a proof, in classical monotonic logic, an argument does not establish warrant for a conclusion in a definite way as it may be defeated by counter-arguments which are defeasible.

4 Agent Communication: Towards a Formalism

Dialogues are assumed to take place between two agents, G and G1, that have Partial Information States (PIS) which are subject to revision. A dialogue consists of a course of successive utterances (moves) made by the dialogue participants. A dialogue system defines the principles of coherent dialogue, i.e., it defines the conditions under which an utterance is appropriate in the sense that it contributes towards achieving the goal of the dialogue in which it is made. We will adopt the notion of a *dialogue game* in which two agents generate moves to pass on *relevant* information with respect to their *goals*. Central to the model is that PISs of the agents change as a result of the interpretation of the moves and that these changes trigger the production of a succeeding move. The interpretation involves some *understanding* of the presented information. It does involve an integration of the offered information with the PIS of the receiver. Consequently, we cast context as a consistent subset of an agent's PIS, namely those propositions which bear on the interpretation of the utterance on hand and on the propositions that are relevant to producing the goal(s). Interpretation relies on maintaining consistency in the context whilst adding the utterance's information. An agent can only interpret an utterance with respect to what it has in its PIS. Therefore, failure to complete the interpretation process will point to those propositions which induce failure. Thus, part of a context is entirely local to an agent and that agents may hold incompatible and inaccurate beliefs.

Dialogue protocols provide a lower bound on the conditions needed for dialogue coherence. We believe that dialogue coherence relations are mainly driven by dialogue history and the dynamics of PIS of the agents regarding the main goal of the dialogue. The coherence of a dialogue moves is tied to local interactions that are dependent on the agent's particular situation reflected in the changes in its PIS and intermediary goals judged to contribute towards the main goal. Thus, the reasoning abilities and specialized knowledge of the agents do play an important role as they do capture the agent's problem-solving and strategic reasoning ability that may affect the selection of the most appropriate legal move.

Definition 4.1. A dialogue system is a triple $D = (L_{COM}, PRO, EFF)$ where L_{COM} is the communication language that specifies the locutions which the participants are able to express. Let L_{COM} = {Assert A, Retract A, Accept A, Reject A, Question A and Challenge A} where $A \in L_{Topic}$ is a proposition of the language of some topic (e.g., topic of the dialogue). PRO is a protocol for L_{COM} and EFF is a set of rules that specify the effects of utterances (locutions in L_{COM}) on the participants' commitments.

Definition 4.2. (Dialogue Move). A dialogue move M is a 5-tuple
$$M = <ID(M), SEND(M), \delta(M), LOC(M), TARG(M)> \text{ where}$$
ID(M) is the identifier of M (i.e.,   ID(M) = i indicates that M is the ith move in the dialogue sequence), SEND(M) is the participant that utters $<\delta(M), TOPIC(M)>$, $\delta(M)$

$\in$ {Assert, Retract, Accept, Reject, Question, Challenge}, TOPIC(M) is the sentence which the sender utters and TARG(M) is the target of M.

Context

The essential use of context, in a dialogue, is for the agents to judge the relevance of moves and the continual change to their PIS throughout the different stages of the dialogue. It can be defined as the set of all conditions that may influence the understanding and generation of appropriate locutions. A model of context should include: (1) information needed for the interpretation (and generation) of appropriate locutions needed to achieve particular goals; (2) information about participants' goals and beliefs; (3) information about the interaction such as information about protocols, about the interpretation, evaluation and application of previous utterances.

Let D_k, where $1 \leq k < \infty$, refer to a finite sequence of moves $M_1, . . , M_k$. It is not possible to give a precise definition of context within the scope of this paper. However, we shall employ Context(D_k, G, G1) to refer to the context of a dialogue between G and G1, at stage k, from the perspective of G. We shall say that A $\in$ Context(D_k, G, G1) if A is not inconsistent with Context(D_k, G, G1) and Relevant(A, G, D_k) to mean that A is judged by G to be relevant to the dialogue D_k regarding the specified criteria.

Context(D_k, G, G1) satisfies the following conditions:

(c1) Context(D_k, G, G1) $\subseteq$ KB(G)

(c2) if A $\notin$ Context(D_k, G, G1) and Relevant(A, G, D_k) then A is not inferable from KB(G).

(c3) it cannot be the case A $\in$ Context(D_k, G, G1) and not Relevant(A, G, D_k).

Effect Rules

Let G be an agent, involved in an On-Going dialogue d at stage i-1 with another agent G1. Let KB(G) (resp. KB(G1)) represents the set of propositions which G (resp. G1) accepts as true. With such background, we may give the update rules that specify how context of agents are modified by the moves. We shall just present the effect rules for "Assert" (For more details cf. [10]).

Let $j < i$, M_j a move made by participant G1, and M is an "Assert A" made by G as a reply to M_j, i,e., M = <i, G, Assert, A, j> then Context(G, D_i, G1)= Context(G, D_{i-1},G1)$\cup$\{A\} and Context(G1, D_i, G) = Context(G1, D_{i-1}, G).

Rules of Protocols of Some Types of Dialogue

We may define the rules of protocols for all the dialogue types presented in Table 3.1. However, due to lack of space, we shall only give these rules for Persuasion.

G is trying to persuade G1 to accept A.

(1) G begins with a move that asserts A.
(2) G1 replies with one of (i), (ii) or (iii)
      (i) accepts A   (ii) asserts ~A   (iii) challenges A.
(3) There are two possibilities:
  (a) If the answer of G1 in the previous step is (ii), then go to step (2) with the roles of the agents reversed and ~A in place of A..
  (b) If the answer of G1 in the previous step is (iii) (challenge), then

(α) G should reply with a move that provide/asserts a proof P of A in KB(G)
(β) go to step (2) for every for every proposition C $\in$ P.

Dialogue Control Using Theorem Proving

In problem-solving contexts, solutions can be developed by one, or more participants, dynamically, on the basis of the previous steps captured in the dialogue history, and the current information states of the participants. This approach differs form the approach adopted in some plan-based systems, where an agent, say G, first develops a complete solution, before it sends it to another other agent, say G1. A theorem prover of the system TFONL [cf. 10] can be used to determine what is accomplishable and the dialogue is used for the acquisition, if at all possible, of what is considered necessary or required to complete a specific task step. Failure to provide the needed/ missing information by one agent may leave no choice but to make assumptions, i.e., by invoking nonmonotonic inference rules. In doing so, dialogue is integrated closely with problem solving and is invoked when an agent is unable to complete a task using its own knowledge and/or resources. The nonmonotonic proof method can be employed to create sub-dialogues to solve sub-tasks needed for the overall task.

5 Single Agent and Multi-agent Cooperative Diagnosis

In single agent diagnosis, the agent has to rely on its diagnostic knowledge, reasoning ability and what has been observed. In the case of multi-agent diagnosis, knowledge has to be distributed among the agents and/or the system has to be decomposed into a set of subsystems where each has some clear form of interactions with other subsystems that must be known to the appropriate agents. Each subsystem or aspect of the system is diagnosed by an appropriate diagnostic agent that has detailed knowledge of it and an abstract view of the subsystems with which it interacts. The distribution of a system's knowledge among a set of cooperating agents could follow various criteria such as physical proximity, functionality and/or knowledge-based. Ideally, the agents cooperate among each other in order to achieve some individual objectives, to handle the dependencies that result from what they are involved with or to reach a consensus regarding a diagnostic goal. When a diagnostic agent receives an alarming message/signal of a possible fault, it investigates if there is one in its subsystem, its cause and whether its effect is local or could spread into other subsystems. In any case, a message is sent to the appropriate diagnostic agents. The other agents could do the same with regard to passing on the appropriate message(s) and thus the process continues until the fault is located and its causes are known.

5.1 Single-Agent Based Diagnosis

The Model-Based Diagnosis (MBD) approach [5] can be characterized as follows:

Definition 5.1. A Diagnosis Problem DP=<SD, CONT, OBS> such that
(a) SD= <COMP, BEHAV> is a System Description where COMP = $\{C_1, ..., C_n\}$ is a list of components (and interconnections between them) and BEHAV = $\{B_1, ..., B_n\}$ specifying the behavioral mode of components.

(b) CONT = contextual data which may include the notion of parameters and inputs in numerical approaches.

(c) OBS = observations which encompasses the notion of outputs in numerical approaches.

Diagnosis amounts to comparing the *predicted behavior* with the *observed behavior* (OBS). Let $\Delta \subseteq$ COMP and $\Delta_{AB} = \{Mode(C, AB): C \in \Delta\} \cup \{\neg Mode(c, AB): C \in COMP-\Delta\}$ where AB(C) stands for "component C has an abnormal behavior".

Definition 5.2

A consistency-based (resp. abduction-based) diagnosis for DP is a *minimal* set $\Delta \subseteq$ COMP such that

SD $\cup$ OBS $\cup$ CONT $\cup \Delta_{AB}$ is consistent (resp. SD $\cup$ CONT $\cup \Delta_{AB} \models$ OBS)

5.2 Multi-agent Based Diagnosis

Whichever distribution of knowledge is employed, some constraints and axioms are needed in order to regulate the subsystem-subsystem hierarchies and relations, component-subsystem relations, abstraction criteria and consistency, intra-subsystems and inter-subsystems behavioral modes.

Definition 5.3. Let $Sub_1, \ldots Sub_n$ be a physical proximity distribution of a system S and let $SD_1, \ldots SD_n$ be their description respectively. An agent G_j where $1 \leq j \leq n$ responsible for sub_j is said to need to cooperate with other agents iff
$SD_j \cup CONT \models \neg Mode(Sub_j, AB)$ and
$SD_j \cup CONT \cup \{Mode(Sub_j, AB)\} \not\models F$ where F stand for falsity.

Suppose that G_j receives a message from G_k reporting an observation OBS such that $SD_j \cup CONTEXT \cup \{Mode(Sub_j, AB)\} \models F$, then there are two possibilities:

(1) Sub_j is faulty and g_j has to investigate what may have caused the fault.
(2) There is a need for cooperation with other agents to obtain more information.

5.3 An Example

Consider a series, S12, of two batteries, B1 and B2, connected to a device D. Suppose that D is not operational. G2's task is to find out which battery is faulty. G1 is in charge of testing the voltage of the batteries and carrying out a repair procedure such as RP1(B,NB) which consists of replacing B with a new one NB from the store, or $RP2(B_1, B_2, NB_1, NB_2)$ which requires the agent to drive to town in order to fetch two new batteries NB_1 and NB_2 and then to replace B_1 with NB_1 and B_2 with NB_2.

Suppose that D is not operational. According to G2's reasoning, there are two possibilities: (1) D is faulty and (2) S12 is faulty. Suppose also that there is available one new battery, New-B, kept in the store.

Let (1) Batt(B) = B is a battery.   (3) Volt(S12, V) = voltage of S12 is V.
    (2) Volt(B,V) = voltage of B is V.   (4) Ok(V) = $1.2 \leq V \leq 1.6$.

G2 and G1 have different diagnostic knowledge and skills:

$KB(G2) \supseteq \{$ $Volt(S12,V_0)$ & $V_0 < 2.4 \leftrightarrow Faulty(S12),$

        $Faulty(C)$&$Volt(S12,V_0)$& $2.8 \leq V_0 \leq 3.2 \rightarrow check(D)),$

        $Ok(V_1)$&$\sim Ok(V_2) \rightarrow RP1(B_2, New\text{-}B),$

        $Ok(V_2)$&$\sim Ok(V_1) \rightarrow RP1(B_1, New\text{-}B),$

        $\sim Ok(V_1)$&$\sim Ok(V_2) \rightarrow RP2(B_1, B_2, NB_1, NB_2)\}$

$KB(G1) \supseteq \{$ $Volt(S12,V_0)$ & $V_0 < 2.4 \leftrightarrow Faulty(S12),$

        $\sim Ok(V_2) \rightarrow RP1(B_2, New\text{-}B),$

        $\sim Ok(V_1) \rightarrow RP1(B_1, New\text{-}B)\}$

We could employ $\Rightarrow$ instead of $\rightarrow$ to express defeasibility.

Suppose that S12 is faulty. G2 has to find out whether both B_1 and B_2 are faulty or just one of them. In the latter case, there is a need to find out exactly which one.

Consider the following sub-dialogues:

(1) Inquiry Sub-dialogue: The aim to find the voltage of the series S12.

\# $M_1 = <1, G2, Question, "Volt(S12,V)", 0>$

\# $M_2 = <2, G1, Assert, "Volt(S12,1.45)", 1>$

\# $M_3 = <3, G2, Accept, "Volt(S12,B_2),1.45)", 2>$

(2) Inquiry Sub-dialogue: The agents have to decide whether to replace B_1.

\# $M_4 = <4, G1, Assert, "RepB_1", 3>$

    where $RepB_1 = (Volt(S12),1.45) \rightarrow RP1(B_1, New\text{-}B))$

\# $M_5 = <5, G2, Reject, "RepB_1", 4>$

\# $M_6 = <6, G1, Challenge, "U(RepB_1)", 5>$

\# $M_7 = <7, G2, Assert, "Proof of S1, Proof of S2 and proof of S3", 6>$

\# $M_8 = <8, G1, Accept, "Proof of S1, Proof of S2 and proof of S3", 7>$ where

(S1): From $Volt(S12,1.45)$, we can infer that $\sim(Ok(V_1)$&$Ok(V_2))$.

(S2): From $\sim Ok(V_1)$ we cannot infer $Ok(V_2)$.

(S3) From $Ok(V_1)$ we can infer $\sim Ok(V_2)$.

The outcome of (2) is that there is a need for further testing of the voltage B_1 or B_2.

(3) Inquiry sub-dialogue: aims to find the voltage of B_1.

\# $M_9 = <9, G2, Question, "Volt(B_1,V_1)", 8>$

\# $M_{10} = <10, G1, Assert, "Volt(B_1,1.3)", 9>$

\# $M_{11} = <11, G2, Accept, "Volt(B_1,1.3)", 10>$

The outcome of (3) is that both G1 and G2 agree that B_1 is ok.

(4) A simple question and answer sub-dialogue: G1 asserts that B_2 is to be replaced and G2 accepts.

\# $M_{12} = <12, G1, Assert, "RP1(B_2, New\text{-}B)", 11>$

\# $M_{13} = <13, G2, Accept, "RP1(B_2, New\text{-}B)", 12>$

6 Previous Work

Little consideration has been given to dialogue in previous Work on distributed systems and multi-agent problem solving (cf. [17, 14]). In [17] an abstract formal model of cooperative problem-solving is presented. In [14] it is shown how the concept agent can be used to realize a multi-agent system for distributed diagnosis.

Most existing dialogue systems focus on simple and constrained tasks [13, 15]. These approaches are not suitable for complex tasks. PIS can be a model of the participant's mental state and its model of other the participant's state. The use of the notion of PIS in our system is compatible with that of information state used in the TRINDI project [7]. However, PIS is partial and supported by a nonmonotonic logic. A method to merge conflicting PIS's based on their preference-based argumentation framework is proposed in [3]. In our proposal, arguments may be built from the PIS of one agent and an appropriate subset of the context set of another.

7 Concluding Remark

We have in this paper made a first step towards a formal model of dialogue and argumentation for a multi-agent (model-based) diagnostic system. We have discussed some of the issues in multi-agent cooperative fault diagnosis, the theories of communicating agents and their reasoning capabilities. We have proposed a PIS based framework for dialogue and argument. We have employed a three-valued based nonmonotonic logic, NML3, for representing and reasoning about PISs. The system can easily be customized to handle distributed problem-solving tasks. On the argumentation side, it is worthwhile investigating further the subtleties of each type of dialogue in relation to different tasks that may be accomplished by an agent. It would be beneficial to further investigate, within the framework, strategic and tactic reasoning for rational agents.

References

1. Amgoud, L., Maudet, N.: Strategical Considerations for Argumentative Agents. In: Proc. of 10th Int. Workshop on Non-Monotonic Reasoning, pp. 409–417 (2002)
2. Colombetti, M.: A Commitment-Based Approach to Agent Speech Acts and Conversations. In: 4th Int. Conf. On Autonomous Agent, pp. 21–29 (2000)
3. Finin, T., Labrou, Y., Mayfield, J.: KQML as an Agent Communication Language. In: Bradshaw, J.M. (ed.) Software Agent, pp. 291–315. AAAI Press, Menlo Park (1995)
4. FIPA-ACL. Communicative act library specification. Technical Report XC00037H, Foundation for Intelligent Physical Agents (2001)
5. Hamscher, W., Console, L., De Kleer, J.: Reading in Model-Based Diagnosis. Morgan Kaufmann Publishers, San Francisco (1992)
6. Kone, M.T., Shimazu, A., Nakajima, T.: The State of the Art in Agent Communication Languages. Knowledge and Information Systems, 259–284 (2000)
7. Larsson, S.S., Traum, D.: Information State and Dialogue Management in the TRINDI Dialogue Move Engine Toolkit. Natural Language Engineering 6, 323–340 (2000)

8. McBurney, P., Parsons, S.: A Denotational Semantics for Deliberation Dialogues. In: Rahwan, I., Moraïtis, P., Reed, C. (eds.) ArgMAS 2004. LNCS (LNAI), vol. 3366, pp. 162–175. Springer, Heidelberg (2005)
9. Moubaiddin, A., Obeid, N.: The Role of Dialogue in Remote Diagnostics. In: 20th Int. Conference on Condition Monitoring & Diagnostic Engineering Management (2007)
10. Moubaiddin, A., Obeid, N.: Partial Information Basis for Agent-Based Collaborative Dialogue. Applied Intelligence (accepted for publication, 2008)
11. Obeid, N.: Three Valued Logic and Nonmonotonic Reasoning. Computers and Artificial Intelligence 15(6), 509–530 (1996)
12. Obeid, N., Rao, R.: Diagnostic Temporal Reasoning in Model-Based Diagnosis (MBD) of Dynamic Systems. International Journal of COMADEM 7(1), 13–28 (2004)
13. Pellom, B., Ward, W., Hansen, J., Hacioglu, K., Zhang, J., Yu, X., Pradhan, S.: Dialog Systems for Travel and Navigation. In: Proc. Human Language Technology Conference (2001)
14. Schroeder, M., Wagner, G.: Distributed Diagnosis by Vivid Agents. In: Proc. of 1st international conference on Autonomous agents, pp. 268–275. ACM Press, New York (1997)
15. Seneff, S., Polifroni, J.: Dialogue Management in the Mercury Flight Reservation System. In: Proceedings of ANLP-NAACL Workshop on Satellite Dialogue, pp. 1–6 (2000)
16. Walton, D.N., Krabbe, E.C.W.: Commitment in Dialogue: Basic Concepts of Interpersonal Reasoning. State University of New York Press, New York (1995)
17. Wooldridge, M., Jennings, N.R.: Cooperative Problem Solving. In: Proc. Modelling Autnomous Agents in Multi-Agent World (1994)

Reinforcement Q-Learning and Neural Networks to Acquire Negotiation Behaviors

Amine Chohra, Kurosh Madani, and Dalel Kanzari

Images, Signals, and Intelligent Systems Laboratory (LISSI / EA 3956), Paris-XII University, Senart Institute of Technology, Avenue Pierre Point, 77127 Lieusaint, France
{chohra,madani,kanzari}@univ-paris12.fr

Abstract. Learning in negotiation is fundamental for understanding human behaviors as well as for developing new solution concepts. Elsewhere, negotiation behaviors, in which the characters such as Conciliatory (Con), Neutral (Neu), or Aggressive (Agg) define a 'psychological' aspect of the negotiator personality, play an important role. In this paper, first, a brief description of SISINE (Integrated System of Simulation for Negotiation) project, which aims to develop innovative teaching methodology of negotiation skills, is given. Second, a negotiation approach essentially based on the escalation level and negotiator personality is suggested for SISINE. In fact, the escalation level defines gradually different negotiation stages from agreement to interruption. Afterwards, negotiation behaviors acquired by reinforcement Q-learning and Neural Networks (NN) under supervised learning are developed. Then, behavior results which display the suggested approach ability to provide negotiators with a first intelligence level are presented. Finally, a discussion is given to evaluate this first intelligence level.

1 Introduction

In a context where agents must reach agreements on matters of mutual interest, *negotiation* techniques for reaching agreements are required. In general, any negotiation settings will have four different components [1]:

- a negotiation set, the space of possible proposals that agents can make,
- a protocol, the legal proposals that agents can make, as a function of prior negotiation history,
- a collection of strategies, one for each agent, which determine what proposals the agents will make,
- an agreement rule that determines the reach agreements and stops the negotiation.

Negotiation usually proceeds in a series of rounds, with every agent making a proposal at every round. The proposals that agents make are defined by their strategy (a mapping from state history to proposal ; a way to use the protocol), must be drawn from the negotiation set, and must be legal, as defined by the protocol (which defines possible proposals at different rounds). If agreement is reached, as defined by the agreement rule, then negotiation terminates with the agreement deal. These four parameters lead to an extremely rich and complex environment for analysis.

Another source of complexity in negotiation is the number of agents involved in the process, and the way in which these agents interact [1]. First possibility is one-to-one

N.T. Nguyen, R. Katarzyniak (Eds.): New Chall. in Appl. Intel. Tech., SCI 134, pp. 23–33, 2008.
springerlink.com

negotiation, in which one agent negotiates with just one another agent, e.g., a particularly case where the agents involved have symmetric preferences with respect to the possible deals, e.g., when discussing terms with a car salesman. Second possibility is many-to-one negotiation. In this setting, a single agent negotiates with a number of other agents, and can often be treated as a number of concurrent one-to-one negotiations. Third possibility is many-to-many negotiation, where, many agents negotiate with many other agents simultaneously. In the worst case, where there are n agents involved in negotiation in total, making such negotiations hard to handle. For these reasons, most attempts to automate the negotiation process have focused on rather single issue settings. Such issue, symmetric, one-to-one negotiation is the most commonly analyzed, and it is on such settings that we will mainly focus in this work.

Finding a shared solution to a problem within a group requires *negotiation*, a potentially exhausting and time-consuming process. To negotiate successfully, members have to involve the whole group, explain their position clearly and do their best to understand those of others [2]. However, in reality, groups often fail to negotiate, even when negotiation would be useful for each part of the group. Sometimes the problem lies in sheer size of the group, or in hierarchical organizational structures or in impediments to communication deriving from language, culture or history. In other cases, the main barriers lie in the individual psychology of specific group members. Typical problems include weak communications skills, lack of empathy with others, and poor control over the emotions arising during prolonged discussion. Such problems can be directly or indirectly related to the personality of each group member participating to the negotiation. Thus, negotiation behaviors, in which the characters such as Conciliatory (Con), Neutral (Neu), or Aggressive (Agg) define a 'psychological' aspect of the personality of a negotiation member (negotiator), play an important role [3], [4].

For this purpose, the suggested negotiation approach, for the Integrated System of Simulation for Negotiation (SISINE), relies on a theoretical model of the negotiation process (between two negotiators) which is mainly based first on the escalation level (defining gradually several negotiation stages from agreement to interruption) of the negotiation. Second it is based on the negotiator personality, i.e., characters Con, Neu, Agg. Elsewhere, learning from interaction in negotiation is fundamental, from embodied cognitive science and understanding natural intelligence perspectives [5], [6], for understanding human behaviors and developing new solution concepts [7].

Learning from interaction is a foundational idea underlying nearly all theories of learning. Indeed, whether a human is learning to drive a car or to hold a conversation (during a negotiation), he is acutely aware of how his environment responds to what he does, and he seeks to influence what happens through his behavior. Elsewhere, reinforcement learning is much more focused on goal-directed learning from interaction than other approaches to machine learning [8], [9], [10]. More, reinforcement learning approaches offer two important advantages over classical dynamic programming [11]. First, they are on-line having capability to take into account dynamics nature of real environments. Second, they can employ function approximation techniques, e.g., Neural Networks (NN) [12], [13], [14], to represent their knowledge, and to generalize so that the learning time scales much better.

The aim of this research work is to develop an adaptive negotiation model, for a first intelligence level, capable of exhibiting a rich set of negotiation behaviors with modest computational effort. In this paper, first, a description of SISINE project is given in Sect. 2. Second, a negotiation approach, essentially based on the escalation level and negotiator personality, is suggested in Sect. 3. Afterwards, for a first intelligence level, negotiation behaviors acquired by reinforcement Q-learning and NN under supervised gradient backpropagation learning are developed and the behaviors results are presented in Sect. 4. Finally, a discussion is given in Sect. 5 to evaluate this first intelligence level in negotiation.

2 Integrated System of Simulation for Negotiation (SISINE)

Humans have developed advanced skills in the intentions and the bodily expressions of the other human being, particularly important in high level communication which is at the basis of any "successful" negotiation (interaction process). Dealing with this, research work, in this paper, is a part of the SOcial and COgnitive SYStem, SOCOSYS, developed for an intelligent human-agent interaction by SCTIC (Complex Systems, and Intelligent Knowledge Processing) team, at Senart Institute of Technology, of LISSI (Images, Signals, and Intelligent Systems Laboratory, Paris-XII University, France). This part has been developed for SISINE (Integrated System of Simulation for Negotiation). SISINE project is funded, by European Union Leonardo Da Vinci Program, to develop an innovative teaching methodology of negotiation skills exploiting an integrated system platform of simulation for negotiation (http://www.sisine.net), [2].

In SISINE project, simulation environments enable a participant to interact with a virtual entity called "bot" (a software agent), through a communicative exchange: texts (one among three), voice (tone and volume), facial expression, and gesture. The objective of such simulation environments is to allow participants to directly experience the basic elements of negotiation relationships, through two kind of agents: - a "standard bot", an agent with reactions from simple rules, and a "smart bot" an agent with reactions from a first intelligence level.

3 Negotiation Approach Based on Escalation Level and Negotiator Personality

The suggested negotiation approach relies on a theoretical model of the negotiation process (between two negotiator agents) which is mainly based first on the escalation level (defining gradually several negotiation stages from agreement to interruption) of the negotiation. Second, it is based on the negotiator personality, i.e., characters Conciliatory (Con), Neutral (Neu), and Aggressive (Agg) which define a 'psychological' aspect of the negotiator agent personality.

In this theoretical model, seven possible escalation level stages are modeled by the variable escalation level EscLevel belonging to the interval [0, 60] and numbered from 0 to 6 as follows: (0). agreement if [0, 10 [; (1). defense of positions [10, 19], where each part defends a position and attempts to persuade the other of its validity ; (2). intermediate level stage [20, 29] ; (3). attack on the other's position [30, 39], where each part do not discuss its position but only seek to attack the other's position ; (4). intermediate level stage [40, 49] ; (5). attack on the other [50, 59], where each part do not discuss the problem but attack each other ; (6). interruption of the negotiation] 59, 60], ending the negotiation process.

Also, in this theoretical model, the character of a negotiator agent is defined by a character vector [Con, Neu, Agg] where each component belongs to the interval [0, 100] in percentage (%) such as the character vector components verify Eq. (1):

$$Con + Neu + Agg = 100 \% \qquad (1)$$

Consequently, during a negotiation round, each negotiator agent is defined by its current character vector **CurrentChar** = [CurrentCon, CurrentNeu, CurrentAgg], then, its new character vector **NewChar** = [NewCon, NewNeu, NewAgg] is updated from the user sentence effect explained below. Indeed, with such model, three negotiation behaviors (conciliatory, neutral, and aggressive behaviors), corresponding to three different agent personalities, arise according to the great value among the three characters, e.g., an agent with [Con = 60.00 %, Neu = 25.00 %, Agg = 15.00 %] is a conciliatory negotiator, whereas an agent with [Con = 30.00 %, Neu = 40.00 %, Agg = 30.00 %] is a neutral negotiator, and so on an aggressive negotiator has a the greater value for the character Agg. More, depending on such value the personality can be gradually weak or strong on that same character.

Thus, during a negotiation round, each negotiator agent is defined by its internal current state CurrentState(CurrentEscLevel, **CurrentChar**, **UserSentence**), having its CurrentEscLevel and **CurrentChar**, and receiving a sentence vector **UserSentence** = [DeltaEscLevel, CharToModify, DeltaChar] from a user where: DeltaEscLevel, an escalation level variation belonging to [-60, +60] ; CharToModify, a character to modify (Con, Neu, or Agg) ; and DeltaChar, a character variation belonging to [-10, +10], with "- large", "- average", "- small", "+ small", "+ average", and "+ large" values corresponding to -6, -4, -1, +1, +4, and +6, respectively.

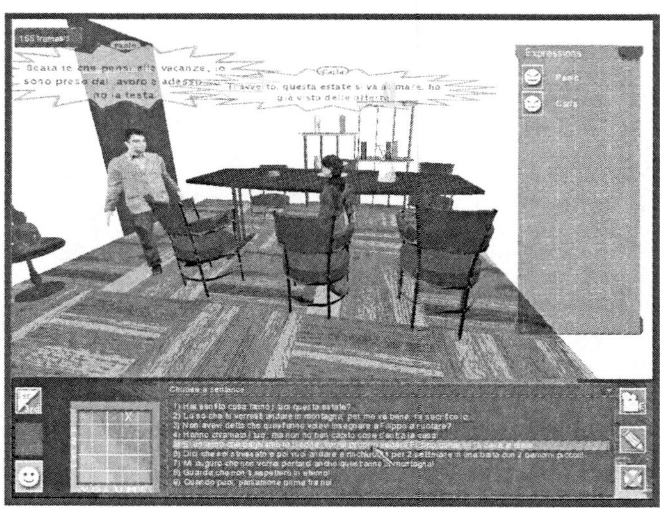

Fig. 1. A negotiation session between two agents from SISINE software

In fact, during a round (a given specific state) of a negotiation session as illustrated in Fig. 1, an agent has a given CurrentEscLevel and a given **CurrentChar**, and receives DeltaEscLevel, CharToModify, and DeltaChar extracted in SISINE software from a user (another agent or a human) sentence. An example of such negotiation session is illustrated in Fig. 1, where a conversation is shown, during a negotiation round, between two agents (woman and man) which are arguing about whether they should spend their holidays at seaside or in mountains. The goal for an agent, is then from a user sentence to update its escalation level (NewEscLevel) and character (**NewChar**), and to choose, based on these new values, an answer.

3.1 Standard Agent

For a *Standard* agent, the reactions are based on an updating, of the escalation level and character, from the simple rules given in Eq. (2) and Eq. (3), respectively.

$$NewEscLevel = CurrentEscLevel + DeltaEscLevel. \tag{2}$$

$$
\begin{aligned}
&\text{IF CharToModify is Con Then DeltaChar} = (DeltaChar/100)*CurrentCon, \\
&\qquad\qquad NewCon = CurrentCon + DeltaChar, \\
&\qquad\qquad NewNeu = CurrentNeu - DeltaChar*0.3, \\
&\qquad\qquad NewAgg = CurrentAgg - DeltaChar*0.7 \ ; \\
&\text{IF CharToModify is Neu Then DeltaChar} = (DeltaChar/100)*CurrentNeu, \\
&\qquad\qquad NewCon = CurrentCon - DeltaChar*0.5, \\
&\qquad\qquad NewNeu = CurrentNeu + DeltaChar, \\
&\qquad\qquad NewAgg = CurrentAgg - DeltaChar*0.5 \ ; \\
&\text{IF CharToModify is Agg Then DeltaChar} = (DeltaChar/100)*CurrentAgg, \\
&\qquad\qquad NewCon = CurrentCon - DeltaChar*0.7, \\
&\qquad\qquad NewNeu = CurrentNeu - DeltaChar*0.3, \\
&\qquad\qquad NewAgg = CurrentAgg + DeltaChar.
\end{aligned}
\tag{3}
$$

3.2 Intelligent Agent (A First Intelligence Level)

The technology of multi-agent systems facilitates the negotiation at operative level of the decision-making [15]. It allows agents to embody a notion of autonomy, in particular, to decide for themselves whether or not to perform an action on request from another agent. More, in order to satisfy their design objectives, agents are designed to be intelligent, i.e., capable of flexible behavior [1], [5], [6]: able to perceive their environment, and respond in a timely fashion to changes that occur in it (reactivity), able to exhibit goal-directed behavior by taking the initiative (proactiveness), and capable of interacting with other agents and possibly humans (social ability) in the sense of cooperation, competition, neutrality, and negotiation.

Thus, for an *intelligent* agent, the reactions are based on the updating of the character from a first intelligence level allowing to acquire negotiation behaviors: using the reinforcement Q-Learning, and using a NN under supervised gradient backpropagation learning. Note that, for an intelligent agent the updating of the escalation level is the same as for standard agent from simple rule given in Eq. (2).

4   Reinforcement Q-Learning and Neural Networks to Acquire Negotiation Behaviors: A First Level of Intelligence

One of the most important breakthroughs in reinforcement learning was the development of an off-policy temporal-difference control algorithm known as Q-learning [8], [9], [10]. Elsewhere, NN implementation of reinforcement Q-learning offers the advantages of the generalization quality and limited memory requirement for storing the knowledge [16], [17]. In addition, NN are characterized by their learning, and generalization capabilities (essential traits of intelligent behaviors), robustness, massively parallel computations and distributed memory [12], [13], [14].

4.1   Reinforcement Learning

Reinforcement learning allows an agent (the learner and decision-maker) to use its experience, from the interaction with an environment, to improve its performance over time [8], [9], [10]. In other words, from the interaction with an environment an agent, can learn, using reinforcement Q-learning, to maximize the reward r leading to an optimal behavior policy. Indeed, in this on-line reinforcement learning, the agent incrementally learns an action/value function Q(s, a) that it uses to evaluate the utility of performing action a while in state s. Q-learning leads to optimal behavior, i.e., behavior that maximizes the overall utility for the agent in a particular task environment [5]. The used Q-learning paradigm [8], [10] is shown in Fig. 2 (a).

The parameter settings of the initial Q values, the constant step-size parameter ($0 < \alpha <= 1$), and the discount rate ($0 < \gamma <= 1$) have been done following the choice approaches given in [8] and [10] resulting in: initial Q values = 0.5, α = 0.1, and γ = 0.01. Q-learning results are given through the example where the input vector is **X** = [CurrentCon = 38.00, CurrentNeu = 31.50, CurrentAgg = 29.50, CharacterToModify = Con, DeltaChar] and the components of the output vector **O** = [NewCon, NewNeu, NewAgg] represented in Fig. 3 (left) for different values of DeltaChar -6, -4, -1, +1, +4, +6. In this example, the character is conciliatory starting at the level 38.00 and decreases at levels 37.60, 36.40, 35.60 when decreasing DeltaChar -1, -4, -6 whereas increases at levels 38.39, 39.59, 40.39 when increasing DeltaChar +1, +4, +6. Another example is given in Fig. 3 (right) where the input vector is **X** = [CurrentCon = 40.00, CurrentNeu = 30.50, CurrentAgg = 28.50, CharacterToModify = Con, DeltaChar].

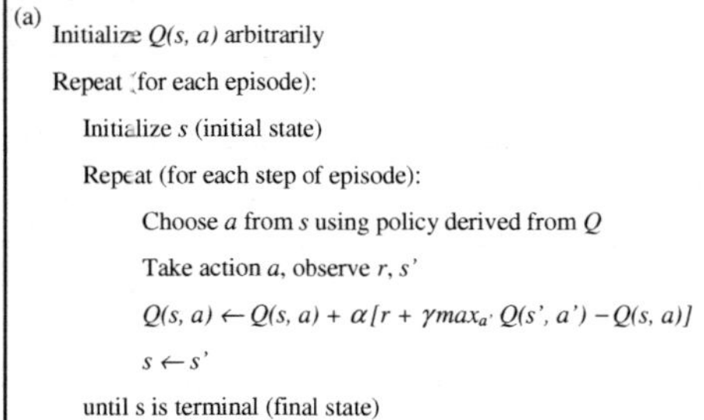

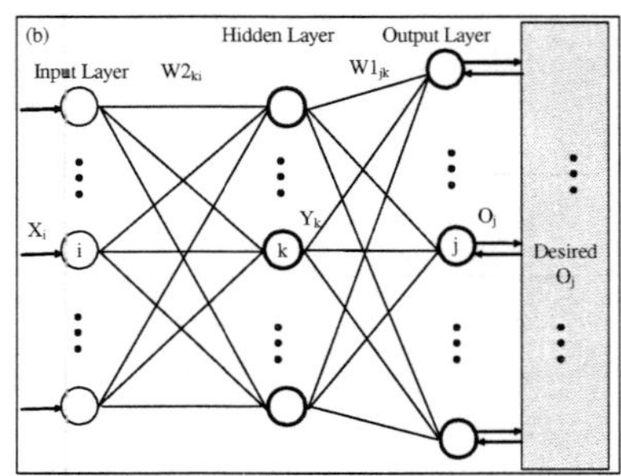

Fig. 2. (a) Q-Learning: an off-policy temporal-difference control learning paradigm. (b) NN architecture where X_i (i = 1, ..., 5), Y_k (k = 1, ..., 11), and O_j (j = 1, ..., 3).

Thus, Q-learning results are satisfying the DeltaChar tendency in decreasing (increasing) the character defined by CharacterToModify while modifying the others verifying Eq. (1).

4.2 Multilayer Feedforward Neural Network (NN) under Supervised Gradient Backpropagation Learning

Multilayer feedforward Neural Networks (NN) are neural global approximators [12], [13], [14]. From this, a NN is suggested for approximation, and trained under the supervised gradient backpropagation learning algorithm [12], [13], [14]. The NN architecture is built of three layers input layer, hidden layer, and output layer as shown in Fig. 2 (b). The input vector is $\mathbf{X}$ = [CurrentCon, CurrentNeu, CurrentAgg, CharacterToModify, DeltaChar]. Note that for CharToModify, the characters Con, Neu, and Agg are coded with 0.1, 0.5, and 0.9, respectively, and for DeltaChar, the values -6, -4, -1, +1, +4, and +6 are coded with 1, 3, 5, 7, 9, and 11, respectively. These components are then pre-processed, in Eq. (4), to constitute input vector $\mathbf{X}$.

$$X_1 = (1/\rho) \exp(-\text{CurrentCon}/a),$$
$$X_2 = (1/\rho) \exp(-\text{CurrentNeu}/a),$$
$$X_3 = (1/\rho) \exp(-\text{CurrentAggt}/a),$$
$$X_4 = (1/\rho) \exp(-\text{CharToModify}/a),$$
$$X_5 = (1/\rho) \exp(-\text{DeltaChar}/a),$$

(4)

where ρ: norm of input vector $\mathbf{X}$ and a: input pre-processing factor with a > 1.

Input Layer: This layer is input layer with i input nodes receiving the components of the input vector $\mathbf{X}$. This layer transmits inputs to all nodes of next layer.

Hidden Layer: This layer is the hidden layer with k hidden nodes. The output of each node is obtained using the output sigmoïd function f as follows:

$$\text{net}_k = \sum_i X_i W2_{ki}, \text{ and } Y_k = f(\text{net}_k), \text{ where } f(x) = \frac{1}{1 + \exp(-x)}.$$

(5)

Output Layer: This layer is the output layer with j linear output nodes obtained by:

$$O_j = \sum_k Y_k W1_{jk}.$$

(6)

The steps in the used learning algorithm are outlined as follows:

1- Random weight ($W2_{ki}$ and $W1_{jk}$) initialization [-1, +1].
2- Apply an input vector $\mathbf{X}$ to the input layer.
3- Compute net_k and outputs Y_k of the hidden layer.
4- Compute outputs O_j of the output layer.
5- Compute the error δ_j for the outputs of the output layer:

$$\delta_j = (\text{DesiredO}_j - O_j).$$

(7)

6- Compute the error δ_k for the outputs of the hidden layer:

$$\delta_k = f'(\text{net}_k) \sum_j \delta_j W1_{jk}, \text{ and } \delta_k = Y_k (1 - Y_k) \sum_j \delta_j W1_{jk},$$

(8)

$$\text{since f: sigmoïd function} \Rightarrow f' = f (1 - f).$$

7- Update the weights of the output layer:

$$W1_{jk}(t+1) = W1_{jk}(t) + \Delta W1_{jk}, \quad \text{with } \Delta W1_{jk} = \eta\, \delta_j\, Y_k. \tag{9}$$

8- Update the weights of the hidden layer:

$$W2_{ki}(t+1) = W2_{ki}(t) + \Delta W2_{ki}, \quad \text{with } \Delta W2_{ki} = \eta\, \delta_k\, X_i. \tag{10}$$

9- Compute the error E:

$$E = (1/2) \sum_j (\text{DesiredO}_j - O_j)^2 . \tag{11}$$

10- Repeat 2- to 9- with the same input vector **X** (the same training example) until the error E is very close to the tolerance.

11- Repeat 2- to 10- for each input vector **X** (each training example).

12- Repeat 2- to 11- under several epochs.

The connection weights are thus updated until the network convergence: a state permitting the coding, i.e., the approximation of all the training examples or input space. This state is reached when the error E is very close of the tolerance, i.e., the error for all training examples is reduced to an acceptable value (preventing inappropriate memorization, also called over-training).

The NN is trained from the training set (learning base built of 144 examples) and tested from the testing set (generalization base built of 144 examples). For an efficient discrimination among input vectors **X**, the input pre-processing factor a = 5 is used. This NN yields convergence to the tolerance T = 0.001 in well under N = 1310 epochs with the learning rate η = 0.04.

NN-learning results are given through an example where the input vector is **X** = [CurrentCon = 38.00, CurrentNeu = 31.50, CurrentAgg = 29.50, CharacterToModify = Con, DeltaChar] and the components of the output vector **O** = [NewCon, NewNeu, NewAgg] represented in Fig. 3 (left), Fig.4 (left), and Fig. 5 (left). for different values of DeltaChar -6, -4, -1, +1, +4, +6. For conciliatory character, the values to learn (from Q-learning) are 35.60, 36.40, 37.60, 38.39, 39.59, 40.39 and the values resulting from NN are 36.19, 37.08, 37.16, 37.61, 39.40, 40.78, respectively. For neutral character, the values to learn are 33.66, 32.94, 31.86, 31.13, 30.05, 29.33 and the values resulting from NN are 33.09, 32.18, 32.49, 31.27, 30.13, 28.96, respectively. For aggressive character, the values to learn are 31.66, 30.94, 29.86, 29.13, 28.05, 27.33 and the values resulting from NN are 31.09, 30.35, 30.45, 29.49, 28.23, 26.84, respectively.

NN-generalization results are given through an example where the input vector **X** = [CurrentCon = 40.00, CurrentNeu = 30.50, CurrentAgg = 28.50, CharacterToModify = Con, DeltaChar] and the components of the output vector **O** = [NewCon, NewNeu, NewAgg] represented in Fig. 3 (right), Fig. 4 (right), and Fig. 5 (right) for different values of DeltaChar -6, -4, -1, +1, +4, +6. For conciliatory character, test values (from Q-learning) are 37.48, 38.32, 39.58, 40.42, 41.68, 42.52 ; resulting values from NN are 37.72, 38.79, 39.02, 39.45, 40.99, 42.42, respectively. For neutral character, test values are 32.78, 32.02, 30.88, 30.12, 28.98, 28.22 and resulting values from NN are 32.70, 31.70, 31.91, 30.75, 29.69, 28.48 respectively. For aggressive character, test values are 30.78, 30.02, 28.88, 28.12, 26.98, 26.22 and resulting values from NN are

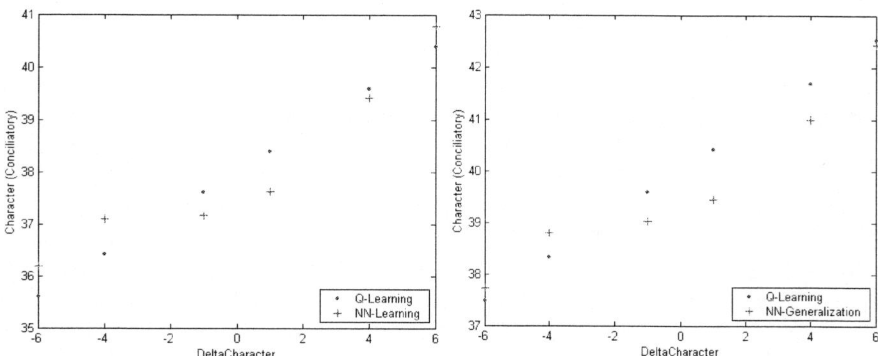

Fig. 3. Q-learning, NN-learning, and NN-generalization results (NewCon)

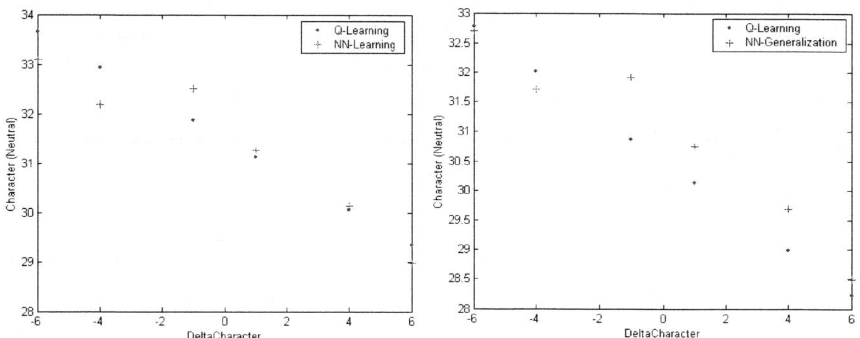

Fig. 4. Q-learning, NN-learning, and NN-generalization results (NewNeu)

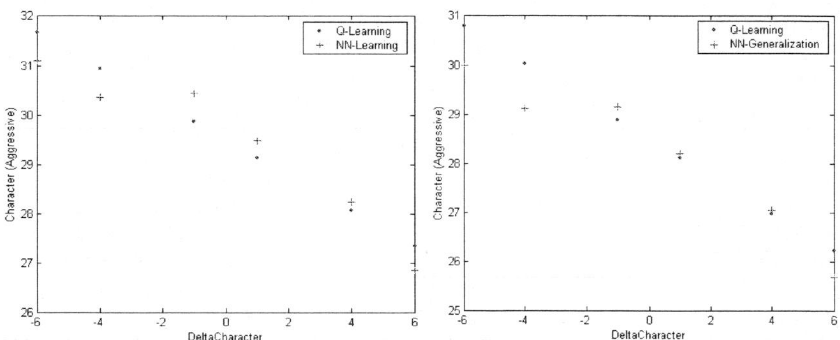

Fig. 5. Q-learning, NN-learning, and NN-generalization results (NewAgg)

29.98, 29.12, 29.14, 28.20, 27.04, 25.67 respectively. Thus, the NN-learning results as well as NN-generalization results (compared to those of Q-learning) are satisfying the tendency of DeltaChar in decreasing (increasing) the character defined by CharacterToModify while modifying the others verifying Eq. (1).

5 Discussion and Conclusion

In this paper, a negotiation approach essentially based on the escalation level and negotiator personality is suggested for SISINE. In fact, negotiation behaviors in which characters Conciliatory, Neutral, or Aggressive define a 'psychological' aspect of the negotiator personality, play an important role in negotiation. Afterwards, such negotiation behaviors acquired by reinforcement Q-learning and Neural Networks (NN) under supervised learning are developed for a first intelligence level. Behavior results of Q-learning as well as for NN learning and generalization are satisfying the tendency of the character variations of the character while modifying the others.

Thus, this first intelligence level provided to an intelligent agent through the negotiation behaviors acquired by Q-learning and NN allows to interact, during a negotiation session, with another agent or with human through SISINE software. More, with such interactions, different training scenarios are possible with the on-line learning (Q-learning) and off-line learning (NN) exploiting one or the other, or exploiting them together. This first intelligence level still not enough to allow an agent to learn a negotiation strategy, for instance, from a human with a high negotiator quality. For this, a second intelligence level is under work handling problems of strategy, cooperation, competition.

Finally, we would like acknowledge **"Leonardo Program and related Authorities"** who have supported this project.

References

1. Wooldridge, M.: An Introduction to MultiAgent Systems. John Wiley & Sons, England (2002)
2. Miglino, O., Di Ferdinando, A., Rega, A., Benincasa, B.: SISINE: Teaching negotiation through a multiplayer online role playing game. In: The 6th European Conference on E-Learning, Copenhague, Danemark, October 04-05 (2007)
3. Bales, R.F.: Interaction Process Analysis: A Method for the Study of Small Groups. Addisson-Wesley, Cambridge, Massachussetts (1950)
4. Rubin, J.Z., Brown, B.R.: The Social Psychology of Bargaining and Negotiation. Academic Press, New York (1975)
5. Pfeifer, R., Scheier, C.: Understanding Intelligence. MIT Press, Cambridge (1999)
6. Chohra, A.: Embodied cognitive science, intelligent behavior control, machine learning, soft computing, and FPGA integration: towards fast, cooperative and adversarial robot team (RoboCup). Technical GMD Report, No. 136, ISSN 1435-2702, Germany (June 2001)
7. Zeng, D., Sycara, K.: Benefits of learning in negotiation. In: Proc. of the 14th National Conference on Artificial Intelligence (AAAI 1997), Providence, RI (July 1997)
8. Watkins, C.J.C.H.: Learning from Delayed Rewards. PhD Thesis, King's College (1989)
9. Whitehead, S.D.: Reinforcement Learning for the Adaptive Control of Perception and Action. Technical Report 406, University of Rochester (February 1992)
10. Sutton, R.S., Barto, A.G.: Reinforcement Learning. MIT Press, Cambridge (1998)
11. Dieterich, T.G.: Hierarchical reinforcement learning with the MAXQ value function decomposition. Journal of Artificial Intelligence Research 13, 227–303 (2000)
12. Anderson, J.A.: An Introduction to Neural Networks. The MIT Press, England (1995)

13. Patterson, D.W.: Artificial Neural Networks: Theory and Applications, Prentice-Hall, Simon & Schuster (Asia) Pte Ltd., Singapore (1996)
14. Haykin, S.: Neural Networks: A Comprehensive Foundation, 2nd edn. Prentice-Hall, Englewood Cliffs (1999)
15. Sandholm, T.W.: Distributed Rational Decision Making. In: Multiagent Systems: A Modern Introduction to Distributed Artificial Intelligence, pp. 201–258. MIT Press, Cambridge (1999)
16. Lin, L.-J.: Self-improving reactive agents based on reinforcement learning, planning and teaching. In: Machine Learning, vol. 8, pp. 293–321. Kluwer Academic Publishers, Dordrecht (1992)
17. Touzet, C.F.: Neural reinforcement learning for behaviour synthesis. Robotics and Autonomous Systems 22, 251–281 (1997)

getALife - An Artificial Life Environment for the Evaluation of Agent-Based Systems and Evolutionary Algorithms for Reinforcement Learning

Daniel Machado and Miguel Rocha

Dep. Informatics / CCTC - University of Minho
Campus de Gualtar, 4710-057 Braga - Portugal
dmachado@di.uminho.pt, mrocha@di.uminho.pt

Abstract. An *Artificial Life* environment - *getALife* - is proposed, whose major aim is to provide a framework to evaluate single and multi-agent systems and evolutionary approaches to the development of reinforcement learning algorithms. The environment is based on a predator-prey scenario, with multiple species and where individuals are mainly characterized by their decision modules and genetic information. The platform is quite powerful, flexible, modular, visually attractive, easy to program and to use, making an interesting tool both to research and teaching. Two applications based on *getALife* are provided: the evaluation of a *Neural Network* based decision module with evolutionary learning and the development of a children's game.

Keywords: Artificial Life simulators, Prey-predator systems, Evolutionary Algorithms for Reinforcement learning.

1 Introduction

In the last decades, a number of *Artificial Life (ALife)* simulators have been proposed, with distinct aims. Two major groups can be identified, the former more concerned with the simulation of natural ecosystems and living beings, while the latter includes systems devoted mainly to the benchmarking of agent systems, decision making strategies and *Machine Learning* algorithms.

In this work, an ALife simulator is proposed, named *getALife*, that can be included in the latter group. The main aim of this work is to develop a framework that enables researchers to test agent-based systems in an environment that is visually attractive, flexible, multi-platform, as well as easy to use, to program and to increment. The system allows the evaluation of *Evolutionary Algorithms for Reinforcement Learning (EARLs)* in quite complex and dynamic environments, that are however, easy to implement. Furthermore, the platform aims to provide analysis tools that make easier the comparison of different approaches.

The basic idea is to create an artificial ecosystem, based on predator-prey systems [2], where a number of beings, from distinct species, interact and struggle for survival. Each living being is mainly characterized by a *genome* that carries

N.T. Nguyen, R. Katarzyniak (Eds.): New Chall. in Appl. Intel. Tech., SCI 134, pp. 35–44, 2008.
springerlink.com   © Springer-Verlag Berlin Heidelberg 2008

its genetic information and by a *decision module* that controls the actions taken by the individuals, based on sensorial information gathered from the environment. The concept of *species* is used to group the individuals and control the reproduction, defining the structure of both the genome and decision modules, although the parameters can change from individual to individual. Each species can also define its own *lifetime learning* algorithm, that can update these parameters given the reward provided by the environment. The species of each individual also defines which species it can fed.

Multiple species can co-evolve in the same simulation and simple or very complex food chains may be defined. The learning task that is implicitly defined for each individual is to be able to eat the maximum number of preys, while avoiding predators (if any). The behaviour of the individuals of a given species in a simulation can be used to evaluate the performance of the strategy it implements in terms of the decision making methods. Also, the evolution of the overall species performance can be used to evaluate *EARLs* [7], namely the encoding scheme of its genetic information and the reproduction operators.

One of the main features of the proposed framework is the ability to be not only a useful tool available to researchers from the *Artificial Life/ Machine Learning (AL/ML)* communities to evaluate algorithms, but also an environment that can be used in teaching, for instance in practical projects within subjects related to *Artificial Intelligence, Multi-agent systems* or *Natural Computation*. This has been the case with several projects at University of Minho, where the software has been used for the last two years. An additional field of application is its use to develop games for elementary levels of education, helping to explain concepts related to complexity and emergence (as shown in Section 5).

In terms of its implementation, the *getALife* system is developed using the Java programming language. Therefore, it is easily portable to the main hardware and operating systems platforms. A GUI is built to allow running the system in a visual mode, but this can be switched off when performing tests.

It is also important to explain what *getALife* is not. Firstly, it does not aim to provide a general purpose tool to develop ALife environments (e.g. such as *Swarm* [6]), but it intends to provide tools to allow the rapid development and testing of new species involving instances of decision modules and/or genome encodings and reproduction operators. Furthermore, it does not aim to provide means for the evolution of the physical features of the individuals/ species, nor does it provide a sophisticated 3D simulation environment (e.g. Breve [4]).

2   Related Work

A number of *ALife* simulators have been proposed in the last decades, from which we would emphasize the following:

- The most popular systems are the ones that create a virtual computer, where computer programs that can mutate, replicate and recombine, compete for CPU time and access to main memory, such as *Tierra* [9] or its close successor

Avida [1]. These environments have been used to explore in computer simulations the processes of evolutionary dynamics.

- *Swarm* [6] is a generic platform for the programming of complex adaptive systems. It provides a sophisticated system for creation of object hierarchies and production of events. Using Swarm involves a lot of programming.
- *Breve* [4] is a software package that allows the building of 3D simulations of decentralized systems and artificial life. It includes articulated bodies and physical simulation with collision detection.
- The *Pursuit Domain Package* [5] is an environment used to test multi-agent system's techniques. It consists on a grid representing a world where predators and preys can move. It allows the testing of strategies within the predator agents and communication, coordination and cooperation issues. It is however very limited to test learning algorithms and does not directly support evolutionary systems.
- *PolyWorld* [11] is a graphical system for artificial life, that consists on a 2-D world with obstacles, where individuals can evolve. The decision module of each individual is restricted to a Hebbian learning neural net.
- The *Artificial Life Environment (ALE)* [3] is a simulator based on building blocks made out of Genetic Algorithms, Neural Networks and Cellular Automata. Individuals live in a discrete world and several types of simulation (e.g. predator-prey, the game of life) can be implemented. The system was developed in 2000 in C++ but no improvements have been made.

3   System Description

The main components of the *getALife* system will be explained in detail in this section. An overview of the system components is given in Figure 1.

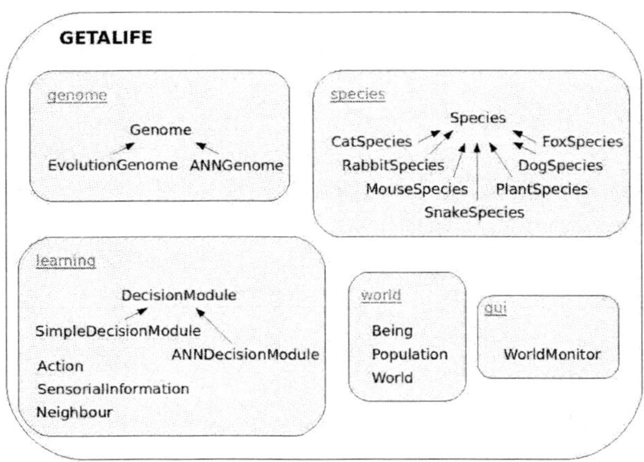

Fig. 1. A scheme showing the main components of the *getALife* system

3.1 World

The core of the *getALife* system is the *world*, the component that controls the whole simulation. It can have an arbitrary number of dimensions, with a specific size for each of the dimensions, and its topology can be closed or toroidal. The world has an internal clock that measures elapsed time or iterations. In each iteration, the world implements the following sequence of steps:

1. Sensorial information is sent to the all the beings according to the world status and the parameters of each being. Each individual will in turn use this information to make a decision of the action to take. At this stage, only visual information is transmitted.
2. The actions from all the beings are applied and their state is updated.
3. If at any time a prey enters inside a predator's attack range, it will die and will be removed from the world. The world rewards the beings, recalculating their energy and fitness.
4. Each individual is given the chance to perform lifetime learning according to its own algorithm, using information from the reward.
5. Finally, the world takes care of invoking reproduction methods for those species where the defined criteria implies the creation of new individuals. Reproduction will try to maintain the population of each species, although it may fail if there aren't enough mature individuals.

3.2 Individuals

In this system, an *individual* (or *being*) is an autonomous agent that has a decision module used to define its behavior and a genetic code, that is inherited from its parents and that is used to initialize the decision module, when it is born. Each individual has also dynamic properties that vary throughout simulation time, such as its position, velocity, acceleration, energy and age. These are updated by the world in every generation.

An individual can sense its surrounding environment and use that information to make decisions. Currently, the being senses visual information, consisting on information about all beings that are closer to the individual than a pre-defined visual radius. The decisions are taken by running a method specified by the decision module. Those decisions will result in changing its energy and fitness. A being will die if its energy reaches zero, or if it is eaten by a predator. The consequences of the actions can be used to perform lifetime learning, i.e. to update the free parameters of its decision module.

3.3 Species

The species of an individual defines its characteristics, such as maximum velocity and acceleration, visual range, attack range, maturity age and initial energy. It also defines on which species the being can feed, the energy reward associated with each one, the energy spent on performing an action and on reproduction.

Furthermore, the species defines the structure of the decision module and of the genome of the individual, as well as the algorithm used for lifetime learning (if any). Therefore, it makes sense to compare the statistics of two species to evaluate each of these components. The basic package has some species already built-in as an example, but more can easily be added.

3.4   Decision Modules

The *decision module* defines the behavior of the individual. After receiving the sensorial information from its surroundings, it is responsible for replying with an action. Currently, the only possible action is to change the absolute value and direction of the acceleration, which will in turn change the velocity vector. New actions will be added in the near future and can also be added by the user with minor programming effort. The sensorial information includes information about the being such as velocity and energy, and information about the beings inside its visual range, such as relative position, velocity and species.

By performing an action, the being will spend energy. A reward is given if a prey is caught, an event that takes place if the prey passes inside the predator's attack range. After making a decision, the individual will be notified (rewarded) with its energy level variation. A decision module may have a static behavior or an adaptive one, evolving the decision module during the individual's lifetime, using *RL* techniques to improve its behaviour. Some examples of decision modules are provided. Most have a static behavior such as not moving, always attacking, running away or moving randomly. In the case studies, species that use *Artificial Neural Networks (ANNs)* to take decisions are also presented.

3.5   Genome

The genome is responsible not only for holding the genetic information of an individual, but also for defining how it evolves through the generations. It defines how genes are created, how parents are selected for reproduction and how new genes can be obtained from the genes of the parents.

This module is closely integrated with a general purpose software environment for the development of *Evolutionary Algorithms* previously developed by one of the authors [8]. In this framework genetic encodings with binary, integer, permutation and real valued representations have been developed, as well as more sophisticated representations such as direct encoding of *ANNs*. Also, for all these, reproduction operators (mutation and crossover operators) are also provided. So, the *getALIfe* takes advantage of this existing software and uses it in the generation of new individuals and also in the definition of selection schemes (e.g. roulette wheel or tournament selection).

3.6   World Monitor

A graphical world monitor was implemented to facilitate the creation and monitoring of simulations, allowing the creation of a bi-dimensional world of a specified size, and the insertion of new beings in runtime. Each species implements a

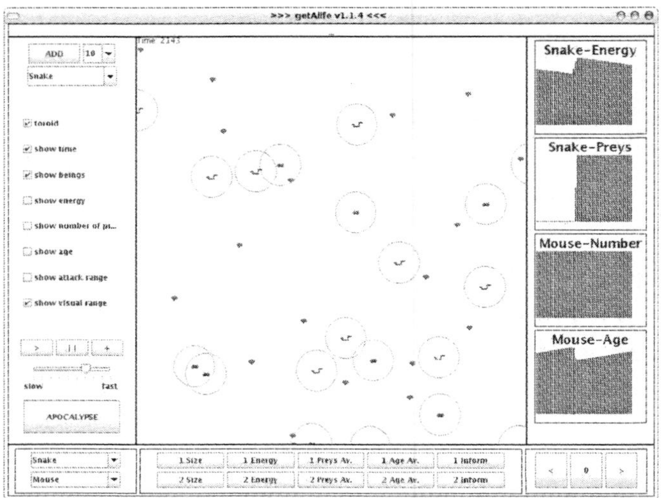

Fig. 2. A screen-shot of the WorldMonitor running a simulation

drawing method, that will be used to draw the individuals of that species, along with some visual aids, such as energy levels, age, caught preys, visual and attack range. The monitor provides several options that allow to enable visual aids, increase/decrease the speed of the world and add new beings. A screen-shot of a simulation is shown in Figure 2.

3.7   Availability

The system and all its source code is available by request to one of the authors. There is a web site (darwin.di.uminho.pt/alife) that makes available an implementation (as a jar file) of the world monitor for demonstration purposes.

4   A Case Study: Evolutionary ANNs

The first application used to illustrate the environment is an example of the definition and empirical evaluation of a decision module based on *ANNs* that are trained based on a evolutionary strategy. A predator species was implemented that uses as its decision module an *ANN*, more specifically a *Multilayer Perceptron (MLP)* with one hidden layer. The *MLP* that is used in the predator's decision module has six input neurons that encode the following information:

- the velocity of the individual;
- distance to the nearest prey;
- direction of the nearest prey;
- the velocity of the prey;
- the angle of the velocity vector of the prey;
- the associated reward.

All the directions are angles relative to the vector of velocity of the individual. On the other hand, the two output neurons encode the absolute value and the angle of the acceleration vector of the individual. All input values were normalized to the range [-1,1] and outputs to [0,1]. Therefore, the final topology of the *MLP* is *6-3-2*.

The *MLPs* are represented using a direct encoding scheme. The only reproduction operator used is the *gaussian* mutation where a noise value is added to a number of weights (randomly selected from 1 to 3)[10]. A new individual is created every time a death occurs, keeping the population size fixed. When this happens, an individual from the population is selected and used as a parent.

An environment was created where a predator (*Snake*) feeds on other two species (*Mouse* and *Plant*), with associated energy rewards (60% and 20%, respectively). The species *Mouse* also feeds on plants, with a 30% energy reward. On Table 1 the features of each species are shown.

Table 1. Main features of the species used in the experiments

Species	Velocity	Acceleration	Visual range	Attack range	Decision
Snake	5	5	25	2	ANN
Mouse	5	5	25	2	Random
Plant	0	0	0	0	Stopped

Two tests were performed using different combinations of preys. For each test, the mean of 30 runs is plotted (each run lasts for 100000 iterations). The world is a two-dimensional 500x500 toroidal square. In Figure 3 the results for a test with 50 snakes and 100 plants are shown; in Figure 4 the world has 50 snakes, 50 mouses and 50 plants. In both cases, the graphs show the number of preys eaten.

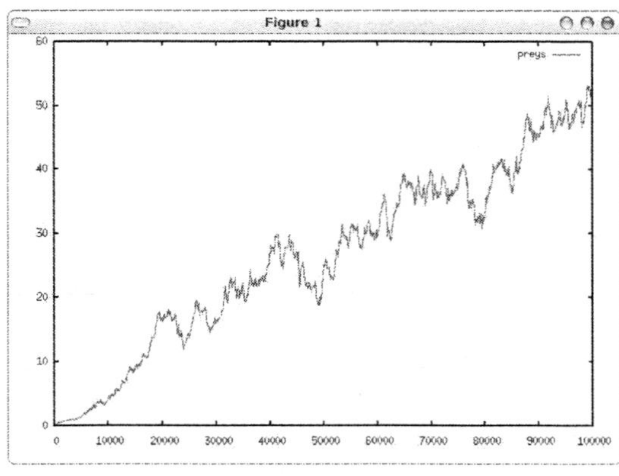

Fig. 3. A graph showing the evolution of the total number of preys (plants) caught by the current generation of snakes. X-axis: number of iterations, Y-axis: average number of preys caught by the predator.

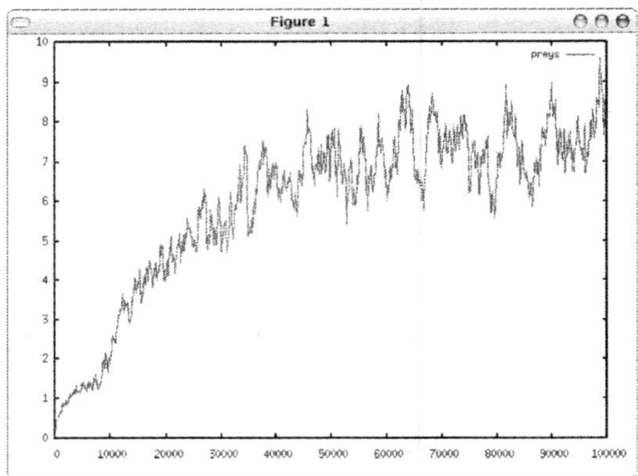

Fig. 4. A graph showing the evolution of the total number of preys (plants and mouses) caught by the current generation of snakes. X-axis: number of iterations, Y-axis: average number of preys caught by the predator.

This simple example shows that the system can be used to evaluate the merits of an evolutionary learning approach. Obviously, other *MLP* encoding schemes and/or reproduction operators can be implemented and compared to this simple strategy. Also, more complex scenarios can easily be considered by making the prey's decision module more complex, both negative or positive rewards (e.g. food/ poison environment), or by having a super-predator that feeds on the predator making the learning task more difficult. Furthermore, environments where two or more species with *MLP* based decision modules can co-evolve as predator and prey or compete as predators for the same preys. These are only a small subset of the experiments that can be conducted using a similar approach to the one described in this case study.

5   A Children's Game

One other quite distinct application is a software project that aims to create a children's game that exploits the concepts of complexity and emergence. Indeed, based on *getALife*, a group of undergraduate students has been developing a children's game named *Species* that allows the users to create predator species by defining a set of IF ... THEN rules in their decision module.

This is done using a visual programming environment where the left side of the rule describes a given scenario and the right side of the rule specifies an action to implement in that case (Figure 5). The species created are evaluated competing against pre-defined preys and getting a score proportional to their energy. The complexity of the game increases with the quality of the opponents prey and also with the degrees of freedom available for the user. In the master levels, the user can even write code to integrate in the decision modules. This code is then compiled dynamically and integrated within the system.

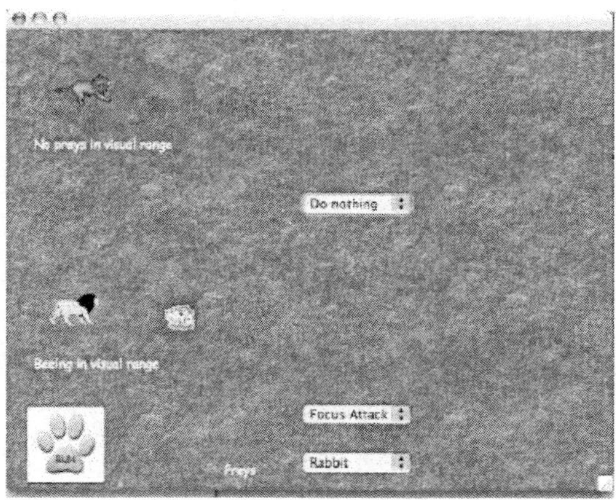

Fig. 5. A screenshot of the *Species* application - selecting a decision module

6   Conclusions and Further Work

This work describes a predator-prey simulator that can be used to evaluate evolutionary approaches to reinforcement learning and agent-based systems in a variety of complex and dynamic environments. The system has the following features, that make it an powerful tool in research and/or teaching scenarios:

- *modularity*: the system is built on a number of modules (e.g. decision modules, genomes, species) that can easily be incremented to achieve distinct environments. This makes it a flexible tool for research.
- *easiness of use*: it is easy to define new species and test them. The need for new code is quite low and the complexity of the software was minimized. This allows its use in teaching scenarios and also for the researcher to concentrate in the algorithms and not on software development.
- *attractiveness*: the system deals with concepts that are inherently interesting and the GUI is attractive to the layman. So, it can be used to exemplify concepts from complex systems and emergence.
- *portability*: it is built in Java, so its portable to the major hardware and operating systems.
- *availability*: the code is available as open-source.

In the future, the system will be developed by adding new functionalities. An important add-on would be to add a communication layer, thus adding a new type of sensorial information, making the tool more powerful to evaluate coordination and cooperation in multi-agent systems.

Acknowledgments

This work was supported by the Portuguese Foundation for Science and Technology under project POSC/EIA/59899/2004, partially funded by FEDER. A special thanks to the students Pedro Silva, Gonçalo Ferreira, Pedro Pinto and Ricardo Lira for their participation in the software development.

References

1. Adami, C., Brown, C.T.: Evolutionary Learning in the 2D Artificial Life System Avida. In: Brooks, R.A., Maes, P. (eds.) Artificial Life IV, pp. 377–381. MIT Press, Cambridge (1994)
2. Benda, M., Jagannathan, V., Dodhiawala, R.: On optimal cooperation of knowledge sources - an experimental investigations. Technical Report BCS-G2010-280, Boeing Advanced Technology Center (1986)
3. Beuster, G.: Artificial Life Environment: A Framework for Artificial LIfe Simulations. In: Proceedings of the 4th German Workshop on Artificial Life (2000)
4. Klein, J.: BREVE: a 3D environment for the simulation of decentralized systemsand Artificial Life. In: Proceedings of Artificial Life VIII, The 8th International Conference on the Simulation and Synthesis of Living Systems, MIT Press, Cambridge (2002)
5. Kok, J., Vlassis, N.: The Pursuit Domain Package. IAS technical report IAS-UVA-03-03, Univ. Amsterdam (2003)
6. Minar, N., Burkhart, R., Langton, C., Askenazi, M.: The Swarm simulation system: a toolkit for building multi-agent simulations. SFI Working Paper 96-06-042, Santa Fe Institute (1996)
7. Moriarty, D., Schultz, A., Grefenstette, J.: Evolutionary algorithms for reinforcement learning. Journal of Artificial Intelligence Research 11, 241–276 (1999)
8. Neves, J., Rocha, M., Rodrigues, H., Biscaia, M., Alves, J.: Adaptive Strategies and the Design of Evolutionary Applications. In: Banzhaf, W., Daida, J., Eiben, A., Garzon, M., Honavar, V., Jakiela, M., Smith, R. (eds.) Proceedings of the Genetic and Evolutionary Computation Conference (GECCO 1999), Orlando, Florida, USA, July 1999, pp. 473–479. Morgan Kaufmann, San Francisco (1999)
9. Ray, T.: Evolution, Ecology and Optimization of Digital Organisms. SFI Working Paper 92-08-042, Santa Fe, NM (1992)
10. Rocha, M., Cortez, P., Neves, J.: Evolutionary Approaches to Neural Network Learning. In: Pires, F., Abreu, S. (eds.) EPIA 2003. LNCS (LNAI), vol. 2902, Springer, Heidelberg (2003)
11. Yaeger, L.: Computational Genetics, Physiology, Metabolism, Neural Systems, Learning, Vision and Behaviour on PolyWorld: Life in a New Context. In: Langton, C.G. (ed.) Artificial Life III, Redwood City, CA, Addison-Wesley, Reading (1993)

An Approach to Efficient Trading Model of Hybrid Traders Based on Volume Discount

Satoshi Takahashi and Tokuro Matsuo

Department of Informatics, Graduate School of Engineering,
Yamagata University, 4-3-16, Jonan, Yonezawa,
Yamagata, 992-8510, Japan
takahashi2007@e-activity.org, matsuo@yz.yamagata-u.ac.jp
http://veritas.yz.yamagata-u.ac.jp/

Abstract. This paper proposes a new cooperation business model in which hybrid traders exist. We define hybrid traders as new traders on the Internet. Hybrid traders can become both buyers and sellers. We assume that hybrid traders do not have enough money. To buy items cheaply, hybrid traders cooperate with other traders. In regard to buying items, we consider a volume discount-based trading. We propose a mechanism in which trader cooperates, buys in a lot of goods, and increases own utility. Our mechanism adopts side payment to promote increasingly cooperation with traders. Cooperative traders commit participation based on a value of side payment. We extend mechanism which hybrid traders deal with multiple items. This mechanism shows new decision of side payment and proposer's strategy.

1   Introduction

In resent years, as e-commerce is developing, researchers regard e-commerce as very important subject of researches, such as, auction [1] and group-buying [2] [3] [4]. End-users can become both buyers and sellers since it is easy for them to open their shops on the web. Such transaction is called B2B(Business to Business)/ B2C(Business to Consumer) [5]. Generally, when end-users open shops on the web, it takes less cost and money. It is easy for consumers to be sellers like a company. In this paper, we call hybrid traders such end-users. When hybrid traders sell items to general consumers, they need purchase items. Hybrid traders do not have enough money to get items. To purchase items at a low price, some traders cooperate with each other on the web since users can communicate with each other easily. Namely, they purchase items as joint capital. In this case, items are sold based on volume discount [6] from sellers such as producers, factories because cooperated traders can purchase a lot of items in one time. If a trader has enough budgets and he/she can purchase a lot of items, price of each item goes down. Although each trader does not enough money, they can purchase in items at a lower price making purchasing community.

When traders cooperate with each other, it is possible to purchase cooperatively cheaper than individually. As the result, each trader's utility increases. If all traders know about types of traders, they make cooperation easily.

We employ side payment mechanism to promote trader's cooperation. If traders are rational, all traders must cooperate with each other. Each trader has a certain participation

N.T. Nguyen, R. Katarzyniak (Eds.): New Chall. in Appl. Intel. Tech., SCI 134, pp. 45–54, 2008.
springerlink.com

incentive based on a valuation of side payment. Side payments are given as cooperation fee from proposing traders to cooperations. If the former's utilities decrease paying side payments to cooperative traders, none search for cooperative agents. This paper proposes a mechanism in which traders' utilities are becoming maximum searching for optimal side payment value. We also propose a decision method of utility maximization and, a mechanism of optimal budget allocation in dealing multiple items.

The rest of this paper consists of the following eight parts. In section 2, we define several terms, assumptions and method of items allocation. In section 3, we define hybrid traders. Then in section 4, we define side payment. In section 5, we propose a mechanism of single item dealing. In section 6, proposing mechanism for expanding multiple items dealing. After that, we discuss about expanding mechanism. Finally, we present our conclusion.

2 Definitions and Assumptions

In this section, we give some definitions of terms and assumptions in our proposed mechanism.

$H = \{h_1, ..., h_i, ..., h_n\}$: A set of hybrid trader with participation of web community.

$A = \{a_1, ..., a_j, ..., a_m\}$: A set of tradable items. All items are sold based on volume discount.

$B = \{b_1, ..., b_i, ..., b_n\}$: Budgets of hybrid traders.

$v_j(\omega)$: A price of item a_j when purchasing with volume discount of the ω phases. ω is parameter that shows number of items which implement discounts.

$p_{i,j}$: A price in which hybrid trader h_i sells item a_j for end-users.

$U_i = p_{i,j} - v_j(\omega)$: A utility when hybrid trader h_i sells item a_j. $(0 \le U)$

$S_{i,n}$: A value of side payment that hybrid trader h_n pays h_i. $0 \le S_{i,n} \le U_n - U'_n$ (U'_n: Utility of h_n with independent transaction)

$Q_{i,j}$: Number of items in which hybrid trader h_i buys in item a_j.

Definition 1. Hybrid traders participate web site community. Hybrid traders deal in this community.

Definition 2. Hybrid trader can purchase all items restricted budgets.

Definition 3. Hybrid traders propose cooperation of buying-in for other hybrid traders.

Assumption 1. Hybrid traders do not have enough money. They do not have enough budgets in which hybrid traders get a grace of volume discount.

Assumption 2. All items are sold as volume discount. Hybrid traders know price and discount ratio of items.

Assumption 3. Items are sold with hopeful price. There are no risk of dealing.

2.1 Volume Discount and Allocation of Items

On above assumption, all items are sold with volume discount. The item price is cheaply by number of items. Figure 1 shows stair-case graph indicating items price in volume discount. It shows prices of items are step function. Increasing number of items, price of each item goes down. Table 1 shows a concrete example of Figure 1. When traders

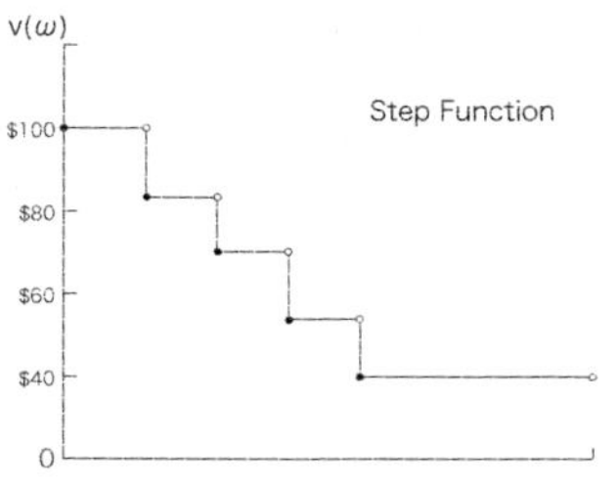

Fig. 1. Volume Discount

Table 1. Value of Item

Quantity	$
1 - 10	100
11 - 50	85
51 -100	70
101 - 200	55
201 -	40

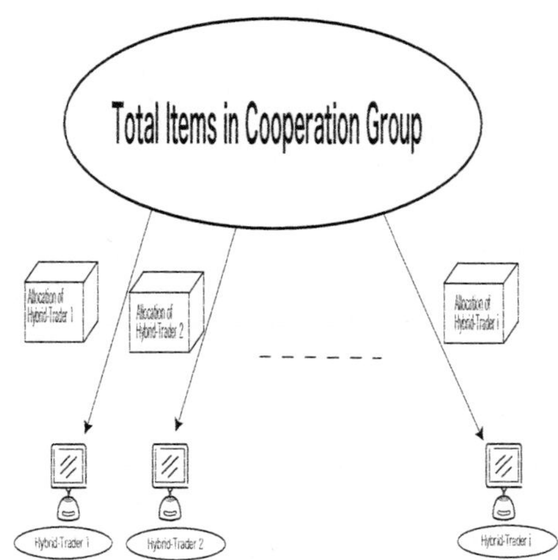

Fig. 2. Allocation of Items

can purchase 11 items, utility calculates on more increasingly about $15 when they purchase only one item.

Hybrid traders sell items with their own gains to end-users. In assumption 1, hybrid traders can not treat a lot of items because they do not have enough money. However, if hybrid traders can cooperate and trade efficiently, they get opportunity of increasing utilities by volume discount. Then, how the items are allocated ?

Hybrid traders allocate the items based on percentage of investing. Total number of allocated items $\sum_{i=1}^{n} Q_i$ equals total number of items bought-in. Figure 2 shows items allocation. Hybrid trader h_i allocates items in total number of items by ratio of investment b_i on total investment $B = \sum_{i=1}^{n} b_i$.

3   Hybrid Trader

In this section, we define hybrid traders. On economic phenomena, we treat dealing between sellers and buyers. But in under continual time, same people sometimes play the seller and the buyer. And the people are end-users who only buy items basically.

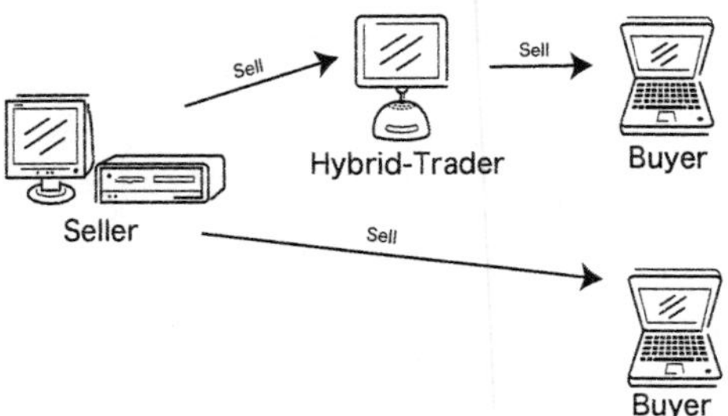

Fig. 3. Hybrid Traders

The sellers are special people who have a certain level of money and procedure. It is difficult to be sellers in which they pay stored cost and advertisement cost without enough money to spend in trading. However, economic activity on the Internet is no cost of their payment. Additionally, users learn indirectly about selling procedure because they use auction and group buying.

End-users do not have enough money for buying-in a lot of items. Traders who have little money can purchase by a pool of capital. One of characteristics of the Internet is that traders can cooperate many and unspecified people. In case of items are sold with volume discount, traders can purchase more items by same budget because a unit price of each item becomes a discounted price.

We define a user who plays seller and buyer as hybrid trader. Figure 3 shows model of transaction environment when hybrid trader stands on their environment. The seller only sells items. The buyer also buys items. Hybrid trader can sell and buy items.

4   Side Payment

In this section, we define side payment institution. We adopt side payment institution for incentive of cooperation to proposer. Side payment is monetary transfer based on some kick-back for cooperation hybrid traders. Incentive of cooperation increases by it. Vendors who pose bid collusion pay kick-back for incentive of cooperation and dropping out of bit collusion. Cooperators can get items cheaply by cooperation of purchasing items. U_i is defined as hybrid trader h_i's utility. U is difference between payments and valuations, such as, $p - v_i(\omega)$. As show in assumption 1, hybrid traders do not have enough money to get items of themselves. In case of existing cooperation traders, a proposer purchases items more cheaply by increasing budget. The proposer's utility is calculated as $U' = p - v_i(\omega)$. His/her utility increases as $U - U' \geq 0$. Consequently, side payment should be paid between 0 to $U - U'$.

5   Single Item Dealing

We handle a situation of single item dealing with value of side payment.

5.1 Cooperation Negotiation Which Depends on Side Payment

All hybrid traders have participation probability depending value of side payment. Hybrid traders are separated into some sets $T_l : \{l = 1, ..., 2, ..., l\}$ by own participation probability. The probability of set T_l is described by function $f_l(s_i)$. s_i is ratio of side payment in which hybrid trader h_i decides. Cooperators decide to participate in purchasing group by that the proposer shows side payment. When value of side payments are increased, proposer h_i's utility reduces due to paying side payment. Instead of this, many cooperators join in the purchasing group. In this case, proposer's utility decreases. If we decide optimal value of side payment, proposer's utility is maximized.

5.2 Dealing Procedure

We shows procedure of single item dealing.

step 1. Hybrid trader h_i is a proposer. h_i proposes about cooperation of purchasing items a_j for other hybrid traders. All traders know discount ratio of items.
step 2. The proposer h_i shows ratio of side payment as s_i.
step 3. Other traders commit participation by side payment.
step 4. The proposer h_i gathers money from purchasing group and purchases the items.
step 5. The items are allocated by each contribution.
step 6. Each trader sells the items by own accountability.
step 7. The proposer pays side payment to all cooperative traders. In this payment, the proposer pays $s_i \cdot (U - U')$ with contribution. A payment $S_n = \{s_i \cdot (U - U')\} \cdot b_n / B$

5.3 Optimization Side Payment

Hybrid traders have participation probability [7] depending on value of side payment. Here, we set up an assumption.

Assumption 4. Proposer knows participation probability.

A proposer can optimize a value of side payment adopting this assumption.

We consider that there are two hybrid traders and a tradable item. Trader h_1 proposes to the other trader cooperating on item a_1. Item a_1 has 3 levels of discount rates. Table 2 shows the item price based on number of items. Trader h_1's budget b_1 is $30. If no traders cooperate, trader h_1 buys 4 items for $28. If trader h_2 who has $3 cooperates with h_1, total budget because $33. Traders h_1 and h_2 buy 11 items and pay for $33. We assume that item's price, where the item is sold to end-users, is $10. Range of side payment is $0 \leq s_1 \leq 4$ per one item.

Table 2. Value of Item 1

Quantity	$
1 - 5	7
6 - 10	5.5
11 -	3

Here, we show a formalization of optimization. Hybrid trader h_1 is the proposing trader. There are n traders in purchasing community and m items. In this condition, trader h_i proposes purchasing item a_j. Items are sold by volume discount. Item's price is $v_j(\omega)\{\omega = 1, ..., \omega', ..., \omega\}$. $v_j(\omega)$ is the cheapest value. Number of l group $T_{\{1,...,2,...,l\}}$ exists with depending probability. All participants reside it. Group $T_{\{1,...,2,...,l\}}$ has participation probability $f_{\{1,2,...,l\}}(s_i)$ depending on side payment. If the proposer gathers n cooperators, he/she decides cooperators based on $n = \sum_{l=1}^{l} f(s_i) \cdot T_l$.

It is possible to calculate the number of traders shown as this formula. The proposing trader uses this formula and decides an optimal value of side payment. When proposer pays side payment, value is $s_i \cdot (U - U')$. U is utility $U_i = \{p_{i,j} - v_j(\omega)\} \cdot Q_{i,j}$ when he/she cooperates. U' is utility of individual dealing.

5.4   Simulation

We simulate based on preceding definition and assumption. We consider hybrid traders who have 3 types of preferences like Figure 4. Figure 4 shows participation probability about changing side payment among $0\% \leq s \leq 100\%$. Type 1 is a group of hybrid traders who has preference which is participation probability rising nonlinearity. Type 2 is a group who has incentive to participate near 0.5. Type 3 is hybrid traders who have participation probability rising linearity. If the value of side payment grows, the cooperator gets less side payment. When the value of side payment is just 0.5, traders who are classified in type 2 participate in cooperation. Table 2 shows concrete values of hybrid traders' utilities. Each trader has budget between \$20,000 and \$200,000. We change ratio of side payment like $0\% \leq s \leq 100\%$. We set up his/her budget based on uniform distribution. Figure 5 shows a visual comparison between proposer's and cooperator's utilities. Table 3, cooperators' utility comes back proposer's utility when investment between $50\% \leq s \leq 60\%$. In this result, the best value of side payment is among $50\% \leq s \leq 60\%$.

Assumption 4 is important condition in this simulation. But, it is difficult to know other traders' type. We propose a method that value of side payment is decided mechanically

Table 3. Utility

side-payment	proposer(\$)	cooperator(\$)
0.0	139	0
0.1	12326	9015
0.2	12631	9281
0.3	12249	9364
0.4	11435	9424
0.5	10274	9478
0.6	8800	9413
0.7	7051	9500
0.8	5054	9482
0.9	2513	9552
1.0	155	9486

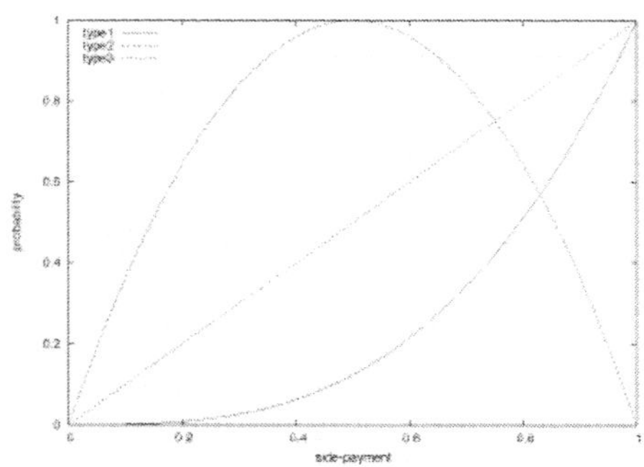

Fig. 4. Type of Hybrid-Traders

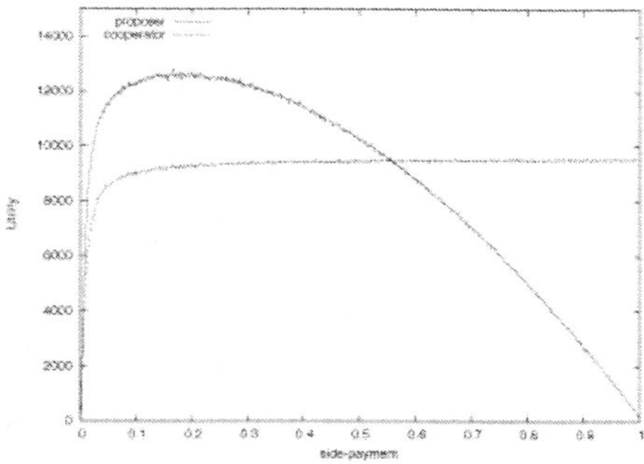

Fig. 5. Result of The Simulation

without assumption 4. Side payment is decided not by depending on side payment but by based on merit of cooperation. Proposer might not gather enough money. But if negotiation communities are made by types of purchasing items, more traders cooperate to purchase items. We separate hybrid traders by their types and create small groups. We define the following definition.

Definition 4. Hybrid traders who use e- commerce site employing our mechanism must propose purchasing items.

Hybrid traders can purchase multiple items on this definition.

6   Multiple Items Dealing

We aspire for fair allocation of payoff by deciding appropriate side payment mechanically. In this section, we consider about a case where a hybrid trader purchases multiple items. Purchasing group should be small set, since dealing in large-scale group is complication and makes computers take a lot of costs to compute the allocation.

6.1   Additional Definition and Protocol

We show an additional definition on dealing multiple items.

$G = \{G_1, ..., G_k, ..., G_l\}$: G is set of hybrid traders. Hybrid traders join a small group. When hybrid trader h_i joins in G_k, h_i do not join in other groups. The following equation shows total sets of groups when there are l groups. $H = \sum_{k=1}^{l} G_k$.

$G' = \{G'_{1,1}, ..., G'_{k,j}, ..., G'_{l,m}\}$: $G'_{k,j}$ is a set in which traders purchases item I_j in small group G_k. $(G'_{k,j} \subseteq G_k)$

Here we show a protocol in trading among multiple hybrid traders on many items.

Protocol
- Hybrid traders are separated by type of purchasing. Hybrid traders have preference about dealing items.

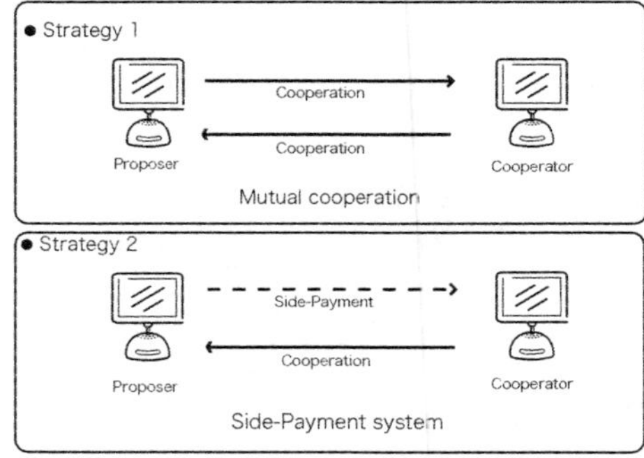

Fig. 6. Strategy of Proposer

- All hybrid traders in a purchasing group must propose purchasing item.
- If there are no cooperator, item are not dealt with traders.
- Side payment is paid after items are allocated.

Further, a proposer has two strategies of cooperation.

1. Cooperating with each other.
2. Using side payment institution.
- In strategy 1, side payments are not paid.

Hybrid traders in group must propose purchasing items. Everyone can cooperate on those items. When a trader does not want to pay side payment, they take cooperation instead of paying it.

We define formula of determine from of side payment mechanically as following formula.

$$S_{i,n} = \{\underset{i \notin G'_j}{v_j(\omega')} - \underset{i \in G'_j}{v_j(\omega)}\} \cdot Q_{i,j}$$

Our mechanism calculates differences between purchasing price in case that when hybrid trader h_i participates cooperation group and purchasing price in case that hybrid trader h_i does not participate cooperation. The mechanism also computes multiplication number of purchasing item by h_i $(\omega' \leq \omega)$. This method can restrain that participants who joins in a cooperating group when purchase price is minimum.

We show proposer's strategy. We adopt not only side payment but also kick-back which increases utility. Figure 6 shows an example of proposer's strategy. $Strategy1$ is a strategy of cooperation with each other. Proposer takes not to pay side payment but to cooperates against cooperator. Side payment is not occur. $Strategy2$ is a side payment institution. Proposer purchases items more cheaply and aspires increasing utility by using two strategies.

It is difficult for free riders to increase utility by deciding side payment automatically. Proposing by all hybrid traders can restrain free rider.

Table 4. Value of Items

Quantity	a_1	a_2	a_3
1 - 10	$100	$80	$50
11 - 20	$70	$60	$40
21 - 50	$50	$45	$30
51 -	$35	$30	$20

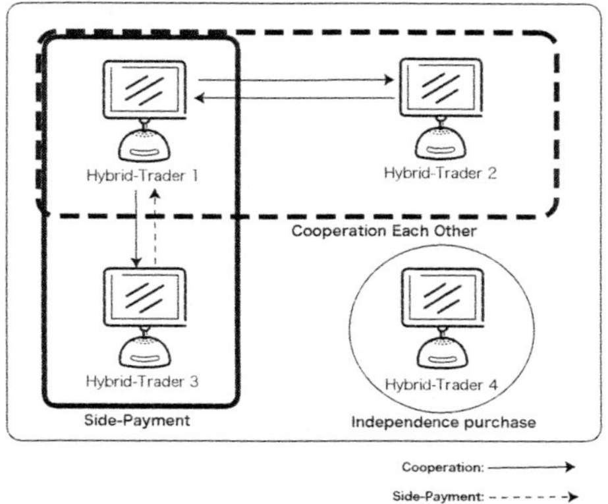

Fig. 7. Cooperation Negotiation with Small Group

7 Discussion

7.1 Example of Multiple Items dealing

We shows an example of multiple items dealing. Four hybrid traders H=h_1, h_2, h_3,h_4 including in group G_1. Three items $A = \{a_1, a_2, a_3\}$ are tradable. Table 3 shows each price and volume discount prices. Each first price of items is $v_1(1) = \$100$, $v_2(1) = \$80$, $v_3(1) = \$50$. Each hybrid traders budget is $b_1 = \$1,000$, $b_2 = \$700$, $b_3 = \$500$, $b_4 = \$600$. Proposing items of each trader are $h_1 : a_1, h_2 : a_1, h_3 : a_3, h_4 : a_2$. Figure 7 shows cooperation and collateral after negotiation among them. Hybrid traders h_1, h_2 propose same items. They cooperate with each other. h_1 cooperates with h_3 and pays $180. h_3 selects side payment against cooperating with h_1. h_4 does not purchase when item's price is more expensive than his/her budget because no one cooperates. Hybrid trader h_1 gets 16 of item a_1 and 6 of item a_3. Hybrid trader h_2 gets 14 of item a_1. Trader h_3 gets 16 of item a_3. Each purchasing prices is $p_1 = \$120, p_2 = \$100, p_3 = \$60$. h_1's utility is $1,100. h_2's utility is $880. h_3's utility is $480. In this transaction, h_1 gets side payment from h_3. Side payment value is $S_{1,3} = (\$50 - \$30) \cdot 3 = \$60$. h_3 pays $60 to h_1. It is trivial to increase hybrid traders' utilities. To buy multiple items are more increasingly than to buy single item. It is important to allocate budget to buy multiple items.

In multiple items dealing, allocation of budget is one of important problems regarding as volume discount-based trading. Purchasing multiple items is more increasingly of utility than purchasing only one item. A simple method for allocation of budgets is that the mechanism divides number of items from total budgets and prepares each divided budget to purchase items. However, the above mechanism sometimes allocates inappropriate items where users' payoffs decrease. Thus, we consider generalized trading of combinatorial items and budgets where users' utilities are maximization/semi-maximization.

8 Conclusion

This paper shows a mechanism of dealing with hybrid traders who play both sellers and buyers. We defined cooperation dealing that hybrid traders cooperate with each other. We also defined side payment as incentive of cooperation. In single item dealing model, we proposed a method of deciding optimal side payment with restricted assumption. In multiple items dealing, we proposed a method of decision side payment mechanically. We showed restraining decreasing proposer's utility and fee rider problem in the cheapest price by decision of kick-back. Our future work includes decision method of the best side payment transfer method without assumptions, combinatorial items and budgets, and a method of restraining free rider completely.

References

1. Leyton-Brown, K., Shoham, Y., Tennenholtz, M.: Bidding clubs: institutionalized collusion in auctions. In: Proceedings of the 2nd ACM conference on Electronic commerce, pp. 253–259 (2000)
2. Yamamoto, J., Sycara, K.: A stable and efficient buyer coalition formation scheme for e-marketplaces. In: Proceedings of the fifth international conference on Autonomous agents, pp. 576–583 (2001)
3. Ito, T., Ochi, H., Shintani, T.: A Group Buying Protocol based on Coalotion Formation for Agent-mediated E-Commerce. International Journal of Computer and Information Science (IJCIS) (2002)
4. Li, C., Sycara, K.: Algorithm for combinatorial coalition formation and payoff division in an electronic marketplace. In: Proceedings of the first international joint conference on Autonomous agents and multiagent systems: part 1, pp. 120–127 (2002)
5. Turban, E., Lee, J., King, D., Chung, M.: Electronic Commerce:A Managerial Perspective. Pearson Education, London (2000)
6. Matsuo, T., Ito, T.: A decision support system for group buying based on buyers' preferences in electronic commerce. In: The proceedings of the Eleventh World Wide Web International Conference (WWW 2002), pp. 84–89 (2002)
7. Varian, R.H.: Intermediate Microeconomics A Modern Approach, 7th edn. W. W. Norton and Company (2005)

Improving the Efficiency of Low-Level Decision Making in Robosoccer Using Boosted SVM

Pravin Chandrasekaran and R. Muthucumaraswamy

Department of Information Technology,
Sri Venkateswara College of Engineering,
Sriperumbudur, India 602105
cp86in@yahoo.co.in, msamy@svce.ac.in

Abstract. Decision making in a multi-agent environment has continued to be one of the most formidable challenges for AI researchers. Increasing the efficiency of predictors is an essential task, especially in a substrate like robosoccer where a misclassification can cost dearly. It is also necessary for the agent to perform well, irrespective of the nature of testing data, generated in a markovian fashion. For this reason, we apply AdaBoost (Adaptive Boosting) algorithm onto support vector machines which helps us achieve a better generalization performance than normal Support Vector Machines (SVM) and a better efficiency when compared to other adaboosted neural networks. To illustrate the concept, we propose a highly efficient decision predictor for low-level behavior in robosoccer using Adaboosted SVM (AdSVM). Through experiments, we have proved that the proposed agent model has outwitted existing neural networks and SVM in classifying two-class data of any nature in a multi-agent environment like robosoccer.

1 Introduction

Robosoccer provides a good platform for developing multi-agent behavior set up. It provides us the freedom of using agents to exercise low-level, high-level, team and collaborative behaviors. In the past, researchers like Stone et al have proposed concepts like 'Keepaway' which helped to explore different kinds of behaviors among the agents. Researchers have also proposed valid concepts for mastery of both high skilled and low skilled behaviors in a multi-agent environment [1] [2] [3] [4] [5]. Unlike in the case of face recognition, data sets in robosoccer cannot be predicted on the nature of testing data as it is generated on the basis of markovian property. A misclassification will have ramifying effects on the proceedings of the game. Imagine a situation wherein an agent is expected to shoot the ball into the goal post. A 'SHOOT' decision would end the episode but on the contrary, a 'PASS' decision will extend the episode further. In any decision making classifier, the testing performance depends on the nature of training data. Therefore, we need a classifier that can exhibit good accuracy irrespective of the nature of training data and testing data. Support vector machines will qualify for the requirement as their performance is outstanding when compared to other neural networks. But these machines show a poor generalization performance when it comes to an imbalanced classification. The solution to this problem would be a boosting algorithm. By using support vector

N.T. Nguyen, R. Katarzyniak (Eds.): New Chall. in Appl. Intel. Tech., SCI 134, pp. 55–64, 2008.

machines as component classifiers, it is possible to show a better generalization performance when compared to normal SVMs in an imbalanced data set and an augmented efficiency in comparison with other Adaboosted neural networks. Since SVM already happens to be a strong classifier, we generate weak SVM classifiers by increasing the value of σ and then linearly summing up their hypothesis to get the final classifier which is a highly efficient one. This methodology provides us with the benefits of both support vector machines and Adaboost algorithm. The proposed model receives input from the physical environment and deduces the available data into feature sets which are then fed to the AdSVM classifier to select between the given two actions i.e pass or shoot. Here it is a simple case of dichotomization. The paper is organized as follows: Section 2 reports the feature representation. Section 3 explains our proposed work. In section 4 we present our algorithm for the proposed predictor. Experiments and results have been discussed in section 5. Finally section 6 concludes the paper with a few suggestions on future work.

2 Feature Representation

In this paper, we have represented data in the form of feature sets, as it is easy to classify data using SVM in this format. In case of robosoccer, low-level decisions are taken based on the situation of their agents; hence it is important to consider a whole lot of factors while selecting the features. For example, an agent decides to pass the ball or shoot, based on factors like the number of opponents, number of teammates around it and their proximity. Now it is important to define a methodology to find out if any agent is within the vision of the ball processing agent A.

2.1 Vision of an Agent

The vision is defined by a triangle that is formed between the agent and any two points (Pt1, Pt2) in the field boundaries including the goal area, with angle $\alpha = 45^{\circ}$ at vertices A. The two vertices are arbitrarily chosen such that the angle at A is 45. Such an arrangement is depicted in Fig 1. In Fig 1 the vision triangle is formed by the vertices A, Pt1, Pt2. Though there are many methodologies proposed for finding a point inside a triangle, we restore ourselves to the Barycentric Technique which is the most efficient and swift one in terms of calculation. In this technique, one of the vertices of the triangle (preferably agent in our case) is chosen and we consider all other locations on the plane as relative to that point. Now we can get to any point on the plane just by starting at A and walking some distance along (Pt1 - A) and then from there walking some more in the direction (Pt2 - A).

Therefore, any point in the plane can be described by the expression.
P = A + u * (Pt1 - A) + v * (Pt2 - A), where u and v are vectors along the line APt1 and APt2.

With the above expression, it is quite easy to find out if the point exists inside or outside the vision of an agent.

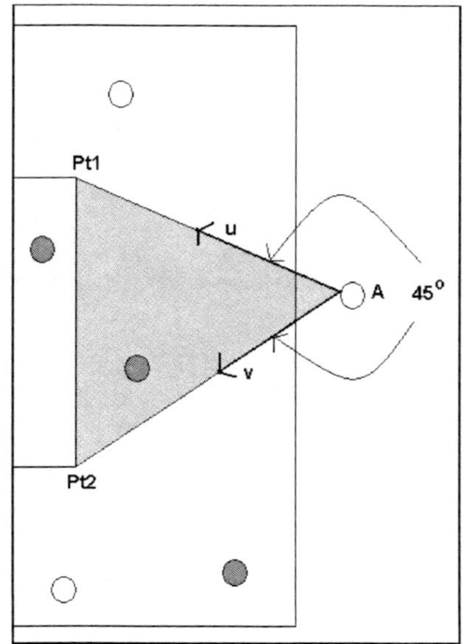

Fig. 1. Vision of an agent

```
// Compute vectors
v0 = Pt2 - A
v1 = Pt1 - A
v2 = P - A

// Compute dot products
d0 = dotproduct (v0, v0)
d1 = dotproduct (v0, v1)
d2 = dotproduct (v0, v2)
d3 = dotproduct (v1, v1)
d4 = dotproduct (v1, v2)

//Compute barycentric coordinates
iD = 1 / (d0 * d3 - d1 * d1)
u = (d3 * d2 - d1 * d4) * iD
v = (d0 * d4 - d1 * d2) * iD

// Check if an agent is inside the
vision triangle of A
return (u>0)&&(v>0)&&(u+v<1)
```

Fig. 2. Pseudo code for Barycentric Technique

The pseudo code is given in Fig 2, which also defines the condition for an agent to be inside the vision of the ball processing agent.

2.2 Feature Set

Now we define the list of features that we have used in the model.

D: The distance from the center of the goal post to the ball processing agent. This feature plays an important role in deciding between the two actions - PASS, SHOOT.

T_V: This feature represents the number of teammates within the vision of the ball possessing agent. It is used to decide on the reliability of a successful pass or to initiate a collaborative formation among its teammates.

O_V: This feature indicates the number of opponents who are inside the vision of the ball processing agent. It greatly influences the decision of the agent, as a wrong pass can overturn the chances of a goal.

T_V_I: It interprets the number of defenders who fall inside the quadrant of the ball processing agent. The number signifies the reliability of its teammates based on their proximity.

O_V_I: This feature indicates the number of offenders in the same quadrant. It expresses the threat level presented by the offenders that lie in the same quadrant as the ball-possessing agent. Though it does not signify the nearest offenders, it gives an overall idea on the formations of opponents near the ball possessing agent's position.

All the above features represent the overall topology of the robosoccer play area and present a clear picture on the threat levels around the ball processing agent for both offensive and defensive behaviors.

3   Adaboosted SVM-Based Decision Classifier Model

In this paper, we propose the model for low-level decision making in robosoccer using an adaboosted SVM. Firstly, we deduce the Feature Vector (FV) from raw data or the input from physical environment. After the creation of FV's, they are fed to the Feature Vector Filter (FVF) module. These two modules have been added for convenience of experimentation. Finally, the FV's obtained from FVF are given as input to the trained AdSVM decision classifier. The hierarchical setup of the decision model is shown in Fig 3.

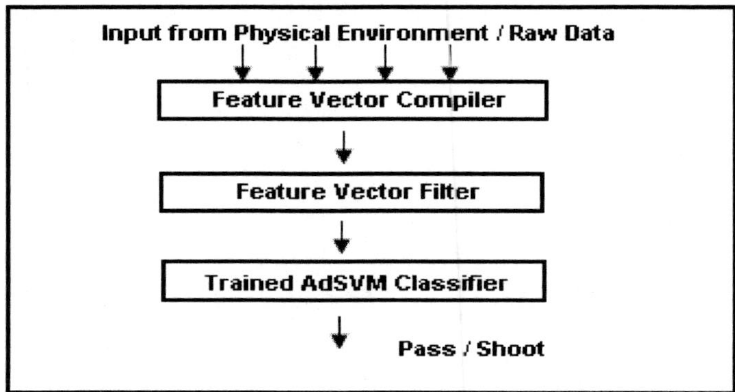

Fig. 3. Architecture of AdSVM Decision Making Model

3.1   Feature Vector Compiler

In this module, we take the sensory input data from the physical environment and extract information like the distance of ball processing agent from center of goal post, number of teammates, opponents and their proximity. As the raw data cannot be used with SVM, we reduce it to a form that is suitable for training and testing these data. In Feature Vector Compiler, we use Barycentric technique to find out if agents are within the vision of the ball possessing agent. The FV is then passed to the next module in pipeline, the Feature Vector Filter. The eligible FV's are filtered and then passed to the AdSVM Decision Classifier.

3.2   Feature Vector Filter

The FV obtained from Feature Vector Compiler is then fed to the FVF, which allows only those FV's that lie within the scope of actions pass and shoot. For example, if there is a situation where the vision area of the ball possessing agent has three or more opponents and no teammates, then neither pass nor shoot will prove fruitful. Since the AdSVM that we use performs dichotomization, we recommend default actions for the filtered FV's. The pseudo code for the FVF module is given in Fig 4.

We use Feature Vector Filter to handle those situations that cannot be resolved through the proposed classifier. When such a rare condition arises, the agent is initially recommended to scan its vicinity, so as to find an environment that is suitable

```
If (T_V>0 || O_V<=2)
          Forward FV to AdSVM Decision Classifier.
Else
   Repeat (for every turn of angle (α+45)°check T_V & O_V)

      If (T_V>0 || O_V<=2)

          Forward FV to AdSVM Decision Classifier.

      Else If (T_V==0 && O_V>2)
           Turn by angle (α+45)°  //action for filtered FVs

Until (one complete rotation)

//If (T_V==0 && O_V>2) after one complete rotation
SHOOT the ball at random    //action for filtered FVs
```

Fig. 4. Pseudo code for Feature Vector Filter

for AdSVM to make a decision. If it fails to meet the condition even after one complete rotation, the agent shoots the ball at random.

3.3 Trained AdSVM Classifier

AdSVM is a classifier that is obtained by applying SVM onto AdaBoost algorithm. In this section we describe a boost based learning or training method to construct a classifier. The aim is to develop a classifier that can perform well irrespective of the nature of training and testing data. It is necessary for the classifier to minimize the error rate and exhibit a good generalization performance.

Boosting Technique: As the name suggests, these techniques are used to enhance the performance of weak classifiers [6] [7] [8]. In the past, boosting algorithms have been successfully applied to neural networks so as to improve their accuracy of classification. Adaboost is a boosting by resampling technique that adjusts adaptively to the errors of the weak hypothesis returned by the weak learning model. An AdaBoost algorithm learns by laying emphasis on the misclassified patterns. Initially, the algorithm sets an equal distribution to all data points in the training set. A subset of that data is selected and is used to train the weak classifier. Now in the forthcoming iterations the algorithm increases the weight of those data points that were misclassified in the previous rounds. This ensures that more attention is given to those data points which are hard to classify. An important property of AdaBoost is that if the weak learners consistently have accuracy only slightly better than half, then the error of the final hypothesis drops to zero exponentially fast. In our model, we use support vector machines as component classifier in AdaBoost algorithm.

Support Vector Machines as Component Classifiers: The idea of support vector machines is to construct a hyper plane as the decision surfaces in such a way that the margin of separation between positive and negative examples is maximized [9] [10]. It is an implementation of the method of structural risk minimization. Even though SVM relies on a linear discriminator, it is not just confined to making linear hypothesis. It is also possible to perform non-linear regression and classification tasks with these machines. In case of non-linear decisions, the data are mapped to a higher dimensional space where they are linearly classified. There is a way to avoid performing these mappings explicitly using a kernel trick. The basic advantage of using kernel functions is to avoid having to perform a mapping and then a linear classification in a higher dimensional feature space. Instead, the required scalar products are calculated directly by computing the kernels which are a function of input vectors. By using the kernel functions, one can avoid the curse of higher dimensionality. For example, if the input dimensionality space is as high as 256, then the decision surface has to be constructed in a billion-dimensional feature space. The computational complexity is a big overhead in such cases. The kernels used primarily depend on the classifier being used. The kernels used are of the following 3 types:

- Polynomial
- Gaussian RBF
- Sigmoidal

The polynomial kernel functions are given by

$$K(x,x_i) = [1+x^T x_i]^d \tag{1}$$

Sigmoidal kernel function for multi layer perceptrons are given by

$$K(x,x_i) = tanh(x^T x_i) + b \tag{2}$$

And finally, the kernel function that is popular and widely used is the Gaussian or ("Radial Basis Function") as it is more adaptive to dimensions when compared to its counterparts.

$$K(x,x_i) = exp(-1/2\sigma^2 ||x-x_i||^2) \tag{3}$$

where x,x_i are feature vectors in the input space while σ is the Gaussian width.
In this paper, we use the Gaussian kernel or Radial Basis Function kernel. Gaussian kernel allows us to control the performance of support vector machines using two parameters namely σ and C, Gaussian width and regularization parameter respectively.

AdSVM: For two class problems, there is a given set of N labeled training examples $(y1, FV_1)$, ..., (y_N, FV_N) , where $y_i \varepsilon$ {PASS,SHOOT} is the class label associated with feature vector FV_i. As we had seen earlier, support vector machines are very strong classifiers. Hence with this level of accuracy, it is not possible to use it in AdaBoost algorithm. For this reason, we weaken the support vector machines by using a relatively large σ value which corresponds to a SVM, with RBF kernel, with relatively weak learning ability. Now the weak SVM is used as a component classifier. During every iteration, if ε_t is greater than 0.5, then the classifier is disregarded, and the value of the Gaussian width is decreased by some fixed constant. It is important to note that the constant, which is used to reduce σ, directly influences the number of iterations (T). During each epoch, we calculate the weight of component classifier (α_t). At the same time, we update the weights of the feature vectors (D_t) in the training data subset. Now we run the algorithm till T times, (i.e T classifiers are constructed) as given in Section 4. Finally, during the testing phase, we feed the FV provided by the Feature Vector Filter module to the AdSVM classifier which is a linear summation of all the hypothesis of component classifiers. The final outcome is an action (PASS/SHOOT) that is recommended by most number of component classifiers. This is computed by summing up the values of α_t (classifier weights) for each action.

4 Algorithm

Training:
   Input:
1.   N feature vectors of form (y_1, FV_1) , ..., (y_N, FV_N) , where y ε {PASS,SHOOT}
2.   Weak SVM as component classifier.
3.   Number of iterations T.
   Initialize D_1 (i) = 1/N., i=1,......, N
   Do for t=1,..., T

    I.   Use SVM with RBF kernel to train a component classifier on the weighted set D_t.

    II. Calculate error rate ε_t

$$\varepsilon_t = \Sigma_{i=1}^{N} D_t(i), \ y_i \neq h_t(x_i)$$

    III. If $\varepsilon_t > 0.5$, then discard classifier, reduce σ by fixed constant and repeat from step I.

    IV. Set component classifier weight : $\alpha_t = \frac{1}{2} (\ln (1 - \varepsilon_t)/ \varepsilon_t)$

    V. Update the weights

$$D_t(i+1) = (D_t(i)*\exp(-\alpha_t y_i h_t(x_i)))/ \ Z_t$$, where Z_t is a normalization constant.

Testing: Feature Vector from Feature Vector Filter.
    Obtain total vote received by actions pass and shoot

$$V_{shoot} = \Sigma_{t:ht(x)=shoot} \ \alpha_t$$
$$V_{pass} = \Sigma_{t:ht(x)=pass} \ \alpha_t$$

    If ($V_{shoot} > V_{pass}$)
        Execute SHOOT
    Else
        Execute PASS

5 Performance Comparisons

We tested our model with data of both balanced and imbalanced nature. We created two training datasets. In one of them, we took more number of PASS samples when compared to SHOOT and trained the AdSVM classifier. In another dataset, actions of equal distribution were taken.

5.1 Imbalanced Classification

In the imbalanced dataset, we tested the generalization performance of the normal SVM and compared it with that of AdSVM classifier. The classification performance of the normal SVM degraded to a random guess, as we reduced the SHOOT keeping the PASS samples constant. On the other hand, AdSVM maintained a substantial generalization performance when compared to the normal SVM. The results are shown in Fig 5.

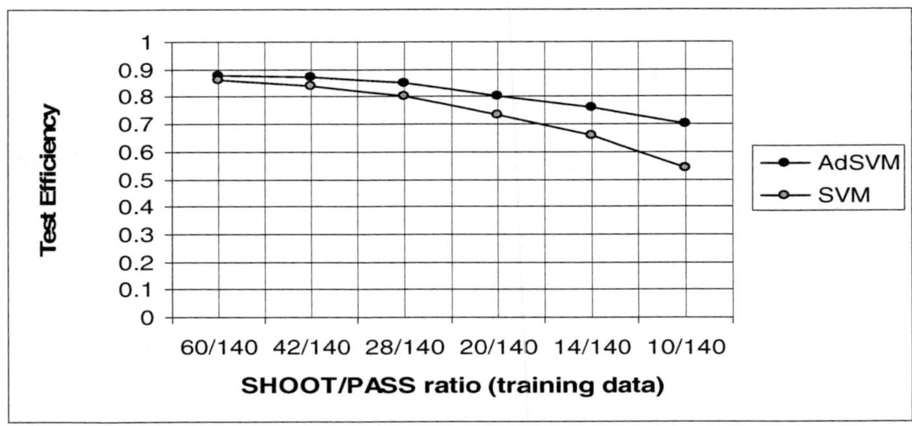

Fig. 5. Generalization performance: normal SVM Vs AdSVM

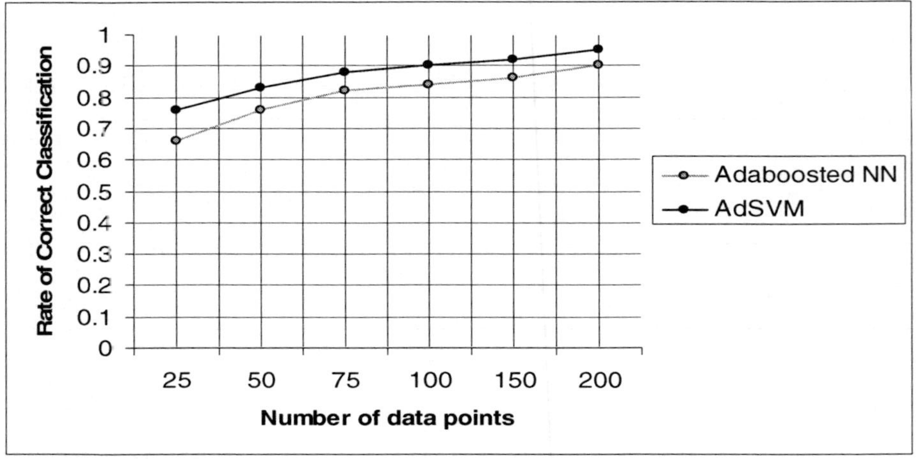

Fig. 6. Comparison of efficiency of AdSVM with Adaboosted Neural Networks

5.2 Normal Dataset

In the normal dataset, we tested our system with both AdSVM and other adaboosted neural networks. The results as shown in Fig 6 clearly indicate that the test efficiency

Table 1. Data Samples from testing data set

D	T_I	O_I	T_I_Q	O_I_Q	Pass/Shoot
10	2	0	1	0	Pass
15	2	1	1	1	Pass
6	0	2	0	1	Shoot
12	3	2	2	1	Pass
2	2	0	0	0	Shoot
18	1	2	1	1	Pass
3	1	0	1	0	Shoot
11	1	1	1	0	Pass
20	2	1	1	1	Pass
16	3	1	2	0	Pass

of AdSVM decision classifier is slightly better than the other adaboosted Neural Networks. It is important to note that the curves shown in the graphs are subjective to value of T, which indicates the number of iterations in the AdSVM algorithm or the number of SVM component classifiers built in the algorithm. Thus from the set of experiments performed, it is quite clear that the proposed AdSVM model exhibits characteristics that are quite required for an ideal decision classifier in a multi-agent environment like robosoccer.

6 Conclusion

In this paper, we have proved that by using SVM as component classifiers in Adaboost algorithm, it is possible to build a highly efficient agent classifier in a multi-agent environment. In the process, we have established the fact that irrespective of the nature of training and test data, the proposed classifier exhibits a better performance when compared to all its counterparts. To the best of our knowledge, it is the first time in the history of a multi-agent environment like robosoccer, a boosted algorithm has been roped in for decision making. Though the classifier has been tested effectively in simulation league variant of robosoccer, it is not clear on how it would adapt itself to other leagues like small, middle and large leagues of robosoccer, where real robots are engaged in play. We intend to test the same set-up for a multi-class decision making. Implementing it in a large scale would be a formidable challenge for researchers in autonomous agents and machine learning.

References

1. Stone, P., Veloso, M.: Using Decision Confidence Factors for Multiagent Control. In: International Conference on Autonomous Agents Proceedings of the second international conference on Autonomous agents (1998)
2. Stone, P., Veloso, M.: A Layered Approach to Learning Client Behaviors in the RoboCup Soccer Server. Applied Artificial Intelligence 12 (1998)

3. Whiteson, S., Stone, P.: Concurrent Layered Learning. In: Proceedings of the Second International Joint Conference on Autonomous Agents and Multi-Agent Systems (AAMAS 2003), Melbourne, Australia, July 2003, pp. 193–200 (2003)
4. Kalyanakrishnan, S., Liu, Y., Stone, P.: Half Field Offense in RoboCupSoccer: A Multiagent Reinforcement Learning CaseStudy. In: Proceedings of the RoboCup International Symposium (2006)
5. Stone, P., Sutton, R.S., Kuhlmann, G.: Reinforcement Learning for RoboCup Soccer Keep away. International Society of Adaptive Behavior (2005)
6. Schapire, R.E., Freund, Y.: Boosting the margin: a new explanation for the effectiveness of voting methods. The Annals of Statistics 26(5), 1651–1686 (1998)
7. Freund, Y., Schapire, R.: A decision theoretic generalization of on-line learning and an application to boosting. Journal of Computer and System Sciences 55(1), 119–139 (1997)
8. Dietterich, T.G.: An experimental comparison of three methods for constructing ensembles of decision trees: Bagging, boosting, and randomization. Machine Learning 40(2), 139–157 (2000)
9. Scholkopf, B., Sung, K.-K., Burges, C., Girosi, F., Niyogi, P., Poggio, T., Vapnik, V.: Comparing support vector machines with Gaussian kernel to radial basis function classifiers. IEEE Transactions on SignalProcessing 45(11), 2758–2765 (1997)
10. Valentini, G., Dietterich, T.G.: Bias-variance analysis of support vector machines for the development of svm-based ensemble methods. Journal of Machine of svm-based ensemble methods (2004)

Part 2
Personal Assistants and Recommender Systems

Closed Pattern Mining for the Discovery of
User Preferences in a Calendar Assistant

Alfred Krzywicki and Wayne Wobcke

School of Computer Science and Engineering
University of New South Wales
Sydney NSW 2052, Australia
{alfredk,wobcke}@cse.unsw.edu.au

Abstract. We use closed pattern mining to discover user preferences in appointments in order to build structured solutions for a calendar assistant. Our choice of closed patterns as a user preference representation is based on both theoretical and practical considerations supported by Formal Concept Analysis. We simulated interaction with a calendar application using 16 months of real data from a user's calendar to evaluate the accuracy and consistency of suggestions, in order to determine the best data mining and solution generation techniques from a range of available methods. The best performing data mining method was then compared with decision tree learning, the best machine learning algorithm in this domain. The results show that our data mining method based on closed patterns converges faster than decision tree learning, whilst generating only consistent solutions. Thus closed pattern mining is a better technique for generating appointment attributes in the calendar domain.

Keywords: Data mining, closed patterns, Formal Concept Analysis, calendar assistants.

1 Introduction

We are interested in the problem of providing automated assistance to the user of a calendar system to help in defining appointments. In a calendar application, the user may initially specify some of an appointment's attributes (e.g. title and day), and the task of the system is to suggest any or all of the remaining attributes (e.g. time and location). What makes this problem difficult is that both the set of attributes given initially by the user and the set of attributes that can be suggested are not fixed; some appointments may contain only basic information such as the title, time, date and duration, while others may have additional attributes specified, such as the location, the attendees, etc. Furthermore the attributes mutually constrain one another.

A calendar appointment can be regarded as a structured solution. A problem requires a *structured solution* if the solution has the form of a set of components that are constrained by other components in the solution. In many practical systems, the challenge of building a consistent solution is solved by defining a set of rules that describe the constraints between components, McDermott [8]. In the calendar domain, the solution "components" are attributes with their values, and the "constraints" are provided by a model of the user's preferences. Additional constraints affecting the solution are the presence of other appointments, dependencies between attributes and other user

N.T. Nguyen, R. Katarzyniak (Eds.): New Chall. in Appl. Intel. Tech., SCI 134, pp. 67–76, 2008.
springerlink.com

knowledge not directly represented in the calendar system. For example, an appointment time and duration may depend on the availability of attendees and the meeting location. These dependencies are not given explicitly, but may be represented in the form of patterns.

In this paper, we investigate the use of closed pattern mining to discover user preferences over calendar appointments and to build structured solutions for a calendar assistant. Traditionally the aim of data mining is to discover association rules, Agrawal and Srikant [1]. We found, however, that mining association rules is not the most suitable method for applications with real-time user interaction, due to the potentially large number of frequent patterns and the number of rules that can be generated from each pattern. In contrast, the number of *closed* frequent patterns can be an order of magnitude smaller than the number of frequent patterns. In fact, all frequent patterns can be generated from a complete set of closed frequent patterns. The data mining algorithm used in this paper is based on the FP-Growth algorithm introduced by Han, Pei and Yin [6] and implemented by Coenen, Goulbourne and Leng [3]. In order to find closed frequent patterns, we filter out all non-closed patterns as they are computed by the FP-Growth method. Details of the pattern mining algorithm can be found in Section 3. Discovered frequent patterns are treated as possibly inconsistent fragments of different solutions that need to be integrated into consistent suggestions before presenting them to the user. We found that it is best to create solutions only from non-conflicting patterns, which makes generated solutions less likely to conflict with user preferences.

The method for pattern selection and the support required for pattern generation were determined from the results of simulated user sessions. The simulation enabled us to compare the accuracy of our appointment prediction method with the best machine learning technique, decision tree learning, on realistic calendar data taken from a user's diary for a 16 month period. We present the results of the comparison and discuss some advantages of our pattern mining approach over decision tree learning

The remainder of this paper is organized as follows. In the next section, we provide the formal framework for the problem of generating structured solutions in the calendar domain. Section 3 describes our data mining and solution generation method, which is evaluated and compared with other methods in Section 4. Section 5 contains a discussion of related research.

2   Formal Problem Statement

This section provides definitions specific to the problem of closed pattern mining for generating structured solutions in the calendar domain.

Definition 1. *Let $A = \{a_1, a_2, \ldots, a_n\}$ be a set of n attributes used in all appointments. Let each attribute a_i have a set of values V_{a_i} specific for the domain of the attribute. For example, $V_{day} = \{Sunday, Monday, \ldots, Saturday\}$. A **feature** is an attribute-value pair $\{a_i, v_{ij}\}$, where v_{ij} is an element of V_{a_i}. The set of all features is denoted $\mathbf{I}$.*

Definition 2. *A **data case** or **case** is a nonempty set of features stored in the database of cases, e.g. $\{\{a_{i_1}, v_{i_1 j_1}\}, \ldots, \{a_{i_m}, v_{i_m j_m}\}\}$. An attribute may appear in a case at most once and may not occur at all.*

For example, a single appointment stored in the calendar database is a data case.

Definition 3. *A **solution** is a potential data case created by the system. A number of solutions can be selected by the system from a set of solutions and presented to the user as suggestions for consideration.*

Definition 4. *A **pattern** is any part of a data case, a set of features, containing at least one feature. Solutions/cases may contain more than one pattern.*

Definition 5. *Two features are **overlapping** if they have the same attribute.*

Definition 6. *Two features are **conflicting** if they have the same attribute with different values.*

Definition 7. *Two patterns are **conflicting** if they contain at least one pair of conflicting features.*

Definition 8. *Two patterns are **overlapping** if they contain overlapping features.*

We also call two conflicting patterns **inconsistent**. It is worth noting that conflicting features/patterns are always overlapping, therefore the "no overlap" condition is stronger than the "no conflict" condition in the solution generation algorithms below.

The underlying theory of closed patterns is based on Formal Concept Analysis, Wille [10]. Pasquier *et al.* [9] extended the theory and introduced the idea of *closed patterns*, applying Formal Concept Analysis to data mining. The key terminology of this theory is summarized below, slightly adjusted for consistency with the above definitions.

Definition 9. *A **data mining context** is a triple $D=\langle O, I, R\rangle$, where O is a set of objects, I is a set of features and $R \subseteq O \times I$ is a binary relation between objects and features. The fact that object o has feature i can be expressed as $(o, i) \in R$.*

Definition 10. *Let $D=\langle O, I, R\rangle$ be a data mining context and let $O \subseteq O$, $I \subseteq I$. The f and g functions map powersets $2^O \rightarrow 2^I$ and $2^I \rightarrow 2^O$ respectively:*

$$f(O) = \{i \in I | \forall o \in O, (o, i) \in R\} \tag{1}$$

$$g(I) = \{o \in O | \forall i \in I, (o, i) \in R\} \tag{2}$$

Less formally, f maps a set of objects into a set of features common to those objects. Similarly, g maps a set of features into a set of objects containing all those features.

Definition 11. *The functions $h = f \circ g$, i.e. $h(I) = f(g(I))$, and $h' = g \circ f$, i.e. $h'(O) = g(f(O))$, are **Galois closure operators**.*

Definition 12. *Let $I \subseteq I$ be a set of features. I is a **closed pattern** iff $h(I) = I$.*

It follows from the last two definitions that a closed pattern is a maximal set of features common to a given set of objects. We regard each mined closed pattern as an implicit user preference. This mapping between closed patterns and user preferences proved to be very useful in data mining for supporting appointment attribute suggestion in the calendar domain.

3   Pattern Mining for Generating Structured Solutions

This section provides a summary of the closed pattern mining method based on the FP-Tree algorithm of Han, Pei and Yin [6], and our approach to generating structured solutions.

3.1   Mining Closed Frequent Patterns

Closed frequent patterns are mined in two steps: 1) build an FP-Tree from the database, and 2) retrieve frequent patterns from the FP-Tree, filtering out all non-closed patterns. In the method implemented by Coenen, Goulbourne and Leng [3], frequent patterns are mined using the FP-Growth algorithm and then stored in a T-Tree structure (Total Support Tree), which also stores the support calculated for all frequent patterns. In our implementation, we store only closed frequent patterns in the T-Tree, which provides fast access to the set of closed frequent patterns. In the first step, an FP-Tree is constructed from the database of past cases using the original FP-Growth method. In the second step, all closed frequent patterns are extracted from the FP-Tree and stored in a T-Tree.

In order to filter out non-closed patterns we use the following property, due to Pasquier *et al.* [9]: if I is any pattern, then $support(I) = support(h(I))$. Thus the support of any pattern is the same as the support of the smallest closed pattern containing it. Therefore any frequent pattern properly contained in the smallest closed pattern containing it is not a closed pattern. This means we can use the following simple algorithm to filter out non-closed patterns.

Algorithm 1 (Finding closed patterns)

```
1      T-Tree = {}
2      while not last frequent pattern
3          FrPat = GetFrPatFromFP-Tree()
4          SmallestClosedFrPat = FindSmallestClosedPatContaining(FrPat, T-Tree)
5          if (SmallestClosedFrPat does not exist)
6          or (SmallestClosedFrPat.Support ≠ FrPat.Support)
7              T-Tree = Add(FrPat, T-Tree)
8          end
9      end
10     Output(T-Tree)
```

The algorithm searches the T-Tree for a smallest closed pattern (line 4) containing the pattern collected from FP-Tree (line 3). If such a pattern is found and it has the same support as the original FP-Tree pattern, it is discarded, otherwise the pattern is stored in the T-Tree (line 7). The original FP-Tree mining algorithm has been modified in such a way that larger patterns are always mined before smaller ones, which enables the above algorithm to discover all closed frequent patterns.

3.2   Generating Solutions

Patterns found in the data mining process are used as building blocks to construct calendar appointment solutions. Individual patterns may complement one another, conflict or overlap (as defined in Section 2). In order to generate useful suggestions, we aim to efficiently find solutions that make use of as many patterns as possible. The algorithm presented below uses the "no conflict" method for pattern selection (for the "no-overlap" method, lines 8 and 17 need to be modified).

The following algorithm is not guaranteed to find all possible solutions, though it has been experimentally verified to provide sufficient time performance and solution quality. The algorithm first computes the set of all patterns that do not conflict with, but have at least one common feature with, the initial user features. The algorithm heuristically finds subsets of these patterns jointly consistent with the initial user features; each such set is heuristically extended to one maximal set of non-conflicting features.

Algorithm 2 (Generating user suggestions)

```
1  Input: InitFeatures, ClosedPatterns
2  Output: Solns
3      Solns = {}
4      InitSoln = MakeInitSoln(InitFeatures)
5      InitSoln.Patterns = {}
6      PatternList = {}
7      for each Pattern in ClosedPatterns
8          if not Conflicting(Pattern, FeatureList)
9              and HasCommonFeature(Pattern, FeatureList)
10             Add(Pattern, PatternList)
11         end
12     end
13     UnusedPatterns = PatternList
14     while UnusedPatterns.Size > 0
15         Soln = InitSoln
16         for each Pattern in UnusedPatterns
17             if not Conflicting(Pattern,Soln.Patterns)
18                 Soln = Update(Soln, Pattern)
19                 Soln.Patterns = Add(Pattern,Soln.Patterns)
20             end
21         end
22         for each Pattern in PatternList
23             if not Conflicting (Pattern,Soln.Patterns)
24                 Soln = Update(Soln, Pattern)
25                 Soln.Patterns = Add(Pattern,Soln.Patterns)
26             end
27         end
28         for each Pattern in Soln.Patterns
29             UnusedPatterns = Delete(Pattern,UnusedPatterns)
30         end
31         Solns = Add(Soln, Solns)
32     end
```

As an example, suppose the initial features are as follows:

Title="Project Meeting", Category="Team Meeting", Period="Semester"

Suppose the existing closed frequent patterns are as follows:

P1. Category="Team Meeting", Period="Semester", AmPm="am", Time=1030
P2. Category="Team Meeting", Period="Break", AmPm="pm"
P3. Category="AI Lecture", Period="Semester", AmPm="pm", Time=1500
P4. AmPm="pm", Day="Wednesday", Attendees="Anna, Alfred, Rita, Wayne"
P5. Period="Semester", AmPm="am", Time=1030, Day="Monday",
    Attendees="Anna, Alfred, Wayne"
P6. Category="Team Meeting", Day="Wednesday"

The initial solution (line 4) is just the initial set of features entered by the user. Since patterns P2 and P3 are conflicting and P4 has no common features with the initial solution, the PatternList and UnusedPatterns sets (line 13) contain only patterns P1, P5 and P6. Solutions always start with an initial user solution. A new solution is generated in lines 16–27. Since the initial solution has no associated patterns, P1 is added and the solution becomes:

Title="Project Meeting", Category="Team Meeting", Period="Semester", AmPm="am", Time=1030

In the next iteration (lines 16–21), P5 is evaluated and, since it is not conflicting, is also added to the solution, which becomes:

Title="Project Meeting", Category="Team Meeting", Period="Semester", AmPm="am", Time=1030, Day="Monday", Attendees="Anna, Alfred, Wayne"

Next P6 is evaluated and rejected as conflicting with this solution. The procedure then continues to add patterns from the PatternsList set (lines 22–27), but there is nothing new to add at this stage. Therefore it becomes the first solution and UnusedPatterns is updated. UnusedPatterns is still not empty, so the solution generation iterates again, this time adding P6 to the initial solution, generating the second solution, as follows:

Title="Project Meeting", Category="Team Meeting", Period="Semester", Day="Wednesday"

4   Experimental Evaluation

In this section, we describe our experimental framework for evaluating the closed pattern mining approach for solution generation using simulations over data extracted from a single user's calendar. This gave us around 1000 cases of real calendar data (about 16 months of appointments).

4.1   Method

The simulation was conducted in two stages. In the first stage, we compared two methods of appointment solution generation used in conjunction with closed pattern mining: the "no conflict" and the "no overlap" methods. This confirmed the superiority of the "no conflict" approach, which is the algorithm presented above. In the second stage, we compared these results with those generated using decision tree learning, the best performing machine learning method.

The simulator runs real calendar data through the solution generation system in a manner resembling interaction with the real user of a calendar system. The calendar data used for the simulation had 8 attributes: Category, Period, Attendees, Location, Duration, AmPm, Day and Time. The simulation was conducted as follows. The "user" (which means "simulated user") enters case n, which is stored in the database and data mining on all past cases is performed. Then the "user" enters the first three attributes of case $n + 1$ as initial attributes, which are always assumed to be the Category, Period and Attendees (this is the typical behaviour of actual users based on our informal observation). The system produces a number of suggestions out of which the "user" selects one closest to case $n + 1$. Differences are then calculated between the best suggestion and the real case $n + 1$. These differences reflect the number of modifications the "user" needs to make to turn the suggestion into case $n + 1$. The "user" needs to either add a missing feature or delete one which is not required, therefore each difference is counted as 1. These differences are then averaged over a number of data cases. For compatibility with the decision tree learning method, as explained further in this section, the simulator produces 32 suggestions.

The machine learning part of the simulation was done using the C4.5 decision tree algorithm implemented in the Weka toolkit [11], called J48. The method was selected by testing the performance of a range of machine learning algorithms on the calendar data for various combinations of attributes. The tested algorithms were rule induction (OneR), decision tree learning (J48), Bayesian methods (Naive Bayes, BayesNet), k-nearest neighbour (IBk) and case based reasoning (KStar). The best performing, J48, was then used on five calendar data sets, each to predict one of the five attributes of case $n + 1$ not specified by the "user" (i.e. the Location, Duration, AmPm, Day and Time). The predicted values were put together to make a set of complete calendar appointments as in the data mining method. So that the decision tree learning methods could generate a number of alternative solutions, we had to combine a number of suggested values for each attribute. This was achieved by modifying the Weka code so that each prediction consisted of two different values rather than one. For the five attributes to be predicted this was equivalent to $2^5 = 32$ possible solutions for each appointment.

4.2   Results

We first present the results comparing the "no conflict" and "no overlap" methods used with closed pattern mining, shown in Figure 1.

The difference between the two methods shows that the "no conflict" method produces significantly better results. This can be explained by the way solutions are created. It is generally easier to find overlapping patterns in a solution than non-overlapping, hence the former method creates a higher number and variety of solutions.

One of our objectives in evaluating machine learning methods for generating solutions was to compare our system with CAP, Dent *et al.* [4]. Although a direct comparison was not possible, the method (ID3 for CAP) and overall results are similar. The accuracy achieved by decision tree learning on our data set is shown in Figure 2. These results broadly confirm those reported in the original experiments with CAP, e.g. accuracy for location is close to 70% after around 150 cases. Note that our experiments compensate for a deficiency in the experimental setup with CAP in that a parameter

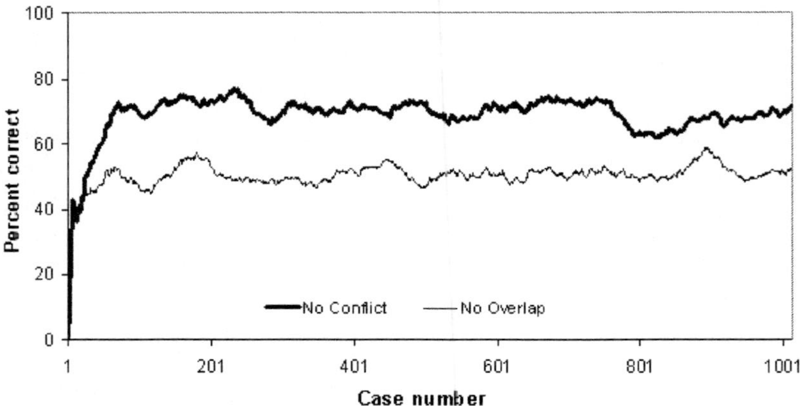

Fig. 1. Average accuracy of appointment prediction for two methods: "no conflict" (*thick line*) and "no overlap" (*normal line*).

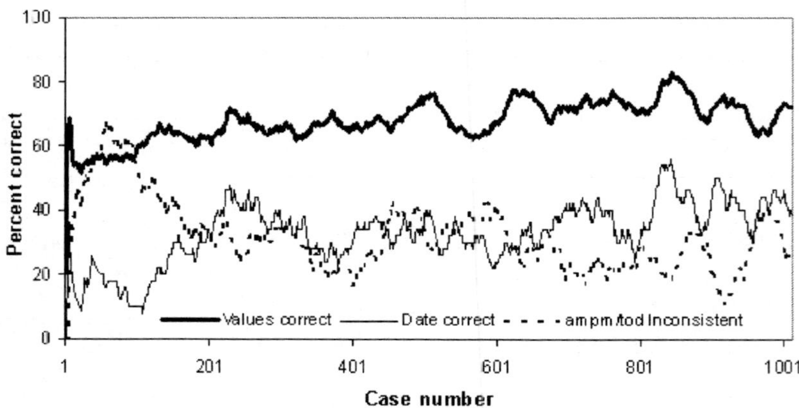

Fig. 2. Decision tree learning prediction results of calendar appointments. The *thick line* shows the overall average accuracy of appointment prediction, the *continuous line* is the average appointment date prediction and the *dotted line* reflects the number of inconsistent values in predicted data (AmPm and Time).

specific to the time period (e.g. semester, break) is included, which means the average accuracy fluctuates much less than in the original CAP evaluation. However, our results also show a greater fluctuation of average accuracy for decision tree learning than with closed pattern mining. We suspect that this could be because at certain points in the simulation, the decision tree is restructured, resulting in a loss of accuracy, whereas the pattern mining approach produces smoother behaviour over time. Additionally, unlike the pattern mining method, where values within one case are created from non-conflicting patterns, the decision tree learning method predicts values separately for each attribute. In consequence, it is possible that some associated attributes may conflict, e.g. AmPm=am and Time=1500. In effect, the system has to choose randomly between the two options to resolve the conflict, meaning that the date suggestion is often wrong (our simulation arbitrarily chooses the value of Time to resolve the conflict, and the date is determined from the next free time slot satisfying the chosen solution). The chart in Figure 2 provides some confirmation of this explanation, where a low average date prediction accuracy corresponds roughly to a high number of inconsistencies between AmPm and Time.

Comparison of Figure 2 with Figure 1 shows that, although the average accuracy of prediction is similar for the two methods (closed pattern mining 69%, decision tree learning 68%), closed pattern mining gives significantly better prediction in the first 200 cases. More specifically, the closed pattern mining method reaches its average accuracy after only 62 cases, whereas the decision tree learning method reaches its average after 224 cases, staying about 10 percent lower in the first 200 cases. This is an important difference for interactive calendar users, who would clearly prefer useful suggestions in a shorter period of time (this corresponds to roughly 1 month for closed pattern mining vs. 3 months for decision tree learning). Moreover, decision tree learning prediction is less stable, showing greater sensitivity to user preference changes in transition periods.

5 Related Work

As far as we know, there are no calendar applications supported by pattern mining, however there are examples of research where some kind of machine learning has been applied. As described above, the CAP system, Dent *et al.* [4], provides suggestions for various appointment attributes. Two methods for predicting attributes were compared: backpropagation neural networks and ID3. Their results showed that for the Location attribute, around 70% accuracy was achieved by both learning methods after sufficient training. As described above, our experiments broadly confirm this result in the case of decision tree learning, though over the whole set of predicted attributes (not only Location). As also mentioned above, CAP predicts each attribute of the appointment separately, which may result in inconsistent appointment solutions when these predictions are combined.

Another preference learning calendar assistant is described by Berry *et al.* [2]. Their system, PCalM, is a framework designed to schedule meetings in the open calendar environment. Instead of learning to predict individual appointment attributes, as in CAP, PCalM learns to rank candidate appointments from pairwise selections provided by the user. Unlike our calendar system, designed to build and present suggestions unobtrusively, PCalM forces the user to choose amongst suggestions in order to generate training data for learning the preference function. Furthermore, similar to our method, PCalM has been evaluated using simulated user interactions, however the data used in the PCalM evaluation is synthetically generated, while we have used appointments from a user's real calendar, providing a more realistic data set for experimentation.

6 Conclusion

We have designed and evaluated a structured solution builder with data mining support for generating suggestions for calendar appointments. Closed patterns proved to be a suitable alternative to association rules due to their compactness and flexibility. Moreover, pattern mining has an advantage over single class machine learning methods in that it better supports creating multiple solutions with consistent structures. We simulated user interaction with real calendar data to configure and tune data mining and appointment solution generation methods. Our results show the superiority of closed pattern mining to decision tree learning, the best performing machine learning algorithm in this domain.

We believe that concept based data mining for building structured solution can be applied to other configuration domains. Due to the fact that cases are added to the system incrementally, it might be possible to use incremental data mining methods in conjunction with the FP-Growth algorithm, similar to the approaches of Ezeife and Su [5] and Koh and Shiehr [7].

Acknowledgments. This work was funded by the CRC for Smart Internet Technology. We would like to thank Paul Compton for making his calendar data available for research and Frans Coenen for his open source implementation of the FP-Growth algorithm.

References

1. Agrawal, R., Srikant, R.: Fast Algorithms for Mining Association Rules. In: Proceedings of the 20th Conference on Very Large Data Bases, pp. 478–499 (1994)
2. Berry, P.M., Gervasio, M., Uribe, T., Myers, K., Nitz, K.: A Personalized Calendar Assistant. In: Proceedings of the AAAI Spring Symposium on Interaction between Humans and Autonomous Systems over Extended Operation (2004)
3. Coenen, F., Goulbourne, G., Leng, P.: Tree Structures for Mining Association Rules. Data Mining and Knowledge Discovery 8, 25–51 (2004)
4. Dent, L., Boticario, J., Mitchell, T.M., Zabowski, D.A.: A Personal Learning Apprentice. In: Proceedings of the Tenth National Conference on Artificial Intelligence (AAAI 1992), pp. 96–103 (1992)
5. Ezeife, C.I., Su, Y.: Mining Incremental Association Rules with Generalized FP-Tree. In: Cohen, R., Spencer, B. (eds.) Advances in Artificial Intelligence, pp. 147–160. Springer, Berlin (2002)
6. Han, J., Pei, J., Yin, Y.: Mining Frequent Patterns without Candidate Generation. In: Proceedings of the 2000 ACM SIGMOD International Conference on Management of Data, pp. 1–12 (2000)
7. Koh, J.-L., Shiehr, S.-F.: An Efficient Approach for Maintaining Association Rules Based on Adjusting FP-Tree Structures. In: Lee, Y., Li, J., Whang, K.-Y., Lee, D. (eds.) Database Systems for Advances Applications, pp. 417–424. Springer, Berlin (2004)
8. McDermott, J.: R1: A Computer-Based Configurer of Computer Systems. Artificial Intelligence 19, 39–88 (1982)
9. Pasquier, N., Bastide, Y., Taouil, R., Lakhal, L.: Efficient Mining of Association Rules Using Closed Itemset Lattices. Information Systems 24, 25–46 (1999)
10. Wille, R.: Formal Concept Analysis as Mathematical Theory of Concepts and Concept Hierarchies. In: Ganter, B., Stumme, G., Wille, R. (eds.) Formal Concept Analysis, pp. 1–23. Springer, Berlin (2005)
11. Witten, I.H., Frank, E.: Data Mining. Morgan Kaufmann, San Francisco (2005)

An Implementation of Goal-Oriented Fashion Recommendation System

Mikito Kobayashi[1], Fumiaki Minami[2], Takayuki Ito[2], and Satoshi Tojo[1]

[1] Japan Advanced Institute of Science and Technology,
   Asahidai, Nomishi, Ishikawa 1-1, Japan
   {m-kobaya,tojo}@jaist.ac.jp
   http://www.jaist.ac.jp
[2] Nagoya Institute of Technology,
   Gokiso, Showaku, Nagoya 466-8555, Japan
   {minami,ito}@itolab.mta.nitech.ac.jp
   http://www-itolab.mta.nitech.ac.jp

Abstract. On the Web, Electronic Commerce is widely thriving with development of the Web technology. However, users still have trouble finding products that will find their desires. In recent researches, they have introduced many kinds of method for Recommender systems, but these system still have problems which are based on concrete attributes of the products and a complex users model. Within this paper, we introduce a new technique, Goal Oriented Recommendation, which works even when users do not want exactly products that they are looking for. Moreover, the system processes users' input (e.g. "I'm going to have dinner with my boss" or "I'm looking for my wife's birthday presents") with a own concept dictionary which contains a occasion word and a person word. The system can recommend items based on users' desire, if users input their desire.

Keywords: Recommender System, Concept Dictionary, Goal Oriented Recommendation.

1 Introduction

On Electronic Commerce, users still have problems finding products which is their desires or goals. In recent years, almost shopping sites employ keyword or category based searching systems[1, 16]. The advantage of keyword based searching, the system provides products for users with a simple keyword for products, but the disadvantage is that users must guess a keyword of products. Similarly, category based searching also obligates users who should guess a category for the products. On the other hands, bulletin board system provide a free input for users, when they want to ask other users, but the disadvantage is that only the limited user will answer the users' question. Hence, even if users get information about their desires, it would not be accurate information or incorrect answers.

In this paper, we introduce a new recommender system, Goal oriented fashion recommendation system. The system provides the input with a sentence, which make user's remove above obligations. In fact, users of the system can input

N.T. Nguyen, R. Katarzyniak (Eds.): New Chall. in Appl. Intel. Tech., SCI 134, pp. 77–86, 2008.
springerlink.com

their desires freely. In addition, we assume users' situation that they want to find items for their occasions, and also have no ideas "what should they wear" for the occasions.

The outline of this paper goes as follows. First, it shows overview of goal oriented fashion recommendation system, second we introduce how such a system may be built, and give an implementation example of the system, third, discusses the implemented system. Fourth, we introduce related works, finally, we conclude this paper.

2 Goal Oriented Fashion Recommendation System(GOFRS)

2.1 System Overview

We implemented a fashion recommendation system. In this system, we assume a situation that each user is looking for clothes for some events but he/she has no idea what should he/she wear. Firstly, Each user registrates his/her clothing items (jackets, shirts, pants). There is an input entry for users to find items for particular occasions, i.e., "I'm going to ...". For example, "I'm going to have dinner" or "I'm going to attend the entrance ceremony", etc. Based on own dictionaries, the system matches the clothes' styles and functions for context. the system returns suggestions for complete outfit with distinguished items (jackets, shirts, pants), and outfits can be selected by users, which the user interested in. Moreover, users can check a pair of clothes from outfits on the avatar space in the system. Additionally, the system discriminates a formal occasion from a informal occasion depend on a person who is together with the user in an occasion, e.g.,"I'm going to have dinner with my family" must be an informal occasion, on the other hand, "I'm going to have dinner with my boss" must be a formal occasion.

2.2 Item Database and Concept Dictionary

Our database of items is based on an item of Marui web channel[9], and the database contains 100 items which include all categories such as jackets, shirts, pants. Moreover, these items has a key which represents features for items, we use a six-tuple to represent dimensions of the concept style:

(1) Formal
(2) Trendy
(3) Sporty
(4) Luxurious
(5) Elegant
(6) Funky

where a key ranges from 0 to 10. An example of keys goes as follow.

$$leather\ jacket(2, 6, 3, 4, 3, 8)$$

when the formal value of an item is high, the item will be used for a formal occasion. These keys are defined in terms of [11], we also use their keys for express the style for any pieces of clothing. Then, GOFRS has two concept dictionaries, one is the dictionary for occasions, the other is the dictionary for persons. The dictionary for occasions contains words which indicates occasions, e.g., dinner, party, ceremony, etc. In addition, each word of these dictionaries contains the formal value of a key of items and also contains a range of the formal value, which is "more than" or "less than" , e.g., if an item has the formal value 8 and "less than", then the system recommends an item which has the formal values less than 8. Next the dictionary for persons contains words which change the range of the formal value for occasions such as boss, supervisor, etc. So each word of the dictionary has "more than" or "less than". When a person word has the range of formal values "more than" and an occasion word has "less than", then the system preferentially uses the range of formal values for a person word. Because an occasion word often changes to a formal or a casual with a person who spends an occasion with users. Therefore, a person word is more preferential than an occasion word.

Using above the database and two concept dictionaries our system can guess the style for any English words, even if the changing occasion depends on a person who spends occasions with a user.

2.3   Attribute Sensing

In users' input and items will be sensed before matched, which are extracted according to tree types of information, including the clothing items, a formal key and words that are related the occasions. All these types of information are processed with an uniform computational representation. We use a formal key for such matching. The other keys are used for users' selecting items, in fact, if an item is recommended by the system, users can decide which items are suited for their desire by the keys to represent the dimensions of the concept "style". Using the formal key, we are able to express the style for any clothes, and for any users' input.

Each item has a formal key and also other keys, if an item has a formal value 8 and a range of value "more than", this item is suited for a formal occasion. Based on the key and the range, the system derives the attribute of items. If users do not agree with the outfit of the system, they can adjust values of all keys on the key form. All possible items and words need to be listed in the occasions dictionary, and the persons dictionary.

2.4   Database Structure

In the system, each database are made by XML type. An example of a data of occasions word dictionary is following as:

```
<word>
<word_name>dinner</word_name>
<range>less than</range>
```

```
<parameter>8</parameter>
<spcial></special>
</word>
```

In the above the XML data, a tag "word" shows a data for one occasion word, "word_name" shows a word, "range" is a range of formal value, "parameter" is a formal value and "special" shows a special attribute respectively. Then, the persons word dictionary is structured as follows:

```
<word>
<word_name>boss</word_name>
<range>more than</range>
</word>
```

Where "word" shows a data for one persons word, "word_name" shows a word and "range" is a range of formal value for the word.

Finally, we would like to show the structure of our item database in the follows:

```
<item>
<name> leather jacket </name>
<category>jacket</category>
<brand_name> VISARUNO </brand_name>
<img>clothes/jacket/jacket10.jpg</img>
<no>WW753-30004</no>
<price> 29400 </price>
<parameter>
<formal>9</formal>
<trendy>4</trendy>
<sporty>1</sporty>
<luxurious>8</luxurious>
<elegant>7</elegant>
<funky>4</funky>
</parameter>
</item>
```

In the above data, "item" shows whole data of an item, "name" is a name of an item, "category" shows a category for items, "brand_name" shows the brand name of items, "price" is a price for items and "parameter" shows six dimensions as formal, trendy, sporty, luxurious, elegant and funky.

2.5   System Architecture

The structure of GOFRS shows as Fig.1. The system has the input form, the output form, user's item form, the key form and the avatar form. Following processes, the system will recommends an item from user's input.

i.   In the input form, a user inputs what he/she is going to do such as "I'm going to have dinner with my family", and the sentence of the user's input must include words which are an occasion, place or a person. Because of the system assumes that a user wants to find items in terms of them.

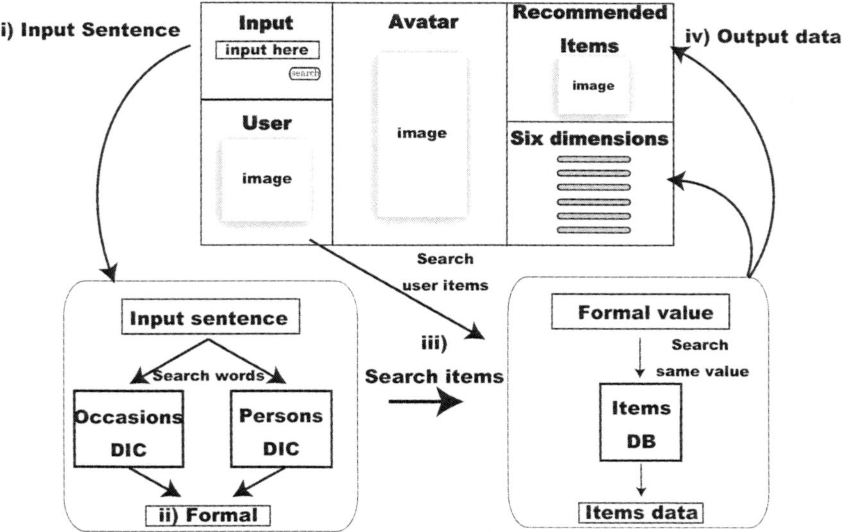

Fig. 1. Interface of GOFRS

ii. When the system gets user's input, the system tries to find an occasion word from the dictionary for occasions. Then, the system extracts parameters of the items such as the formal value, the range of the formal value "more than" (or "less than") and a special attribute "strong"(or "null").

Fig.2 shows an example of a table for a formal value, a range of values and a special attribute. In the figure, "dinner" and "birthday party" have the

Event	Formal value	Range of fomal value	Special attribute
dinner	8	less than	null
birthday party	7	less than	null
conference	8	more than	strong

Fig. 2. Example of a table for formal values, ranges and special attributes

Person	Range of formal value
boss	more than
family	no change
president	more than

Fig. 3. Example of a table for person words

formal value and the range of the value, but do not have a special attribute which represent a strong occasion. On the other hands, "conference" has a special attribute "strong", the meaning of the attribute is that "conference" does not depend even if some persons spend the occasion with the user. In fact, a special attribute "strong" keeps the range of value for each occasion word.

In this process, the system searches a word that shown a person in the dictionary for persons. Then, when the range is different from the occasion word, the range will be changed, e.g, the occasion word "dinner" has a formal value 8 and a range of the formal value "less than", but the word "boss" has a range of the formal value "more than", finally, a range of the formal value is changed as "more than". In addition, If there is no the word in the user's input, then the system move to the process iv, and an occasion word which has a special attribute, also move to the next process.

iii. Based on the process ii , the system searches items which has values in range of the value. Then, matched items are categorized as jacket, shirts, pants, furthermore, six dimensions of each item is including the output data.

iv. Finally, recommended items and six dimensions are displayed in the interface of our system.

3 Implementation Example

In this section, we show an example of implementation based on 2.5. First, we assume that the user tries to find items for dinner with his boss. If a person will have dinner with his/her boss, he/she should wear a formal clothes. However, the user would not wear too formal wears such as a suit and a tie. Then, the user inputs "I'm going to have dinner with my boss.", in the user's input the system divides this sentence per word. After that, the system searches a word which matches an occasion word in the occasion dictionary. In this input, "dinner" should be found by the system. Additionally, the formal value and the range of value be extracted from the word "dinner". In this time, the system keeps these values and the range. Next, The system searches a person word in the person

Jacket	
shirts	
Trousers	

Fig. 4. Output of the input "to have dinner with my boss"

history, etc. In recent years, many shopping sites such as Amazon.com, yahoo, etc, employ Collaborative Filtering[10, 5], which works by clustering users' purchase history. However, many of these systems do not consider users' desires in search activity. Furthermore, Collaborative Filtering has The First-Rater problem and Cold Start Problem[7]. The First-Rater problem is caused by new items in the system that understandably have not yet received any ratings from any users. Similarly, The Cold Start Problem is cased caused by new users in the system which have not submitted any ratings. Without any information about the user, these systems can not guess items which is similar to users' preferences.

The other systems is [3, 4, 12]. These systems take into account users' desire such as preferences or even lifestyles, and try to capture users' desire by applying critiquing interaction and complex user models. However, they either require users to provide specific descriptions in terms of product attributes, or require that the correspondence between scenarios and product attributes be explicitly coded. In addition, [11] introduces Scenario-Oriented Recommendation System with common sense reasoning. Reasoning is based on an 800,000-sentence Common Sense Knowledge base, and spreading activation inference. They put a key(same as us) for each brand of items, but each brand has a lot of varieties, in fact all brands would have a formal item and a casual item. Although they have shown a brand a case of Levis in their paper, Levis's items are quite wide range of varieties of the items. So if a user has found an item which is suited for their desire, the system would miss actually there are other suited items. Our approach differs from above systems. GOFRS does not cause such problems because of the system focuses on users' situations and desires, moreover, our system does not need such a complex users' model.

6   Conclusion

In this paper, we introduce a new recommender technique, Goal Oriented Recommendation. Based on an own concept dictionary, the system can recognize users' desire and situations with the attribute sensing and the item network. With the system, even when users do not necessarily know exactly what product characteristics they are looking for, the system can find users' desire from their input. Moreover, the system can apply for any other products such as DVDs, CDs, etc, if each item of them has the keys and the range of the keys. Hence, the system could be an important recommendar system on the Web. However, we have a technical issues in 4 and we also need to an evaluation experiment with examinees in the future. Additionally, our concept dictionary does not have enough words for users' input, so we plan to use OMCS or any other concept dictionaries, and also we need to do an evaluation experiment with examinees in future.

References

1. Amazon.com, `http://www.Amazon.com/`
2. Bruke, R., Hammond, K., Young, B.: The FindMe Approach to Assisted Browsing. IEEE Expert 12(4), 32–40 (1997)

3. Fano, A., Kurth, S.W.: Personal Choice Point: Helping users visualie what it means to buy a BMW. In: Proc of IUI, pp. 32–46 (2003)
4. Ha, V., Haddawy, P.: Problem-Focused International Elicitation of Multi-Attribute Utility Models. In: Proc of UAI, pp. 215–222 (1997)
5. Herlocker, J.L., Konstan, J.A., Terveen, L.G., Riedl, J.T.: Evaluating collaborative filtering recommender systems. In: ACM Transactions on Information Systems 22, 5–53 (2004)
6. Goker, M.H., Thompson, C.A.: Personal Conversational Case-Based Recommendation. Springer, Heidelberg (2000)
7. Guo, H.S.: Live recommendations through social agents. In: Fifth DELOS Workshop on Filtering and Collaborative Filtering, Budapest (1997)
8. Liu, H., Singh, P.: ConceptNet: a practical commonsense reasoning toolkit. BT Technology Journal 22(4), 211–226 (2004)
9. Marui web channel, http://voi.0101.co.jp/voi/index.jsp
10. Resnick, P., Varian, H.R.: Recommender Systems. Communications of the ACM 40(3), 56–58 (1997)
11. Shen, E., Lieberman, H., Lam, F.: What am I gonna wear?: Scenatio-Oriented Recommendation. In: Proceedings of the International Conference on Intelligent User Interface, IUI 2007, Honolulu, Hawaii, USA, January 28–31 (2007)
12. Shearin, S., Lieberman, H.: Intelligent Profiling by Example. In: Proc of IUI, pp. 145–151 (2001)
13. Shumida, A., Torisawa, K.: Hacking Wikipedia for Hyponymy Relation Acquisition. In: Proceedings of the Third International Joint Conference on Natural Language Processing (IJCNLP) (2008)
14. Singh, P., Lin, T., Mueller, E.T., Lim, G., Perkins, T., Zhu, W.L.: Open Mind Common Sense: Knowledge acquisition from the general public. In: Proc. of the 1st International Conference on Ontologies, Databases, and Applications of Semantics for Large Scale Information Systems (2002)
15. Wikipedia, http://en.wikipedia.org/
16. Yahoo.com, http://www.yahoo.com/

A Proposal on Recommender System Based on Observing Web-Chatting

Fumiaki Minami[1], Mikito Kobayashi[2], and Takayuki Ito[1]

[1] Nagoya Institute of Technology,
   Gokiso, Showaku, Nagoya 466-8555, Japan
   {minami,ito}@itolab.mta.nitech.ac.jp
   http://www-itolab.mta.nitech.ac.jp
[2] Japan Advanced Institute of Science and Technology,
   Asahidai, Nomi, Ishikawa 923-1292, Japan
   m-kobaya@jaist.ac.jp
   http://www.jaist.ac.jp

Abstract. As electronic commerce thrives, products on the Web are increasing and consumers are having trouble finding products that meet their desires. Many recommender systems help users find desired products, but not all use communication among users positively for recommendations. We propose a recommender system with a web-chat interface to make an environment where users can talk with each other. By observing conversation in a web-chat interface in real-time and always recognizing user interests, the system can recommend products based on conversation contents.

Keywords: Recommender System, Web Chat.

1 Introduction

In recent years, the increase of products on the Web has complicated the location of products for consumers, who then have trouble finding products that meet their desires. Many shopping sites have keyword- or category-based searching systems to help users find desired items. But users cannot find items with these searching systems when they are not sure about the related keywords or categories. To resolve this problem, we need recommender systems that can suggest items for users on Web shopping sites. AI has offered several ways to make recommender systems[1], but not all adequately use communication among users for recommendations.

Imagine searching for items in a shopping site without having specific interests. Since keyword- or category-based searching systems need specific keywords or categories, they offer little assistance. In this situation, talking with someone, for example, a friend, can clarify one's interest. Talking with others can help search for items.

We think web-chat communication among users can be used for recommendations and propose a recommender system with a web-chat interface to create an environment where users can talk with themselves. By constantly observing conversations in the web-chat interface and detecting changes of user interests,

N.T. Nguyen, R. Katarzyniak (Eds.): New Chall. in Appl. Intel. Tech., SCI 134, pp. 87–96, 2008.
springerlink.com

the system can renew user profiles in real-time and always recognize what kind of things users like. The recommended items can be changed based on the real-time changes of user profiles.

With the recommender system, users can talk with others in a web-chat interface and clarify their own interests. Then the system recommends items that match conversation contents. The system is useful because users can get recommendations by only talking with others.

The rest of this paper is organized as follows. We give a detailed explanation of the recommender system and show its architecture in Section 2. Next, in Section 3, we show an example when a user uses the system. In Section 4, we discuss its advantages and applicability. In Section 5, we show related work. Finally, we conclude this paper with comments on future work in Section 6.

2   Recommender System Based on Observing Web-Chatting

2.1   System Outline

This system can recommend items to users by observing conversation among users. We introduce a web-chat interface to our system and allow users to talk with others. The system has user profiles that show what kind of things users like and renews them by analyzing the contents of conversations. To analyze the conversations, we make a dictionary that contains several words and phrases. By getting meaning every time users say something, the system can recognize the latest user interests and renew user profiles in real-time. Recommended items can be changed based on real-time changes of user profiles.

2.2   User Profile and Item Data

This system has user profiles that show what kind of things users like. We used five parameters to represent a user profile: "unusual," "cute," "cool," "simple," and "luxurious." "Unusual" means interesting, "cute" means pretty, "cool" means stylish or neat, "simple" means basic, and "luxurious" means classy in other words. Each parameter ranges from 1 to 5 and higher values show that users like things with strong features at the corresponding parameters. Default values of all parameters are 3, and they are renewed based on contents of conversations among users. For example, a user whose profile is [5, 3, 3, 3, 1] likes unusual or interesting things, but not luxurious or classy ones.

The recommended items are from our original database made from Amazon [10] to be shown now in detail. Each item in the database has its own parameters that represent features. We handcrafted values of the parameters for each item and made the database in XML structure as Fig. 1. In a tag called "Item," we put tags "ASIN," "Title," and "URL". They respectively represent an item's ID, its name, and an URL for an image of it. The rest of the five tags from "unusual" to "luxurious" show an item's features. The meaning of the five tags is identical as the five parameters in a user profile above explained.

```
<Items>
    <Item>
        <ASIN>B000PC1IWM</ASIN>
        <Title>VAGARY IB0-410-94</Title>
        <URL>http://g-ec2.images-amazon.com/images/I/01zwN-akMFL.jpg</URL>
        <Unusual>5</Unusual>
        <Cute>2</Cute>
        <Cool>3</Cool>
        <Simple>1</Simple>
        <Luxurious>2</Luxurious>
    </Item>
</Items>
```

Fig. 1. Example of item data

The system recommends items with more than four points at parameters that have more than four points in the user profile.

2.3 System Architecture

We show the system architecture in Fig. 2. The system interface is composed of a chat space where users can chat in real-time and a showcase where recommended items are displayed. We explain how the system works in the following.

i) In chat space, a user can talk in real-time with others. We introduce a chat server to the system that allows users to use a web-chat interface. The system observes conversations among users in chat space. Every time they say something in the chat space, the system catches what they say and recognizes the speaker. In the following processes, the system behavior changes depending on the speaker.

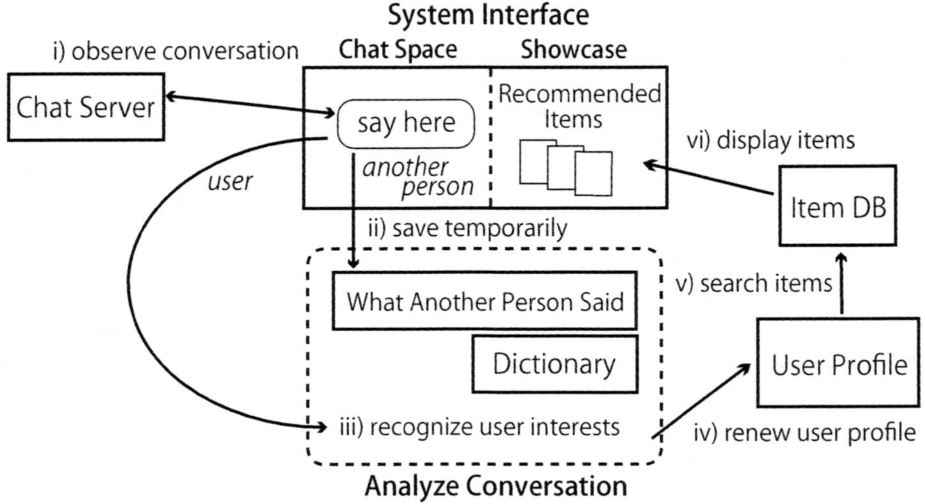

Fig. 2. System architecture

ii) When the system recognizes the speaker as another person, the system temporarily saves what that person said. The system refers to the saved information in the next process.

iii) When the system recognizes the speaker as the user, it tries to recognize the user's interest with a dictionary in the system with two processes. One process decides whether the user likes (or does not like) something, and the other decides what kind of items the user likes (or does not like). We show the processes in Fig. 3.

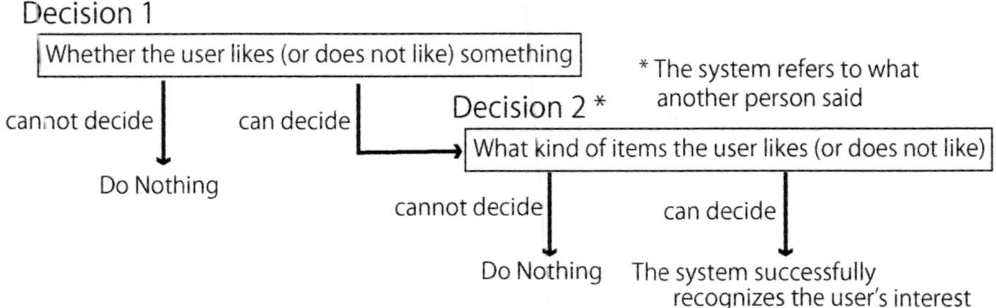

Fig. 3. Two processes to recognize user interest

In Decision 1, the system decides whether the user likes (or does not like) something. For the system's decision, the dictionary includes several words or phrases such as "want," "good," "like," etc. For example, if the word "want" is in what the user says, the system decides that the user likes something. Here, we should pay attention to negative expressions, because we can't decide that the user likes something when he/she says, "I don't like" something. We put several words such as "don't," "won't," "isn't," etc. in the dictionary to cope with negative expressions. As a result, even when the user says "I don't like...," the system can decide that the user does not like something. We should also consider such special expressions as "not bad," "sounds good," etc. The system contains these special expressions in its dictionary and can deal with them.

In Decision 2, the system decides what kind of items the user likes (or does not like). The dictionary has several words related to the five parameters shown in 2.2. For example, "classy" are related to the parameter "luxurious." When the word "classy" is contained in what the user says, the system decides that the user likes (or does not like) luxurious things. With only this decision, if the system cannot decide what kind of items the user likes (or does not like) by only what the user says, the system refers to what another person said just before the user.

Here, we show some examples to explain how the system behaves in Decisions 1 and 2. If the system can make decisions in both Decisions 1 and 2, it successfully recognizes the user's interest.

In Table 1, we show the system behavior when it only uses what a user says to recognize user interest. In the four examples from the top, the system can

Table 1. System behavior when only using what a user says

Examples (What a user says)	Decision 1 (LIKE or NOT LIKE)	Decision 2 (Corresponding Parameter)
I want an interestingly designed one.	LIKE	Unusual
Basic design might be good.	LIKE	Simple
I don't like classy design.	NOT LIKE	Luxurious
I won't buy an unusual one.	NOT LIKE	Unusual
Not bad.	LIKE	cannot decide
I don't like something like that.	NOT LIKE	cannot decide

recognize the user's interest by using only what the user says. But, for the last two examples, the system cannot recognize the user's interest because there is no keyword in what the user says to decide what kind of items the user likes (or does not like).

In Table 2, we show the system behavior when it uses what a user (U) and another person (A) say to recognize the user's interest. Here, remember that the system refers to what another person says only in Decision 2. Even if the system cannot recognize the user's interest with only what the user says, the system refers to what another person said and successfully recognizes the user's interest as readers can see in the two examples.

Table 2. System behavior when referring to what others said

Examples (What a user & another person says)	Decisions 1 (LIKE or NOT LIKE)	Decisions 2 (Corresponding Parameter)
A: How about a simple one? U: Not bad.	- LIKE	Simple cannot decide
A: Do you like luxurious design? U: I don't like something like that.	- NOT LIKE	Luxurious cannot decide

iv) When the system successfully recognizes the user's interest in the previous process, the system renews the user profile shown in 2.2. For example, if the system decides that the user likes "unusual" things, it adds one point to the value of the parameter "unusual."

v) Using the renewed user profile, the system searches for suitable items in the item database. The system searches for items with more than four points at parameters that have more than four points in the user profile.

vi) The suitable items are shown in the showcase of the system interface. The recommended items change in real-time based on the real-time changes of the user profile.

3   Example

We here explain the system behavior in detail by showing an example of a conversation between a user and a friend. Suppose the user is searching for

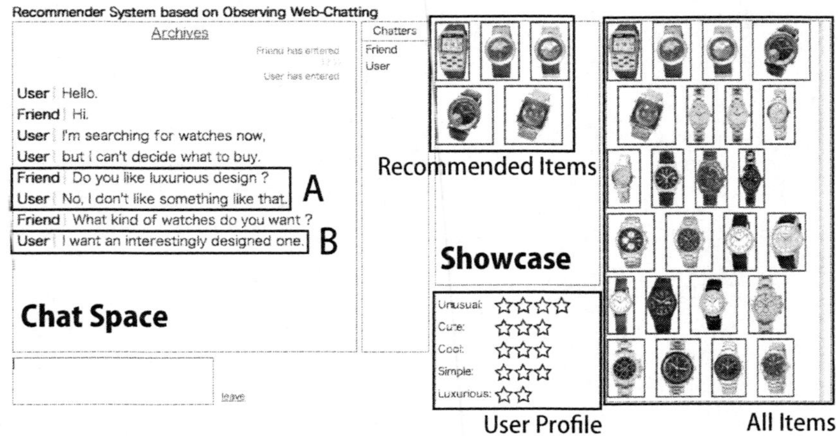

Fig. 4. Example of system 1

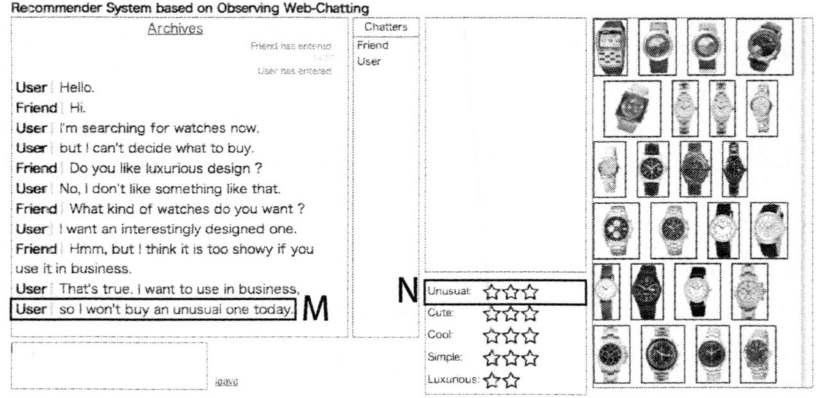

Fig. 5. Example of system 2

watches in the system and asking the friend what to buy. Fig. 4 shows the user talking with the friend in chat space. At A in Fig. 4, the user reacts to "Do you like luxurious design?" from the friend and says "I don't like something like that." As explained in 2.3 iii, the system decides that the user does not like "luxurious" items by referring to what the friend said. Then the value of the parameter "luxurious" changes to two points from three points. We show the state of the user profile with stars and set the default value at three points. At B in Fig. 4, when the user says "I want an interestingly designed one," the system decides that the user likes items related to the parameter "unusual" by only using what the user says. At this time, the value of the parameter "unusual" is changed to four points and the system recommends items in a showcase. We show all items in the system on the right side of Fig. 4.

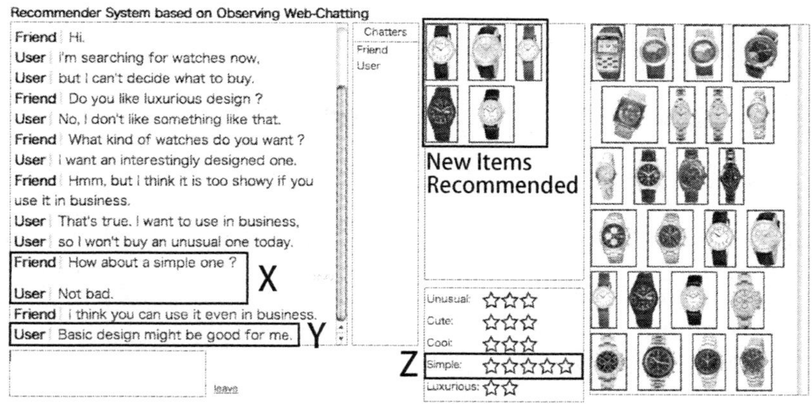

Fig. 6. Example of system - 3

The conversation continues and the system decides that the user does not want to buy the items related to the parameter "unusual" at M in Fig. 5. At this time, the system subtracts one point from the value of the parameter "unusual" (shown at N in Fig. 5) and the recommended items disappear from the showcase.

At X in Fig. 6, the system decides that the user likes "simple" items by analyzing the user's reaction "Not bad" to the friend's question "How about a simple one?" and recommends new items. As shown in 2.3 iii, the system understands such special expressions as "Not bad". Finally, the system decides that the user likes "simple" items again at Y in Fig. 6, and the value of the parameter "simple" becomes five points (shown at Z in Fig. 6).

4 Discussion

4.1 Advantages

Using the system and talking with others, users can get opinions and advice. Getting such opinions and advice can help users change or clarify their interests. In Section 3, we showed an example where a user is sharpening his/her interest during conversation with a friend. The system solves problems well when users are searching for items without clear interests. Advantage when using communication in a recommender system is that users can be affected by others and get more suitable recommendations.

User interests can change in real-time, so a recommender system must always deal with them. Since we introduce a web-chat interface into our system, users can communicate interactively with others and the system always recognizes changes of user interests. An example in Section 3 shows that the system changes recommended items based on real-time changes of a user profile.

To simplify understanding of the system behavior, we show a state of a user profile in the system interface. But users do not have to be concerned with the state of their profile; they just chat with others and get recommendations from the system. A web-chat interface in the system allows users to have a natural conversation with others. Natural conversation facilitates communication among users, and users can easily express their own interests.

4.2 Applicability

We made a system with a web-chat interface that can be applied to shopping sites on the Web. Like keyword-based search or category-based search, users can use the web-chat interface to find items. Users can talk to others in shopping sites with the interface. Then the users get recommendations from the system.

As for others with whom users talk, we imagine not only friends but also sellers in shopping sites. We believe that the sellers can give good advice to users, because they have a lot of information about items in their shopping sites. The users can clarify their interests by asking the sellers about items. The system has a suitable interface for users who want to talk with sellers to get advice. The system is beneficial for the sellers too, because they can directly communicate with users in their shopping sites.

As for measures to deal with spam on the Web, shopping sites manage user accounts or have social network functions to allow people use the system. Users in the shopping sites must register their own information to use the system. Even if spam exists in the shopping sites, administrators can detect offenders and ban their accounts. With such measures, we believe the system can be applied to shopping sites.

5 Related Work

The term "recommender system" refers to systems that suggest music CDs, books, movies, restaurants, news, etc. [1]. In recent years, collaborative filtering techniques are often used to develop recommender systems [2]. These systems recommend items by clustering user purchase histories. The clustering techniques identify groups of users who appear to have similar preferences. However, they have several problems. Depending on user purchase histories to recommend items, they have the cold-start problem when they do not have enough histories[3]. And there are shilling acts that have specific products recommended more often than others by taking advantage of the techniques[4]. Another problem is that they do not consider user goals in searching activities.

More advanced techniques, which consider user goals, have been applied to recommender systems. In these systems, user goals such as preferences or lifestyles are considered [5, 6]. Many systems have been implemented with various approaches such as using commonsense reasoning [7]. However the interfaces of these systems are not interactive enough to decide what users think.

Some researchers have been developing conversational recommender systems to make interactive systems [8]. The interaction supported by conversational systems has a sequence of questions to narrow down recommended items. Users in the systems are requested to answer questions to get recommendations from the systems. Such interactive systems are good approaches to recognize user interests well, but they do not always provide suitable questions for the users. Even though one research is using grammatical analysis to develop a conversational recommender system [9], it is difficult to prepare the rules and questions for all situations in conversations with users.

Our approach differs from all the above techniques, because it has a web-chat interface and users can talk with others. Since others answer the users, our system does not have to answer what users say. Our system just analyzes conversations among users and recommends items. For ways to show items to users, most systems need user requests. But our system does not need requests to show items. When the system sufficiently recognizes user interests while analyzing conversations among users, recommended items are shown automatically and changed in real-time based on changes of user interests.

6   Conclusion and Future Work

In this paper, we use communication among users to develop a recommender system with a web-chat interface in which users can talk with others. By observing conversation in the web-chat interface in real-time and always recognizing user interests, the system can recommend items that reflect the contents of conversations. As a result, our system can recommend suitable items for the latest user interests. With the recommender system, users can talk with others and make their own interests clear. Users simply talk with others through the web-chat interface to get recommendations.

The system always observes conversations among users and analyzes their conversation contents to renew user profiles. The recommended items depend on how the system renews user profiles by recognizing user interests. For good recommendations, this process must increase its accuracy. We believe that it is beneficial for the system to increase the words in its dictionary. If it does so, the system can recognize user interests more accurately. We should introduce a more complicated algorithm to analyze conversations among users for developing more intelligent recommender systems. For example, if a system can catch topics of conversations, it can recommend more suitable items.

As we explained in Section 2, we handcrafted an item database for recommended items, defined parameters for items in it, and matched items with user interests. The system is not linked to any shopping sites, so it cannot recommend items from big databases. However, we believe our approach can be extended to big databases by automatically setting values of parameters with various kinds of information such as brand, color, material, etc. of each item.

In the example of Section 3, we experimentally implemented a system that only recommends watches. We must also apply our system to various products and estimate its advantages more in detail.

References

1. Resnick, P., Varian, H.R.: Recommender Systems. Communications of the ACM 40(3), 56–58 (1997)
2. Miller, B., Riedl, J., and Konstan, J.: Experiences with GroupLens: Making Usenet useful again. In: Proceedings of the 1997 Usenix Winter Technical Conference (January 1997)
3. Guo, H.: Soap: Live recommendations through social agents. In: Fifth DELOS Workshop on Filtering and Collaborative Filtering, Budapest (1997)
4. Lam, S.K., Riedl, J.: Shilling recommender systems for fun and profit. In: Proceedings of the 13th International World Wide Web Conference (WWW), pp. 393–402. ACM Press, New York (2004)
5. Burke, R., Hammond, K., Young, B.: The FindMe Approach to Assisted Browsing. IEEE Expert 12(4), 32–40 (1997)
6. Fano, A., Kurth, S.W.: Personal Choice Point: Helping users visualize what it means to buy a BMW. In: Proc of IUI 2003, pp. 46–52 (2003)
7. Shen, E., Lieberman, H., Lam, F.: What am I gonna wear?: Scenatio-Oriented Recommendation. In: The proceedings of the International Conference on Intelligent User Interface, IUI 2007, Honolulu, Hawaii, USA, January 28-31 (2007)
8. Goker, M.H., Thompson, C.A.: Personalized Conversational Case-Based Recommendation. Springer, Heidelberg (2000)
9. Bridge, D.: Towards Conversational Recommender Systems: A Dialogue Grammar Approach. In: Procs. of the Workshop in Mixed-Initiative Case-Based Reasoning, Workshop
10. Amazon.co.jp, http://www.amazon.co.jp/

Ontological Query Processing and Maintaining Techniques for FAQ Systems

Sheng-Yuan Yang

Department of Computer and Communication Engineering, St. John's University,
499, Sec. 4, TamKing Rd., Tamsui, Taipei County 25135, Taiwan
ysy@mail.sju.edu.tw

Abstract. This paper illustrated an Interface Agent which works as an assistant between the users and FAQ systems to retrieve FAQs on the domain of Personal Computer. It can effectively tackle the problems associated with traditional FAQ retrieval systems. Specifically, we addressed how ontology helps interface agents to provide better FAQ services and related algorithms described in details. Our preliminary experimentation demonstrated that user intention and focus of up to eighty percent of the user queries can be correctly understood by the system, and accordingly provided the query solutions with higher user satisfaction.

Keywords: Ontology, Template-based Processing, User Modeling, Interface Agents, FAQ Systems.

1 Introduction

With increasing popularity of the Internet, it affects people's life style in terms of how people acquire, present, and exchange information. Especially the use of the World Wide Web has been leading to a large increase in the number of people who access FAQ knowledge bases to find answers to their questions [9]. As the techniques of Information Retrieval [5,6] matured, a variety of information retrieval systems have been developed, e.g., Search engines, Web portals, etc., to help search on the Web. How to search is no longer a problem. The problem now comes from the results from these information retrieval systems which contain some much information that overwhelms the users. Therefore, how to improve traditional information retrieval systems to provide search results which can better meet the user requirements so as to reduce his/her cognitive loading is an important issue in current research [1].

We have proposed an FAQ-master as an intelligent Web information aggregation system, which provides intelligent information retrieval, filtering, and aggregation services [11,14]. It contains four agents, namely, Interface Agent [15], Search Agent [16], Answerer Agent [17,18], and Proxy Agent [12,13,17], supported by an ontology-supported Content Base, which in turn contains a User Model Base, Template Base, Domain Ontology, Website Model Base, Ontological Database, Solution Library, and Rule Base. This paper discussed the Interface Agent focusing on how it captures true user's intention and accordingly provides high-quality FAQ answers. The agent features ontology-based representation of domain knowledge, flexible interaction interface, and personalized information filtering and display. Our

N.T. Nguyen, R. Katarzyniak (Eds.): New Chall. in Appl. Intel. Tech., SCI 134, pp. 97–109, 2008.

preliminary experimentation demonstrated that the intention and focus of up to eighty percent of the users' queries can be correctly understood by the system, and accordingly provided the query solutions with higher user satisfaction. The Personal Computer (PC) domain is chosen as the target application of our Interface Agent and will be used for explanation in the remaining sections.

2   Fundamental Techniques

2.1   Domain Ontology and Services

We have outlined an engineering procedure for developing an ontology in [10]. By following the procedure we developed an ontology for the PC domain in Chinese using Protégé 2000 [3], but was changed to English here for easy explanation, as the fundamental background knowledge for the system. Fig. 1 shows part of the ontology taxonomy, which represents relevant PC concepts as classes and their parent-child relationships as *isa* links and the nodes at the bottom level represent various concept instances which capture real world data. The complete PC ontology can be referenced from the Protégé Ontology Library at Stanford Website (http://protege.stanford.edu/ download/download.html). We also developed a Problem ontology to deal with query questions. Fig. 2 illustrates part of the Problem ontology, which contains query type and operation type. Together they imply the semantics of a question. Finally, we use Protégé's APIs to develop a set of ontology services, which provide primitive functions to support the application of the ontologies. The ontology services currently available include transforming query terms into canonical ontology terms, finding definitions of specific terms in ontology, finding relationships among terms, finding compatible and/or conflicting terms against a specific term, etc.

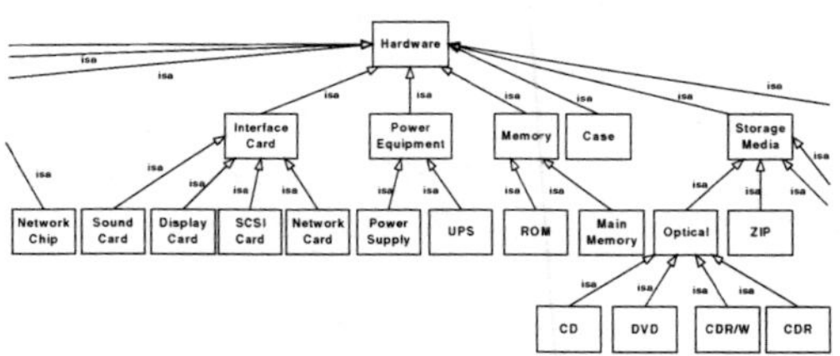

Fig. 1. Part of PC ontology taxonomy

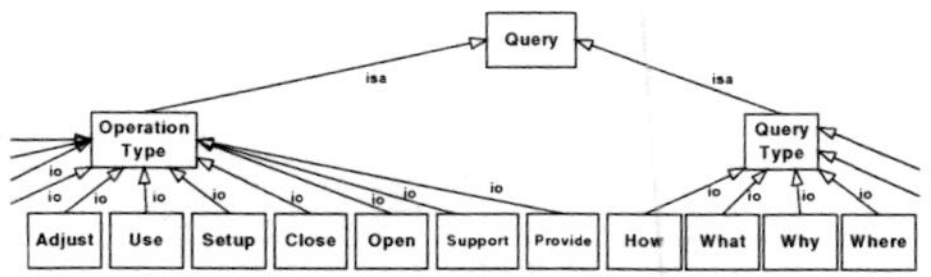

Fig. 2. Part of problem ontology taxonomy

Table 1. Examples of query patterns

Question Type	Operation Type	Intention Type	Query Pattern
是否 (If)	支援 (Support)	ANA_CAN_SUPPORT	<S1 是否 支援 S2>
		GA-7VRX 這塊主機板是否支援 KINGMAX DDR-400？ (Could the GA-7VRX motherboard support the KNIGMAX DDR-400 memory type?)	
如何 (How)	安裝 (Setup)	HOW_SETUP	<如何 在 S1><安裝 S2>
		如何在 Windows 98SE 下，安裝 8RDA 的音效驅動程式？ (How to setup the 8RDA sound driver on a Windows 98SE platform?)	
什麼 (What)	是 (Is)	WHAT_IS	<S1 是 什麼>
		AUX power connector是什麼？ (What is an AUX power connector?)	
何時 (When)	支援 (Support)	WHEN_SUPPORT	<S1 何時 支援 S2>
		P4T何時才能支援 32-bit 512 MB RDRAM 記憶體規格？ (When can the P4T support the 32-bit 512 MB RDRAM memory specification?)	
哪裡 (Where)	下載 (Download)	WHERE_DOWNLOAD	<S1><哪裡 可以 下載 S2>
		CUA 的 Driver CD 遺失，請問哪裡可以下載音效驅動程式？ (Where can I download the sound driver of CUA whose Driver CD was lost?)	
為什麼 (Why)	列印 (Print)	WHY_PRINT	[S1]<S2 無法 列印>
		為什麼在Win ME底下，從休眠狀態中回復後，印表機無法列印。 (Why can I not print after coming back from dormancy on a Win ME platform?)	

2.2 Ontological Query Templates

To build the query templates, we have collected in total 1215 FAQs from the FAQ website of six famous motherboard factories in Taiwan and used them as the reference materials for query template construction. Currently, we only take care of the user query with one intention word and at most three sentences. These FAQs were analyzed and categorized into six types of questions. For each type of question, we further identified several intention types according to its operations. Finally, we define a query pattern for each intention type. Table 1 illustrates the defined query patterns for the intention types. Now all information for constructing a query template is ready, and we can formally define a query template [2,7]. According to the generalization relationships among intention types, we can form a hierarchy of intention types to organize all FAQs. Currently, the hierarchy contains two levels as shown in Fig. 3. Now, the system can employ the intention type hierarchy to reduce the search scope during the retrieval of FAQs after the intention of a user query is identified.

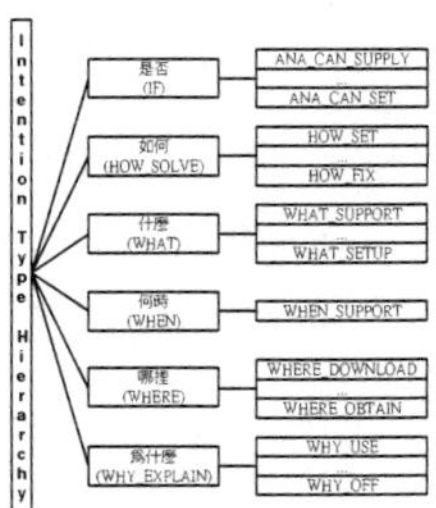

Fig. 3. Intention type hierarchy

3 System Architecture

3.1 User Modeling

A user model [15] contains interaction preference, solution presentation, domain proficiency, terminology table, query history, selection history, and user feedback. The interaction preference is responsible for recording user's preferred interface.

When the user logs on the system, the system can select a proper user interface according to this preference. The solution presentation is responsible for recording solution ranking preferences of the user. In addition, we use a Show_Rate parameter to control how many items of solutions for display each time, in order to reduce information overloading problem. The domain proficiency factor describes how familiar the user is with the domain. By associating a proficiency degree with each ontology concept, we can construct a table, which contains a set of <concept proficiency-degree> pairs, as his/her domain proficiency. Thus, during the decision of solution representation, we can calculate the user's proficiency degree on solutions using the table, and accordingly only show his/her most familiar part of solutions and hide the rest for advanced requests. To solve the problem of different terminologies to be used by different users, we include a terminology table to record this terminology difference. We can use the table to replace the terms used in the proposed solutions with the user favorite terms during solutions representation to help him better comprehend the solutions. Finally, we record the user's query history as well as FAQ selection history and corresponding user feedback in each query session in the Interaction history, in order to support collaborative recommendation.

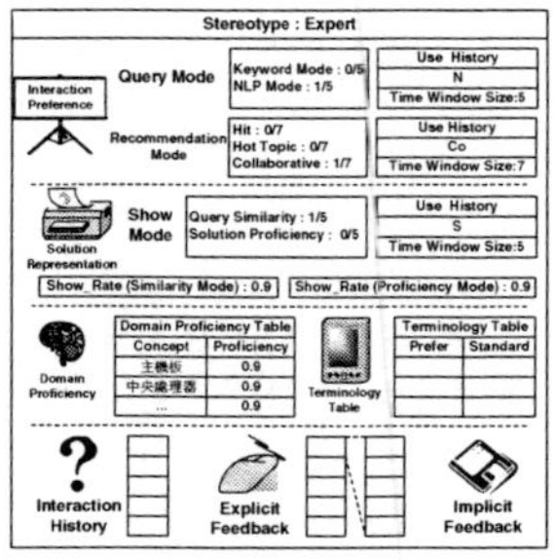

Fig. 4. Example of expert stereotype

In order to quickly build an initial user model for a new user, we pre-defined five stereotypes [4], namely, expert, senior, junior, novice, and amateur [10], to represent different user group's characteristics. This approach is based on the idea that the same group of user tends to exhibit the same behavior and requires the same information. Fig. 4 illustrates an example user stereotype. When a new user enters the system, he is asked to complete a questionnaire, which is used by the system to determine his/her domain proficiency, and accordingly select a user stereotype to generate an initial user model to him. However, the initial user model constructed from the stereotype may be too generic or imprecise. It will be refined to reflect the specific user's real intent after the system has experiences with his/her query history, FAQ-selection history and feedback, and implicit feedback [1].

3.2 System Architecture and Related Techniques

Fig. 5 shows the architecture of the Interface Agent. The Interaction Agent consists of the following three components: Adapter, Observer and Assistant. First, the Adapter constructs best interaction interfaces according to user's favorite query and

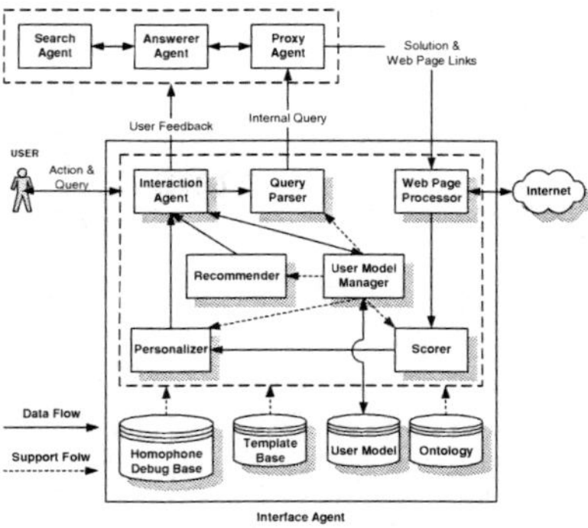

Fig. 5. Interface agent architecture

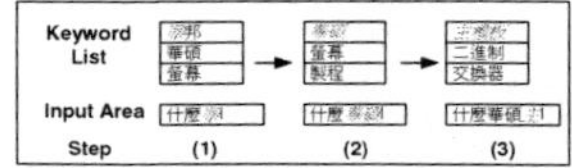

Fig. 6. Examples of automatic keyword scrolling mechanism

recommendation modes. It is also responsible for presenting to the user the list of FAQ solutions (from the Personalizer) or recommendation information (from the Recommender). During solution representation, it arranges the solutions in terms of the user's preferred style (query similarity or solution proficiency) and displays the solutions according to the "Show_Rate." Second, the Observer passes the user query to the Query Parser, and simultaneously collects the interaction information and related feedback from the user. The interaction information contains user preferred query mode, recommendation mode, solution presentation mode, and FAQ clicking behavior, while the related feedback contains user satisfaction degree and comprehension degree about each FAQ solution. The User Model Manager needs both interaction information and related feedback to properly update user models and stereotypes. The satisfaction degree in related feedback can also be passed to the Proxy Agent for tuning the solution search mechanism [17]. Finally, the Assistant provides proper assistance and guidance to help the user query process. First, the ontology concepts are structured and presented as a tree so that the users who are not familiar with the domain can check on the tree and learn proper terms to enter their queries. We also rank all ontology concepts by their probabilities and display them in a keyword list. When the user enters a query at the input area, the Assistant will automatically "scroll" the content of the keyword list to those terms related to the input keywords. Fig. 6 illustrates an example of this automatic keyword scrolling mechanism. In addition to the keyword-oriented query mode, the Assistant also provides lists of question types and operation types to help question type-oriented or operation type-oriented search. The user can use one, two, or all of these three mechanisms to help form his/her query in order to convey his/her intention to the system.

The Query Parser pre-processes the user query by performing Chinese word segmentation, correction on word segmentation, fault correction on homophonous or multiple words, and term standardization. It then employs template-based pattern matching to analyze the user query and extract the user intention and focus. Finally, it transforms the user query into the internal query format and then passes the query to the Proxy Agent for retrieving proper solutions [17]. Fig. 7 shows the flow chart of the Query Parser. Given a user query in Chinese, we segment the query using MMSEG [8]. The results of segmentation were not good, for the predefined MMSEG word corpus contains insufficient terms of the PC domain. For example, it does not contain the keywords "華碩" or "AGP4X", and returns wrong word segmentation like "華", "碩", "AGP", and "4X". The step of query pruning can easily fix this by using the ontology as a second word corpus to bring those mis-segmented words back. It also performs fault correction on homophonous or multiple words using the ontology and homophone debug base [1]. The step of query standardization is responsible for replacing the terms used in the user query with the canonical terms in the ontology and intention word base. The original terms and the corresponding canonical terms will then be stored in the terminology table for solution presentation personalization. Finally, we label those recognized keywords by symbol "K" and intention words by symbol "I." The rest are regarded as stop words and removed from the query. Now, if the user is using the keyword mode, we directly jump to the step of query formulation. Otherwise, we use template-based pattern matching to analyze the natural language input. The step of pattern match is responsible for identifying the semantic pattern associated with the user query. Using the pre-constructed query templates in the template base, we can compare the user query with the query templates and select the best-matched one to identify user intention and focus. Fig. 8 shows the algorithm of fast selecting possibly matched templates, Fig. 9 describes the algorithm which finds out all patterns matched with the user query, and Fig. 10 removes those matched patterns that are generalization of some other matched patterns.

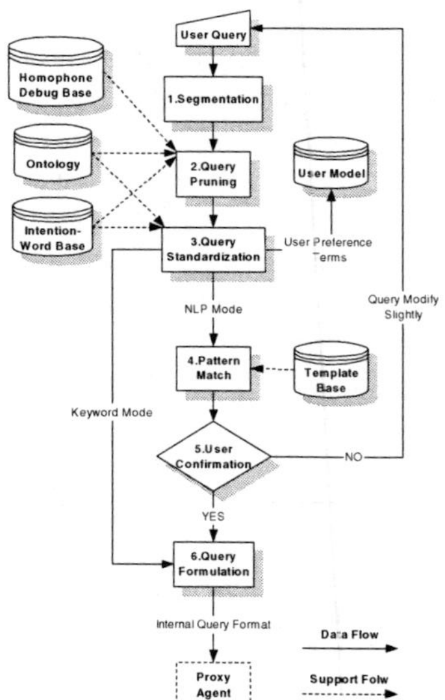

Fig. 7. Flow chart of the Query Parser

```
Template Selection :
Q : User Query.
Q.Intention_Word = {I₁, I₂,..., Iₙ}, Intention words in Q.
Q.Sentence : Number of sentences in Q.
Template Base = {T₁, T₂,..., Tₘ}, M : Number of templates.
For each template Tⱼ in Template Base
{
    If Tⱼ conforms to the follow rules, then select Tⱼ into C.
        1. Tⱼ.Sentence = Q.Sentence.
        2. Tⱼ.Intention_Word ⊆ Q.Intention_Word.
}
return C : Candidate Templates.
```

Fig. 8. Query template selection algorithm

```
Pattern Match :
For each template Tⱼ in C, candidate templates
{
    Tⱼ.Pattern = {P₁, P₂,...}, Pₖ : pattern k in template j.
    For each Pₖ in Tⱼ.Pattern
    {
        If Pₖ match Q, the user query, then
            Pₖ.Intention_Word = Tⱼ.Intention_Word,
            Pₖ.Intention_Type = Tⱼ.Intention_Type,
            Pₖ.Quertion_Type = Tⱼ.Question_Type,
            Pₖ.Operation = Tⱼ.Operation,
            Pₖ.Focus = Tⱼ.Focus,
            and put Pₖ in M and break this inner loop.
    }
}
return M : Patterns matching Q.
```

Fig. 9. Pattern match algorithm

```
Pattern Removal :
For each pattern Pₖ in M, matched patterns
{
    If Pₖ conforms to follow rule, then remove Pₖ from M.
    {
        ∃Pₗ ∈ M, Pₗ.Intention_Type = Pₖ.Intention_Type and
                  Pₖ.Intention_Word ⊂ Pₗ.Intention_Word
    }
}
```

Fig. 10. Query pattern removal algorithm

Table 2. Format of retrieved FAQ

Field	Description
FAQ_No.	FAQ's identification
FAQ_Question	Question part of FAQ
FAQ_Answer	Answer part of FAQ
FAQ_Similarity	Similarity degree of the FAQ met with the user query
FAQ_URL	Source or related URL of the FAQ

```
A : Answer FAQs List, A = {F₁, F₂,..., Fₙ}, A ⊂ C, FAQ Collection.
A : FAQs found by the system that best match user query.
Fᵢ.Q : The Query part of FAQ Fᵢ.
For each Fᵢ ∈ A, we calculate each Fᵢ's Proficiency Score for user k.
```

$$Fi = \frac{1}{\text{Number of Concept appering in Fi.Q}} * \sum_{j} \text{Appearance(Concept } j) * \text{Proficiency(Concept } j),$$

```
where Concept j ∈ Ontology,
```

$$\text{Appearance(Concept } j): \begin{cases} 1, \text{Concept } j \text{ appears in Fᵢ.Q} \\ 0, \text{Concept } j \text{ doesn't appear in Fᵢ.Q,} \end{cases}$$

```
Proficiency(Concept j): The drgree of user k's proficiency in Concept j.
```

Fig. 11. Proficiency degree calculation algorithm

The Web Page Processor receives a list of retrieved solutions, which contains one or more FAQs matched with the user query from the Proxy Agent, each represented as Table 2, and retrieves and caches the solution webpages according to the

FAQ_URLs. It follows to pre-process those webpages for subsequent customization process, including URL transformation, keyword standardization, and keyword marking. The URL transformation changes all hyperlinks to point toward the cached server. The keyword standardization transforms all terms in the webpage content into ontology vocabularies. The keyword labeling marks the keywords appearing in the webpages by boldfaced <B>Keyword<\B> to facilitate subsequent keywords processing webpage readability.

Each FAQ is a short document; concepts involved in FAQs are in general more focused. In other words, the topic (or concept) is much clearer and professional. The question part of an FAQ is even more pointed about what concepts are involved. Knowing this property, we can use the keywords appearing in the question part of an FAQ to represent its topic. Basically, we use the table of domain proficiency to calculate a proficiency degree for each FAQ by calculating the proficient concepts appearing in the question part of the FAQ, detailed as shown in Fig. 11.

The Personalizer replaces the terms used in the solution FAQs with the terms in the user's terminology table, collected by the Query Parser, for improving the solution readability.

The first task of the User Model Manager is to create an initial user model for a new user. To do this, we pre-defined several questions for each concept in the domain ontology, for example, "Do you know a CPU contains a floating co-processor?", "Do you know the concept of 1GB=1000MB in specifying the capacity of a hard disk?", etc. The difficulty degrees of the questions are proportional to the hierarchy depth of the concepts in the ontology. When a new user logs on the system, the Manager randomly selects questions from the ontology. The user either answers an YES or NO to each question. The answers are collected and weighted according to the respective degrees and passed to the Manager, which then calculates a proficiency score for the user according to the percentage of correctness of his/her responses to the questions and accordingly instantiates a proper user stereotype as the user model for the user.

The second task is to update user models. Here we use the interaction information and user feedback collected by the Interaction Agent in each interaction session or query session. An interaction session is defined as the time period from the time point the user logs in up to when he logs out, while a query session is defined as the time period from when the user gives a query up to when he gets the answers and completes the feedback. An interaction session may contain several query sessions. After a query session is completed, we immediately update the interaction preference and solution presentation of the user model. Specifically, the user's query mode and solution presentation mode in this query session are remembered in both time windows, and the statistics of the preference change for each mode is calculated accordingly, which will be used to adapt the Interaction Agent on the next query session. Fig. 12 illustrates the algorithm to update the Show_Rate of the similarity mode. The algorithm uses the ratio of the number of user selected FAQs and that of the displayed FAQs to update the show rate; the algorithm to update the Show_Rate of the proficiency mode is similar.

In addition, each user will be asked to evaluate each solution FAQ in terms of the following five levels of understanding, namely, very familiar, familiar, average, not familiar, and very not familiar. This provides an explicit feedback and we can use it to update his/her domain proficiency table. Fig. 13 shows the updating algorithm.

$$N_s : \text{Number of FAQ in Solution FAQ List.}$$
$$N_{(Similarity\,Mode)} : \text{Number of FAQ shown to user in Similarity Mode.}$$
$$N_{(Similarity\,Mode)} = \lceil N_s * Show\_Rate_{(Similarity\,Mode)_{old}} \rceil$$
$$N_{Hide} = N_s - N_{(Similarity\,Mode)}, \text{Number of hidden FAQ.}$$
$$N_{Select} : \text{Number of FAQ selected by user in the query session.}$$
$$Show\_Rate_{(Similarity\,Mode)} = Show\_Rate_{(Similarity\,Mode)_{old}} + Variation$$
$$Variation = \begin{cases} \left(\frac{N_{Select}}{N_s}-0.7\right)*\left(1-\exp\left(-\frac{N_{(Similarity\,Mode)}}{\alpha}\right)\right), \text{if } \frac{N_{Select}}{N_s} \geq 0.7 \\ 0, \text{if } 0.3 < \frac{N_{Select}}{N_s} < 0.7 \\ \left(\frac{N_{Select}}{N_s}-0.3\right)*\left(1-\exp\left(-\frac{N_{(Similarity\,Mode)}}{\alpha}\right)\right), \text{if } \frac{N_{Select}}{N_s} \leq 0.3, \end{cases}$$
$$\text{where } \alpha : \text{weight change rate,}$$
$$Show\_Rate_{(Similarity\,Mode)_{new}} = Max(Min(Show\_Rate_{(Similarity\,Mode)}, 1), 0.01)$$

Fig. 12. Algorithm to update show rate in similarity mode

$$F = \{F_1, F_2, ...\} \subseteq C, \text{FAQ Collection.}$$
$$F_i \in F : \text{FAQs selected and rated by user.}$$
$$\text{For each } F_i, \text{update user proficiency of each Concept j.}$$
$$Proficiency\,(Concept\ j) =$$
$$Proficiency_{old}\,(Concept\ j) + \alpha * \left(\frac{T(Concept\ j)}{Number\ of\ Concept\ in\ F_i} * Understanding\ Level\right)$$
$$Proficiency_{new}\,(Concept\ j) = Max(Min(Proficiency\,(Concept\ j), 1), 0),$$
$$\text{where } \alpha : \text{Learning rate,}$$
$$\text{Concept } j \in \text{Ontology,}$$
$$T(Concept\ j) : \text{The times Concept j appears in } F_i,$$
$$Understanding\ Level : \begin{cases} +2, \text{user rates "very familiar" for } F_i. \\ +1, \text{user rates "familiar" for } F_i. \\ 0, \text{user rates "average" for } F_i. \\ -1, \text{user rates "not familiar" for } F_i. \\ -2, \text{user rates "very not familiar" for } F_i. \end{cases}$$

Fig. 13. Algorithm to update the domain proficiency table

$$\text{For each user}_i \text{ in each user proficiency group}$$
$$\{$$
$$Proficiency_{avg}(user_i) = \frac{\sum_j Proficiency(Concept\ j)}{Number\ of\ concepts\ in\ Domain\ Proficiency\ Table},$$
$$\text{where Concept } j : jth \text{ Concept in user}_i\text{'s Domain Proficiency Table,}$$
$$\text{if}\,(0.8 \leq Proficiency_{avg}(user_i) \leq 1.0), \text{user i reassigned to Expert group.}$$
$$\text{if}\,(0.6 \leq Proficiency_{avg}(user_i) < 0.8), \text{user i reassigned to Senior group.}$$
$$\text{if}\,(0.4 \leq Proficiency_{avg}(user_i) < 0.6), \text{user i reassigned to Junior group.}$$
$$\text{if}\,(0.2 \leq Proficiency_{avg}(user_i) < 0.4), \text{user i reassigned to Novice group.}$$
$$\text{if}\,(0.0 \leq Proficiency_{avg}(user_i) < 0.2), \text{user i reassigned to Amateur group.}$$
$$\}$$

Fig. 14. Algorithm to re-cluster all user groups

Finally, after each interaction session, we can update the user's recommendation mode in this session in the respective time window. At the same time, we add the query and FAQ-selection records of the user into the query history and selection history of his/her user model.

The third task of the User Model Manager is to update user stereotypes. This happens when a sufficient number of user models in a stereotype has undergone changes. First, we need to reflect these changes to stereotypes by re-clustering all affected user models, as shown in Fig. 14, and then re-calculates all parameters in each stereotype, an example as shown in Fig. 15.

The Recommender uses the following three policies to recommend information. 1) High hit FAQs. It recommends the first N solution FAQs according to their selection counts from all users in the same group within a time window. 2) Hot topic FAQs. It recommends the first N solution FAQs according to their popularity, calculated as statistics on keywords appearing in the query histories of the same group users within a time window. The algorithm does the hot degree calculation as shown in Fig. 16. 3)

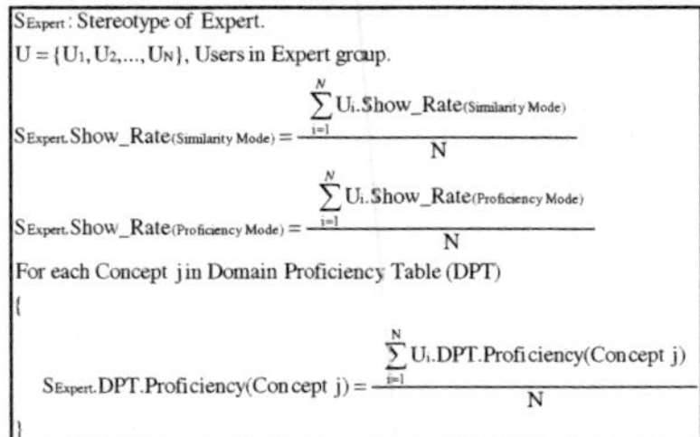

Fig. 15. Example to update the stereotype of Expert

For each $faq_i \in C$, FAQ Collection.

HOT Score $faq_i = $

$$\frac{1}{\text{Number of Concepts appearing in } faq_i} * \sum_j \text{Appearance(Concept } j) * \text{Weight(Concept } j),$$

where $\text{Concept } j \in$ Ontology,

$$\text{Appearance(Concept } j): \begin{cases} 1, \text{Concept } j \text{ appears in } faq_i \\ 0, \text{Concept } j \text{ doesn't appear in } faq_i, \end{cases}$$

Weight(Concept j) : Within a time window, the times that the Concept j appears in user queries of the same user group,

Fig. 16. Algorithm to calculate the hot degree

$G = \{U_1, U_2, ..., U_m\}$, Users in the same group G.

$Y = \{U_j \in G \mid U_j \neq U_i\}$.

FAQ Collection $= \{F_1, F_2, ..., F_n\}$.

$U_i = \{S_{i,1}, S_{i,2}, ..., S_{i,x}\}$, Query sessions of user i.

$S_{i,x} = \{R_{i,x,1}, R_{i,x,2}, ..., R_{i,x,n}\}$, FAQs rated and/or selected by User i in session x.

$$R_{i,x,y}: \begin{cases} 1, \text{If User i select } F_y \text{ in session x, and rate more than satisfying.} \\ 0.8, \text{If User i select } F_y \text{ in session x, and rate satisfying.} \\ 0.6, \text{If User i select } F_y \text{ in session x, and rate average.} \\ 0.6, \text{If User i select } F_y \text{ in session x, but no rating.} \\ 0.4, \text{If User i select } F_y \text{ in session x, and rate dissatisfying.} \\ 0.2, \text{If User i select } F_y \text{ in session x, and rate less than dissatisfying.} \\ 0, \text{If User i doesn't select } F_y \text{ in session x.} \end{cases}$$

$$\text{Similarity}(S_{i,x}, S_{j,k}) = \arg\max_{S_{j,k}} \left(\sum_{U_j \in Y} \sum_{S_{j,k} \in U_j} (1 - \text{Distance}(S_{i,x}, S_{j,k})) \right),$$

$$\text{where Distance}(S_{i,x}, S_{j,k}) = \frac{\sqrt{\sum_{z=1}^{n} (R_{i,x,z} - R_{j,k,z})^2}}{n}.$$

Let $S_{j,k}$ be the most similar to $S_{i,x}$, and $S_{i,x+1}$ be user i in current query session.

Recommend following FAQs for $S_{i,x+1} = $

$\{F_y \mid R_{j,k,y} > 0.5 \text{ and } R_{i,x,y} = 0\} \cup \{F_y \mid R_{j,k+1,y} > 0.5 \text{ and } R_{i,x,y} = 0\}$

Fig. 17. Algorithm to do the collaborative recommendation

Collaborative recommendation. It refers to the user's selection histories of the same group to provide solution recommendation. The basic idea is this. If user A and user B are in the same group and the first n interaction sessions of user A are the same as those of user B, then we can recommend the highest-rated FAQs in the (n+1)th session of user A for user B, detailed algorithm as shown in Fig. 17.

4 System Experiments

Firstly, we focus on the performance evaluation of the most important module, i.e., the Query Parser. Our philosophy is that if it can precisely parse user queries and extract both true query intention and focus from them, then we can effectively improve the quality of the retrieved. Recall that the Query Parser employs the technique of template-based pattern matching mechanism to understand user queries and the templates were manually constructed from 1215 FAQs. In the first experiment, we use this same FAQs for testing queries, in order to verify whether any conflicts exist within the query. Table 3 illustrates the experimental results, where only 33 queries match with more than one query patterns and result in confusion of query intention, called "error" in the table. These errors may be corrected by the user. The experiment shows the effectiveness rate of the constructed query templates reaches 97.28%, which implies the template base can be used as an effective knowledge base to do natural language query processing. Our second experiment is to learn how well the Parser understands new queries. First, we collected in total 143 new FAQs, different from the FAQs collected for constructing the query templates, from four famous motherboard factories in Taiwan, including ASUS, GIGABYTE, MSI, and SIS. We then used the question parts of those FAQs for testing queries, which test how well the Parser performs. Our experiments show that we can precisely extract true query intentions and focuses from 112 FAQs. The rest of 31 FAQs contain up to three or more sentences in queries, which explain why we failed to understand them. In summary, 78.3% (112/143) of the new queries can be successfully understood.

Finally, Table 4 shows the comparison of user satisfaction of our systemic prototype (Fig. 18) against other search engines. In the table, S_T, for Satisfaction of testers, represents the average of satisfaction responses from 10 ordinary users, while S_E, for Satisfaction of experts, represents that of satisfaction responses from 10

Table 3. Effectiveness of constructed query patterns

#Testing	#Correct	#Error	Precision Rate (%)
1215	1182	33	97.28 %

Table 4. User satisfaction evaluation

KEYWORD METHOD	CPU (S_E / S_T)	MOTHERBOARD (S_E / S_T)	MEMORY (S_E / S_T)	AVERAGE (S_E / S_T)
Alta Vista	63% / 61%	77% / 78%	30% / 21%	57% / 53%
Excite	66% / 62%	81% / 81%	50% / 24%	66% / 56%
Google	66% / 64%	81% / 80%	38% / 21%	62% / 55%
HotBot	69% / 63%	78% / 76%	62% / 31%	70% / 57%
InfoSeek	69% / 70%	71% / 70%	49% / 28%	63% / 56%
Lycos	64% / 67%	77% / 76%	36% / 20%	59% / 54%
Yahoo	67% / 61%	77% / 78%	38% / 17%	61% / 52%
Our approach	78% / 69%	84% / 78%	45% / 32%	69% / 60%

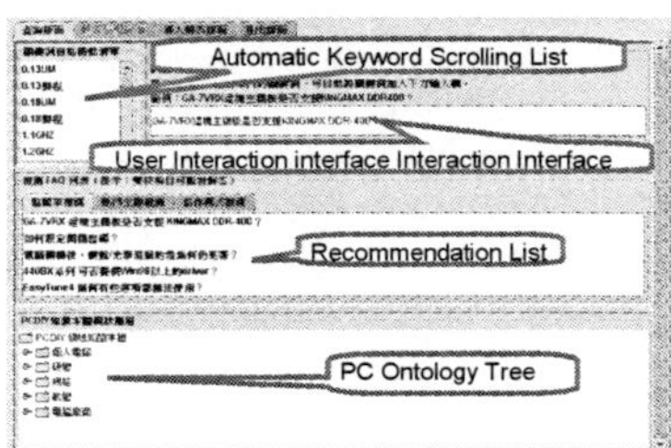

Fig. 18. Main tableau of our systemic prototype

experts. Basically, each search engine receives 100 queries and returns the first 100 webpages for evaluation of satisfaction by both experts and non-experts. The table shows that our approach, the last row, enjoys the highest satisfaction in all classes.

5 Discussions and Future Work

We have developed an Interface Agent to work as an assistant between the users and systems, which is different from system architecture and implementation over our previous work [11]. It is also used to retrieve FAQs on the domain of PC. We integrated several interesting techniques including user modeling, domain ontology, and template-based linguistic processing to effectively tackle the problems associated with traditional FAQ retrieval systems. In short, our work featured an ontology-supported, template-based user modeling technique for developing interface agents; a nature language query mode, along with an improved keyword-based query mode; and an assistance and guidance for human-machine interaction. In other words, we addressed how ontology helps interface agents to provide better FAQ services and related algorithms described in details. Our preliminary experimentation demonstrated that user intention and focus of up to eighty percent of the user queries can be correctly understood by the system, and accordingly provided the query solutions with higher user satisfaction. In the future, we are planning to employ the techniques of machine learning and data mining to automate the construction of the template base. As to the allover system evaluation, we are planning to employ the concept of usability evaluation on the domain of human factor engineering to evaluate the performance of the user interface.

Acknowledgments. The author would like to thank Yai-Hui Chang and Ying-Hao Chiu for their assistance in system implementation. This work was supported by the National Science Council, R.O.C., under Grants NSC-89-2213-E-011-059, NSC-89-2218-E-011-014, and NSC-95-2221-E-129-019.

References

1. Chiu, Y.H.: An Interface Agent with Ontology-Supported User Models. Master Thesis, Department of Electronic Engineering, National Taiwan University of Science and Technology, Taiwan, R.O.C (2003)
2. Hovy, E., Hermjakob, U., Ravichandran, D.: A Question/Answer Typology with Surface Text Patterns. In: Proc. of the DARPA Human Language Technology conference, San Diego, CA, USA, pp. 247–250 (2002)
3. Noy, N.F., McGuinness, D.L.: Ontology Development 101: A Guide to Creating Your First Ontology. Stanford Knowledge Systems Laboratory Technical Report KSL-01-05 and Stanford Medical Informatics Tech. Rep. SMI-2001-0880 (2001)
4. Rich, E.: User Modeling via Stereotypes. Cognitive Science 3, 329–354 (1979)
5. Salton, G., Wong, A., Yang, C.S.: A Vector Space Model for Automatic Indexing. Communications of ACM 18(11), 613–620 (1975)
6. Salton, G., McGill, M.J.: Introduction to Modern Information Retrieval. McGraw-Hill Book Company, New York (1983)

7. Soubbotin, M.M., Soubbotin, S.M.: Patterns of Potential Answer Expressions as Clues to the Right Answer. In: Proc. of the TREC-10 Conference, NIST, Gaithersburg, MD, USA, pp. 293–302 (2001)
8. Tsai, C.H.: MMSEG: A word identification system for Mandarin Chinese text based on two variants of the maximum matching algorithm (2000), `http://technology.chtsai.org/mmseg/`
9. Winiwarter, W.: Adaptive Natural Language Interface to FAQ Knowledge Bases. International Journal on Data and Knowledge Engineering 35, 181–199 (2000)
10. Yang, S.Y., Ho, C.S.: Ontology-Supported User Models for Interface Agents. In: Proc. of the 4th Conference on Artificial Intelligence and Applications, Chang-Hwa, Taiwan, pp. 248–253 (1999)
11. Yang, S.Y., Ho, C.S.: An Intelligent Web Information Aggregation System Based upon Intelligent Retrieval, Filtering and Integration. In: Proc. of the 2004 International Workshop on Distance Education Technologies, Hotel Sofitel, San Francisco Bay, CA, USA, pp. 451–456 (2004)
12. Yang, S.Y., Liao, P.C., Ho, C.S.: A User-Oriented Query Prediction and Cache Technique for FAQ Proxy Service. In: Proc. of the 2005 International Workshop on Distance Education Technologies, Banff, Canada, pp. 411–416 (2005)
13. Yang, S.Y., Liao, P.C., Ho, C.S.: An Ontology-Supported Case-Based Reasoning Technique for FAQ Proxy Service. In: Proc. of the 17th International Conference on Software Engineering and Knowledge Engineering, Taipei, Taiwan, pp. 639–644 (2005)
14. Yang, S.Y.: FAQ-master: A New Intelligent Web Information Aggregation System. In: International Academic Conference 2006 Special Session on Artificial Intelligence Theory and Application, Tao-Yuan, Taiwan, pp. 2–12 (2006)
15. Yang, S.Y.: An Ontology-Supported and Query Template-Based User Modeling Technique for Interface Agents. In: 2006 Symposium on Application and Development of Management Information System, Taipei, Taiwan, pp. 168–173 (2006)
16. Yang, S.Y.: An Ontology-Supported Website Model for Web Search Agents. In: Proc. of the 2006 International Computer Symposium, Taipei, Taiwan, pp. 874–879 (2006)
17. Yang, S.Y.: How Does Ontology Help Information Management Processing. WSEAS Transactions on Computers 5(9), 1843–1850 (2006)
18. Yang, S.Y., Chuang, F.C., Ho, C.S.: Ontology-Supported FAQ Processing and Ranking Techniques. International Journal of Intelligent Information Systems 28(3), 233–251 (2007)

On Human Resource Adaptability in an Agent-Based Virtual Organization*

Costin Bădică[1], Elvira Popescu[1], Grzegorz Frackowiak[2], Maria Ganzha[2], Marcin Paprzycki[2], Michal Szymczak[2], and Myon-Woong Park[3]

[1] University of Craiova, Software Engineering Department
   Bvd.Decebal 107, Craiova, RO-200440, Romania
   {badica_costin,popescu_elvira}@software.ucv.ro
[2] Systems Research Institute, Polish Academy of Science, Warsaw, Poland
   {maria.ganzha,marcin.paprzycki}@ibspan.waw.pl
[3] Korea Institute of Science and Technology, Seoul, Korea
   myon@kistmail.kist.re.kr

Summary. This paper discusses subsystem responsible for providing human resource adaptation through software-supported training in an agent-based virtual organization. Attention is focused on the requirements, functionalities and components of this subsystem and its interactions with other parts of the system.

1 Introduction

In our recent work on agent-based virtual organizations ([3, 4, 12, 13]) we have argued that support for collaborative work in a project-oriented organization must be *adaptable* on a number of levels. Basically, this means that as projects carried within the organization evolve, the behavior of the support system (including usage and access to resources) should evolve as well. Consequently, we have setup the goal of developing a system to meet this requirement.

In our earlier work we have argued that emergent software technologies such as software agents [16] and the Semantic Web [11] should be the base around which the proposed system is conceptualized. In particular: (i) organizational structure consisting of specific "roles" and interactions between them should be represented by software agents and their interactions, and (ii) domain knowledge, resource profiles and resource matching have to be represented using ontologies and semantic reasoning.

Separately, adaptability within the organization was sub-divided into:

- *System adaptability* obtained through: adapting various "structures" within the agent system; and adapting resource profiles.
- *Human resources adaptability* achieved by (e-)learning.

---

* Work of G. Frackowiak, M. Ganzha, M. Paprzycki. M. Szymczak and M.-W. Park was partially sponsored by by the KIST-SRI PAS "Agent Technology for Adaptive Information Provisioning" grant.

In our previous work we have outlined processes involved when a task/project is introduced into an organization (approached from the point of view of resource management) [12]; in [4] we have approached the proposed system from the point of view of roles played by various entities identified in [12]; while in [13], we have outlined how ontologies are going to be used in the proposed system. This allowed us to conceptualize, in [4], which roles can be played by software agents alone, by human(s), and by human-agent team(s). Separately, focusing on human resources, in [3] we have sketched initial ideas of how e-learning can be introduced into the system to support adaptability.

In this paper we further explore adaptability of human resources, by naturally enhancing the task oriented view of the work with introduction of *training tasks*. Observe that as work carried out by the organization is focused on tasks, this naturally leads to the idea of training as "workplace learning"—i.e. learning taking place in close relation with usual activities performed in the workplace. In this context, we introduce two approaches to training in the system: *reactive*, and *proactive*.

We start the paper by briefly summarizing main features of the system. Then we detail our proposal for achieving human resource adaptability by introducing training tasks covering both reactive and proactive approaches. During this analysis we identify two specialized units that are needed in the system: the *Competence Management Unit* and the *Training Management Unit* and outline their main functionalities.

2   System Overview

Our system is conceived as an agent-based virtual organization, which provides adaptive support for project-based collaborative work [3]. Structure of the organization and interactions between participants are represented using software agents and their interactions. Each human participant (member of the organization) has an associated *Personal Agent—(PA)*. Domain knowledge and resource (human and non-human) profiles are overlaid on top of ontologies [3, 4, 13].

We assumed that work carried out within the organization is project-driven (however, the notion of the project is very broad and includes installation of Cable TV as well as design and implementation of an intranet based information system for a corporation). Based on analysis performed in [4] several components of the system were identified (note that these are "roles," rather than "individuals"):

- *Project Manager (PM)* is created whenever a project proposal is submitted to the organization. Its main duties cover: formulation of project requirements, if project is accepted formulation of project schedule, assignment of resources to project activities, supervising project's progress and assuring its completion.
- *Analysis Manager (AM)* analyzes project requirements and formulates documents that is used to support the decision if the project should be accepted or not.
- *Organization Provisioning Manager (OPM)* is responsible for management of resources of the organization.
- *Resource Procurement Unit (RPU)* represents an interface between the organization and the "outside world." Its role is to seek and potentially deliver resources requested by the *OPM*.

- *Task Monitoring Agent* (*TMA*) is responsible for monitoring a given task according to its schedule and informing the *PM* in the case of any problems.
- *Quality of Service Management Unit* (*QoS*) is responsible with quality control of tasks completed by workers.

3 Conceptualizing Training Tasks

Following [14], we understand *training* as "acquisition of knowledge, skills, and competencies as a result of the teaching of vocational or practical skills and knowledge that relates to specific useful skills". In the context of our work we refer to training as "workplace learning," i.e. training that is closely related to the needs of improving performance of performing tasks in an organization. Clearly, this type of training can and should be closely related to projects carried out by the organization.

First, note that when conceptualizing a training task, three aspects need to be taken into account: (i) *timing*, i.e. when training should be started (possibly also: when it should end) (ii) *goals*, i.e. what should be goals of each specific training activity and, (iii) *trainees*, i.e. who should be enrolled in a given training task.

In our system we found the timing issue crucial for distinguishing between the *reactive* approach, and the *proactive* approach. Basically, the *reactive approach* may occur in two situations. First, when a new project is introduced in the organization, and consequently defined and represented within the support system. Here, based on the project requirements and available resources, management of the organization may decide to enroll selected human resources into training activities, carried out "within the project" (i.e. within the project budget and time). Second, while the project is running either the *Personal Agent* of an employee or the project manager (*PM*) may decide to enroll an employee or group of employees in ad-hoc training tasksto acquire specialized knowledge increments, to solve specific problems.

The *proactive approach* occurs when the management within the organization, based on current market conditions, history of the interactions between the organization and the external environment, specific regulations, expected projects, etc., decides to enroll selected employees into training tasks. Let us now look into these two approaches in more detail.

4 Reactive Approach

4.1 First Case—Project Level

The first case of the reactive approach follows the process of introducing a task into the system outlined in [12] and enhances it with activities and roles required for training. Decision to follow this approach is taken at the project level.

When a new project request is submitted to an organization, a new *Project Manager* (*PM*) is created. The first responsibility of the *PM* is to assure that project requirements are carefully analyzed and, on the basis of this analysis, a decision is made to accept or reject the job.

Analysis and suggestion as to what the decision should be are made by the *Analysis Manager (AM)*. The *AM* produces the *Requirements Analysis Document* containing a detailed specification of the project together with resources required and their associated desired competencies. Some details of this process are given in [12]. Note that the final acceptance/rejection decision is made by the *PM* (i.e. the project *PM*, or an appropriate *PM* on a higher hierarchical position in the organization).

Next, assuming that the job is accepted, differently from [12], *PM* creates an *Abstract Project Schedule* based on the general competencies available in the organization that she is aware of. The initial project schedule is called "abstract" because it contains only refined descriptions of necessary resources from the *Requirements Analysis Document*, i.e. the *Resource Request Descriptions*, rather then the resources themselves. Actual assignment of resources can be fixed later based on matching of required competencies from the *Resource Request Description* with available competencies of existing resources. Note that resources can be either human and non-human. As focus of this paper is on human resource adaptability, here we consider only this situation. For details of the general case please consult [12].

The next step of the process is to transform the *Abstract Project Schedule* into an actual *Project Schedule*. In this process, the *PM* can propose utilization of resources that she knows (see [12]). However, resources that *PM* knows might not be sufficient, i.e. either they are unavailable or they do not exist. In this case the *PM* contacts the *Organization Provisioning Manager (OPM)*. The *OPM* is the general resource manager of the organization. The process of seeking missing resources, conducted collaboratively by the *PM* and the *OPM* can produce the following results:

- A matching resource exists and is currently available. This is the simplest situation, i.e. the resource will be assigned to the project.
- A matching resource exists in the organization but it is not currently available. Here we have two cases: (i) the *Project Schedule* can be updated by re-scheduling tasks requiring this resource such that he/she will be available (obviously, this is possible only if project constraints, like deadlines, are not violated), and (ii) if the *Project Schedule* cannot be updated to accommodate the available resource then we treat this situation as the one in the next point.
- A matching resource does not exist in the organization. Here there are two cases: (i) an external resource is found by the *Resource Procurement Unit* (in our case someone with adequate skills is hired), or (ii) a decision is made to train an available resource that has some but not all of the required skills for the job.

Note that, additionally to [12], we propose that human resources should be handled in such a way to accommodate adaptability by interleaving work with training. Thus, we provide an overview of the process of training an available human resource that has some but not all of the required skills for the job. Here, note also that training decision depends on the following factors: (i) current level of competence of available resources; (ii) competence increment that represents the gap between available and required competencies for the job; (iii) project constraints. Furthermore, introduction of the training tasks may require also the update of the *Project Schedule* to accommodate the new training activities within the project timing and costs.

4.2   Second Case—Individual or Group Level

The second case of reactive training occurs after the project has already started (so either all the necessary resources were found within the organization, or/and training sessions were carried out at the beginning of the project, or/and external resources were hired). Usually decision to apply additional training tasks can be undertaken either at the individual or group (part of the team) level. Note however that such decision may be also made at the project level on the basis of the *PM*'s observations of what is happening within the project while it is running.

We will now illustrate this case in the framework of a software and services company. Let us consider an example of a customer requesting creation of an intranet and a company knowledge portal.

Most often, in the case of IT projects, the decision to start a project and to assign human resources to tasks is taken even if there is no perfect match between the competencies of the available resources and needs of required tasks. As a result human resource adaptability issues may arise during the unfolding of the project (e.g. finding tips on how to overcome the vulnerabilities of the MD5). In this case, programmer informs her *PA* about the missing information that she needs in order to carry on her task. It is the job of the *PA* to provide the human with the needed resource—either non-human (a manual, a tutorial, a book excerpt, etc.) or human (a peer who possesses the needed information and is able to share it). Note that each such resource request from the part of a programmer may and usually does represent an interrupt in his or her current task. However, this is usually tackled locally, without the intervention of the *PM* (unlike the case of organizing training activities, which involves the decision of the *PM* of a given project or another, even higher, level authority within the organization). However such request involves approval of the contacted peer.

In this example, these actions pertain to the "Direct support provided by the *PA* to facilitate the master needs", more specifically "searching for a resource" [4]. In the search of the needed information, the *PA* will query other *PA*s, profiles of which indicate that they represent humans interested in the given subject (and thus store knowledge about pertinent resources); this scenario is similar to well-known cases of collaborative filtering (see, for instance [9]). As pointed out in [3], a non-human resource (e.g. an educational material) is considered appropriate to a learner if there is a correspondence between the following characteristics of the non-human and human resource profiles, as described in their respective ontologies: (i) the prerequisites level and the knowledge level; (ii) the intended purpose and the actual learning goal; (iii) the most appropriate learning style and the recorded cognitive characteristics; (iv) the desirable hardware and software features of the used device and the actual platform available. It is the role of the *PA* to apply ontological matchmaking realizing these criteria (recall that *all* resources are described as ontological instances) and to judge suitability of a particular resource for its user-master. Upon receiving responses, on the basis of implicit and explicit feedback, the *PA* will adjust its "trust values" that it applies to recommendations provided by other *PA*s. This step is needed to rank responses in the case when a large number of them is obtained from peer-*PA*s.

In the case when the needed resource is human (peer help), the *PA* will contact appropriate *PA*s, based on their associated profiles. For example, whether a particular programmer possesses the needed information on MD5 can be easily seen from her profile, since an overlay model on domain ontologies is used to represent human resource profiles [5, 6]. In case there are several peers that possess the required information, a "near-peer-matching" rule can be applied—i.e. directing the user to the peer with a slightly higher knowledge and skills level. This insures a fair distribution of help demands, avoiding the situation that most skilled programmers will be overburdened. The matching will be done by means of negotiations between the *PA*s of programmers and will be based on their profiles and schedules.

To illustrate this approach, let us assume that a senior programmer has the information on MD5 required by a junior programmer. However, if the same information is available from a junior programmer, then the *PA* will assign the help task to the latter. Besides insuring a fair load, this approach could also provide the most efficient training, since the trainer's competence level will closely match the trainee's level.

In case the same help request appears several times from the part of different programmers (and each such request is stored in the project *log*), the *PM* (that analyzes the project *log*) might consider organizing an ad-hoc training on the topic, eventually involving only a subgroup of the team interested in that specific topic, in order to optimize the time spent by the programmer who plays the role of the trainer.

Similar situation takes place when the *Quality of Service* module reports that a task has not been carried correctly by one or more team members (all reports from the *QoS* are also collected in the project *log*). Analyzing the project *log* the *PM* may decide that a just-in-time training is needed for one or more team members to improve their skills and reduce number of incorrectly completed tasks.

Once the ad-hoc training is carried out, each *PA* adjusts the profile of its user.

5   Proactive Approach

Consider the situation when a new project request is received and, for various reasons (which might include, among others, that required resources are missing and/or requested competencies are unavailable), the *AM* determines that it should be rejected. Moreover, assuming that situation like this repeats, the management is faced with deciding: (i) to continuously reject similar project proposals; (ii) proactively involve available human resources in training tasks; or (iii) hiring new staff. While situations (ii) and (iii) are instances of human resource adaptation at the organizational level, clearly only option (ii) is within the scope of this paper. Note that other scenarios pertinent to the proactive approach include the organizational management that expects a certain set of projects to materialize within short or mid-range perspective, an expansion or a change in direction of the organization, or more generally long-term and semi-long-term goals and strategies of the organization. Here, the same three possibilities of dealing with availability of human resource competencies arise.

Note that while the *reactive approach* involves mainly decisions at the project level, the *proactive approach* involves mostly decisions at the higher organizational level. These decisions are based on conditions like: recurring competence and expertise needs of incoming project requests, availability of time and financial resources, specific regulations at the national and/or regional level, corporate strategy, etc. Separately, note that granularity of training tasks (and consequently costs, time, and effort) in the reactive approach are expected to be substantially smaller than in the case of proactive approach (i.e. individual and/or small group training, focused training in the reactive approach; larger groups, broader training scope in the proactive approach). For example, proactive training can include continuing professional education, initial training for new employees (e.g. "school to work transition"), coaching and motivational seminars, group/team building activities, etc.

6   Competence and Training Management Units

Based on the material presented thus far, as well as on ideas found in related works [10, 15], two specialized units are going to be added to the proposed system (following [12], we use the term *unit* with specific roles for describing these entities):

- *Competence Management Unit* (*CMU*)—responsible for management of competencies;
- *Training Management Unit* (*TMU*)—responsible for management of training activities.

In what follows we outline the main functionalities of these units and their interactions with existing units in the system (see also Figure 1).

6.1   Competence Management Unit

The *CMU* is responsible for management of competencies within the organization. Representation of competencies will use a competence ontology described in [1, 7, 10], and associated reasoning mechanisms proposed in [8].

Functionalities of the *CMU* comprise: (i) management of individual competencies of available human resources; this requires the ability to represent, record and update competencies at an individual level; (ii) provisioning of a global view of competencies available at the organizational level; this facility is required for example by the *AM* to be able to asses if the organization has competencies "good enough" to accept a given project; (iii) qualitative and quantitative reasoning about matchings between available and required competencies; this functionality is needed to help decide to hire new staff [2, 8], assign human resources to tasks or enroll human resources into training. Note that the *PM* and the *OPM* will have to interact with the *CMU* during the process of fixing the problem of missing resources. Furthermore, the *CMU* will utilize information from the *Quality of Service* unit that assess work done by individuals and teams (each time a task is completed the *QoS* checks the result). This being the case the *QoS* can provide the *CMU* with information which tasks have been successfully or unsuccessfully completed. This information, in turn can be used to assess which individuals, or teams need extra training (i.e. training needs can be assessed directly on the basis of on-the-job performance).

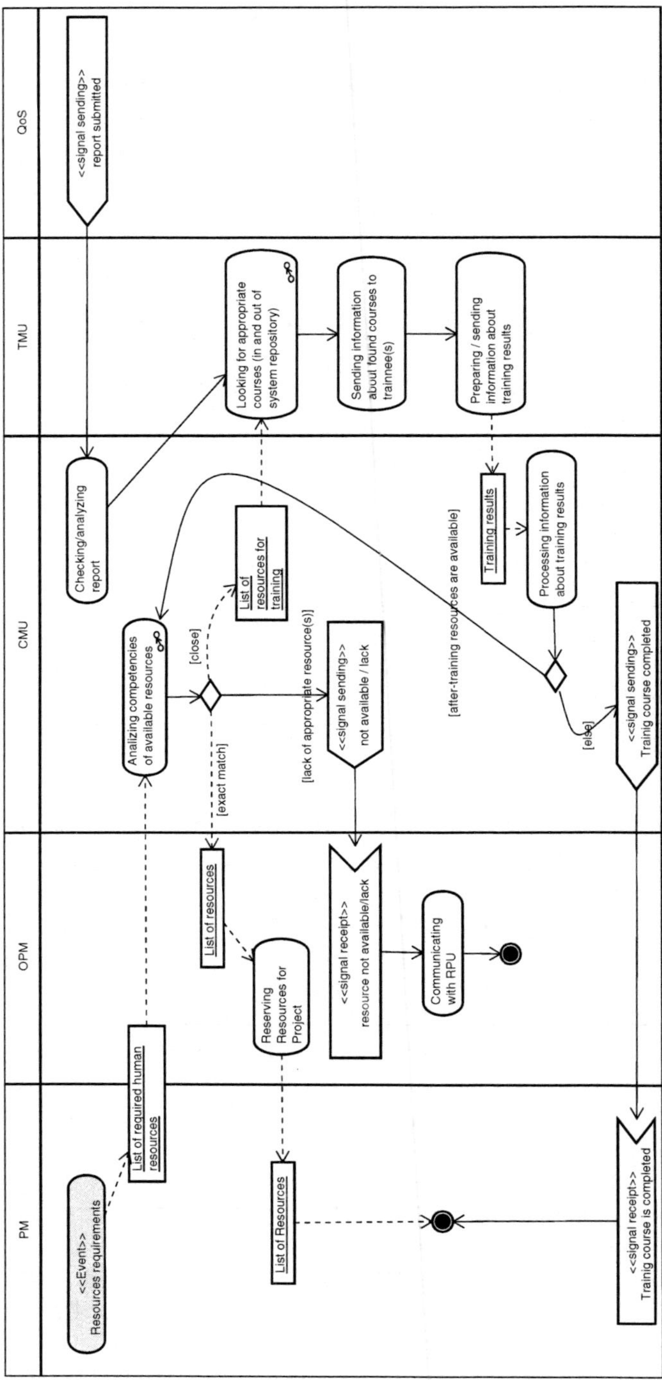

Fig. 1. Interactions of *CMU* and *TMU* with other units in the system

6.2   Training Management Unit

The introduction of the *TMU* is motivated by the need of a specialized unit that is capable of formulating training goals for employees engaged in training activities based on the contextual conditions that resulted in training being requested at various levels within the organization: individual, group, project and organization.

Following [15], the main functionalities of the *TMU* are defined as follows: (i) identification of training goals by analyzing individual, project and business needs, available competencies, and contextual conditions when the training occurs, i.e. reactive (both cases) or proactive approach; (ii) choice of learning objects and selection of a learning strategy.

Note that function (i) requires the interaction with the unit responsible for deciding of the actual assignment of the training task (responsibility of the *PM* or other higher level authority) and with the *CMU* to evaluate the gap between existing and required knowledge. Function (i) requires interaction with *Resource Procurement Unit* (*RPU*) in the case a suitable learning object could not be located at the level of *TMU*. Obviously, work of the *TMU* involves interaction with the actual training unit (structure and functioning of which are out of scope of this paper). However, we can specify that the role of the *TMU* is to provide input specifying: (i) who needs training, (ii) which area needs to be trained, (iii) what training method should be applied, and (iv) when training should take place. The output of the training unit is certification of the the completed training and an assessment of trainee(s), which will be send to the *CMU* (and to appropriate *PA*s) to update profile(s) of trainee(s).

7   Concluding Remarks

In this paper we have conceptualized training-related processes involved in human resource adaptability in an agent-based virtual organization. The main accomplishments of this work are: (i) identification of three approaches for introducing training tasks into the proposed system, two approaches based on the idea of reactive training and one approach based on the idea of proactive training; (ii) identification of additional specialized units that are required to be included in the system: *Competence Management Unit* and *Training Management Unit*. Future work will be targeted on providing more details of interactions between units in the system and development and evaluation of a competencies ontology and associated reasoning mechanisms.

References

1. Biesalski, E., Abecker, A.: Human Resource Management with Ontologies. In: Althoff, K.-D., Dengel, A., Bergmann, R., Nick, M., Roth-Berghofer, T.R. (eds.) WM 2005. LNCS (LNAI), vol. 3782, pp. 499–507. Springer, Heidelberg (2005)
2. Bizer, C., Heese, R., Mochol, M., Oldakowski, R., Tolksdorf, R., Eckstein, R.: The Impact of Semantic Web Technologies on Job Recruitment Processes. In: Proc. International Conference Wirtschaftsinformatik (WI 2005), Bamberg, Germany (2005)
3. Ganzha, M., Paprzycki, M., Popescu, E., Bădică, C., Gawinecki, M.: Agent-Based Adaptive Learning Provisioning in a Virtual Organization. In: Advances in Intelligent Web Mastering.

Proc. AWIC 2007, Fontainebleu, France. Advances in Soft Computing, vol. 43, pp. 25–40. Springer, Heidelberg (2007)
4. Ganzha, M., Paprzycki, M., Gawinecki, M., Szymczak, M., Frackowiak, G., Bădică, C., Popescu, E., Park, M.-W.: Adaptive Information Provisioning in an Agent-Based Virtual Organization—Preliminary Considerations. In: Proceedings of the SYNASC 2007, pp. 235–241. IEEE CS Press, Los Alamitos (2007)
5. Gawinecki, M., Gordon, M., Paprzycki, M., Vetulani, Z.: Representing Users in a Travel Support System. In: Kwasnicka, H., et al. (eds.) Proceedings of the ISDA 2005 Conference, pp. 393–398. IEEE Press, Los Alamitos (2005)
6. Gawinecki, M., Gordon, M., Nguyen, N.T., Paprzycki, M., Zygmunt Vetulani, Z.: Ontologically Demarcated Resources in an Agent Based Travel Support System. In: Katarzyniak, R.K. (ed.) Ontologies and Soft Methods in Knowledge Management, Adelaide, Australia. Advanced Knowledge International, pp. 219–240 (2005)
7. HR-XML Consortium, http://www.hr-xml.org/
8. Mochol, M., Wache, H., Nixon, L.: Improving the Accuracy of Job Search with Semantic Techniques. In: Abramowicz, W. (ed.) BIS 2007. LNCS, vol. 4439, pp. 301–313. Springer, Heidelberg (2007)
9. Montaner, M., López, B., de la Rosa, J.L.: A taxonomy of recommender agents on the Internet. Artif. Intell. Rev. 19(4), 285–330 (2003)
10. Schmidt, A., Kunzmann, C.: Towards a Human Resource Development Ontology for Combining Competence Management and Technology-Enhanced Workplace Learning. In: Meersman, R., Tari, Z., Herrero, P. (eds.) OTM 2006 Workshops. LNCS, vol. 4278, pp. 1078–1087. Springer, Heidelberg (2006)
11. Semantic Web, http://www.w3.org/2001/sw/
12. Szymczak, M., Frackowiak, G., Ganzha, M., Gawinecki, M., Paprzycki, M., Park, M.-W.: Resource Management in an Agent-based Virtual Organization—Introducing a Task Into the System. In: Proceedings of the MaSeB Workshop, pp. 458–462. IEEE CS Press, Los Alamitos (2007)
13. Szymczak, M., Frackowiak, G., Ganzha, M., Gawinecki, M., Paprzycki, M., Park, M.-W.: Adaptive Information Provisioning in an Agent-Based Virtual Organization—Ontologies in the System. In: Proceedings of the AMSTA-KES Conference (to appear)
14. Training, http://en.wikipedia.org/wiki/Training
15. Tzelepis, S., Stephanides, G.: A conceptual model for developing a personalized adaptive elearning system in a business environment. In: Current Developments in Technology-Assisted Education, vol. III, pp. 2003–2006. Formatex Publishing House (2006)
16. Wooldridge, M.: An Introduction to MultiAgent Systems. John Wiley & Sons, Chichester (2002)

Part 3
Knowledge Modelling and Processing

On the Pursuit of a Standard Language for Object-Oriented Constraint Modeling

Ricardo Soto[1,2] and Laurent Granvilliers[1]

[1] LINA, CNRS, Université de Nantes, France
[2] Escuela de Ingeniería Informática
Pontificia Universidad Católica de Valparaíso, Chile
{ricardo.soto,laurent.granvilliers}@univ-nantes.fr

Abstract. A main trend in CP is to define a standard modeling language. This challenge is not a minor matter whose success may depend on many years of experimental steps. Several concerns must be studied such as the simplicity, the level of expressiveness and a suitable solver-independent architecture. In this paper we introduce the s-COMMA modeling language and its execution platform. In this approach a constraint language including extension mechanisms has been carefully fused with object-oriented capabilities in order to provide a considerable level of expressiveness and simplicity. The system is supported by a solver-independent three-layered architecture where models can be mapped to many solvers. We believe the work done on s-COMMA represents a concrete step on the pursuit of a standard constraint modeling language.

Keywords: Constraint Programming, Constraint Satisfaction Problems, Constraint Modeling, Languages.

1 Introduction

In the last decade, CP (Constraint programming) has emerged as one of the most efficient technologies for solving combinatorial NP-hard problems. Under this paradigm, problems are formalized as CSP (Constraint Satisfaction Problems). Such a representation describes a system in terms of variables and constraints. Variables include a set of allowed values, namely its domain; and constraints restrict the possible values that variables can take. The goal is to find the variable-value assignments which satisfies the whole set of constraints.

The CP community has done a considerable effort on the development of systems for using this technology. Several programming languages and libraries were designed, for instance, ECLiPSe [7], ILOG SOLVER [14] and Gecode [2]. In these approaches, a host language is used to state control operations, a constraint language is used for modeling in terms of variables and constraints. The search process may also be tuned using customized strategies.

In this field, it is recognized that constraint programming languages and constraint libraries are mostly oriented to CP specialists. Their constraint vocabulary is very extensive. A significant background in CP techniques is needed. Advanced programming skills are in general required to deal with such systems.

N.T. Nguyen, R. Katarzyniak (Eds.): New Chall. in Appl. Intel. Tech., SCI 134, pp. 123–133, 2008.
springerlink.com © Springer-Verlag Berlin Heidelberg 2008

This complexity of use, of course has not improved the acceptance of CP in the "average-engineer" audience.

In order to fix this problem, some years ago an important research direction has emerged to simplify the use of CP. The aim was to implement simple but at the same time expressive modeling languages. Some examples are OPL [17], Essence [6] and Zinc [9]. In general, modeling languages succeed to be simpler than programming languages and libraries thanks to these three important features:

- Modeling languages provide a higher-level of abstraction which allows to represent problems in a more human-comprehensible way.
- There is no need to make complex modeling decisions or to know some particular search technique. In general, no CP expertise is required: The user state the model and the system solve it [15].
- There is no need to deal with operational concerns of the host language such as C++ in ILOG SOLVER [14] or Prolog in ECLiPSe [7].

Now, after some years of experience in CSP modeling, a main challenge is to define a standard modeling language [15], such as AMPL [11] or GAMS [4] in the mathematical programming area. This challenge is not a minor matter, several aspects must be considered:

- Simplicity, to be accepted by a big audience.
- Expressiveness, to support a wide range of combinatorial problems.
- Solver-independence, to facilitate experimentations with different solvers.

In order to follow this research direction, we introduce s-COMMA [3], a new language for modeling CSP. s-COMMA can be regarded as an hybrid built from a combination of an object-oriented language and a constraint language. The s-COMMA constraint language provides typical data structures, control operations, and first-order logic to define constraint-based formulas. The object-oriented part is a simplification of the Java programming style. This framework clearly provides model structures using composition and inheritance.

The s-COMMA system is written in Java (22000 lines) and it is supported by a solver-independent execution platform, where models can be solved by four well-known CP solvers: Gecode/J [2], ECLiPSe [7], RealPaver [13] and GNU Prolog [5].

We believe s-COMMA is in compliance with the requirements mentioned above. Their simplicity is similar to the state-of-the-art modeling languages [9, 8, 6]. The expressiveness provided is considerable and even it can be increased with extension mechanisms (see Section 3.2). The solver-independence is the base of the platform which allows experimentations with many solvers.

The definition of a standard or universal language is an evident hard task which may require many years and several experimental steps. We believe that the work done on s-COMMA is one of the steps towards the achievement of this important goal.

This paper is organized as follows. Section 2 presents the related work, an overview of the s-COMMA language is given in Section 3. The implementation of the system is explained in Section 4, followed by the conclusions.

2   Related Work

The definition of a standard constraint modeling language is a recent trend. First encouragements on this issue were done by J-F. Puget in [15]. He suggested to develop a "model and run" paradigm such as in Math Programming. The paradigm involved a standard file format for expressing models and a CP library to solve them. Then, at *The Next 10 Years of CP* [10], this challenge was confirmed as an important research direction. Recently, at CP 2007 Conference, MiniZinc [8] was proposed as a standard modeling language. MiniZinc can be seen as a subset of elements provided by Zinc [9]. The syntax is closely related to OPL [17] and its solver-independent platform allows to translate MiniZinc models into Gecode and ECLiPSe solver models. Essence [8] is another good basis to define such a standard. This core is focused on users with a background in discrete mathematics, this style makes Essence a specification language rather than a modeling language. The Essence execution platform allows to map specifications into the ECLiPSe solver.

We believe s-COMMA may be a good starting point too, the constraint language of s-COMMA is closer to OPL and Zinc. The solver-independent platform is an adequate support to map models to three different solvers; and in addition we provide two important features not present in the aforementioned proposals:

- The constraint language of s-COMMA is extensible. This is a useful capability that allows one to increase the expressiveness of s-COMMA by means of a simple term-rewriting mechanism (see Section 3.2).
- The object-oriented framework, which has demonstrated to be useful for modeling composite problems. These kind of problems are widely present in application areas such as design, biology, computer graphics and engineering (see Section 3.3).

3   s-COMMA Overview

In order to describe the main features of s-COMMA let us first give an overview of the elements of s-COMMA models and then, illustrate these elements by means of two examples: The Packing Squares Problem and the Engine Problem.

The s-COMMA language is an hybrid built from a combination of a constraint language with an object-oriented framework. This framework allows to represent a problem through a model including one or more classes. Each of these classes is composed by attributes and constraints zones. The attributes may represent decision variables or objects. Decision variables must be declared with a type (Integer, Real or Boolean). We provide independence between model and data. Therefore, constants are given in a separate data file.

Constraint zones are stated with a name and allows one to group and organize the constraints of the problem. A constraint zone can contain constraints, loops, conditional statements, optimization statements, and global constraints. Loops can use loop-variables which do not need to be declared (i and j in the example). Associations between classes such as composition and inheritance are allowed. Multiple inheritance is not allowed. Let us notice that a full specification of the language can be found at [3].

3.1   The Packing Squares Problem

The goal of the Packing Squares Problem is to place a given set of squares in a square area. Squares may have different sizes and they must be placed in the square area without overlapping each other.

```
//Data file
1. int sideSize := 5;
2. int squares   := 8;
3. Square PackingSquares.s :=
       [{_,_,3},{_,_,2},{_,_,2},{_,_,2},
        {_,_,1},{_,_,1},{_,_,1},{_,_,1}];


//Model file
1. import PackingSquares.dat;
2.
3. class PackingSquares {
4.
5.    Square s[squares];
6.
7.    constraint inside {
8.       forall(i in 1..squares) {
9.          s[i].x <= sideSize - s[i].size + 1;
10.         s[i].y <= sideSize - s[i].size + 1;
11.      }
12.   }
13.
```

```
14.    constraint noOverlap {
15.       forall(i in 1..squares) {
16.          forall(j in i+1..squares) {
17.             s[i].x + s[i].size <= s[j].x or
18.             s[j].x + s[j].size <= s[i].x or
19.             s[i].y + s[i].size <= s[j].y or
20.             s[j].y + s[j].size <= s[i].y;
21.          }
22.       }
23.    }
24.
25.    constraint fitArea {
26.       (sum(i in 1..squares)
             (s[i].size*s[i].size))
                = sideSize*sideSize;
27.    }
28. }
29.
30. class Square {
31.    int x in [1,sideSize];
32.    int y in [1,sideSize];
33.    int size;
34. }
```

Fig. 1. An s-COMMA model for the Packing Squares Problem

Figure 1 shows an s-COMMA model for the Packing Squares Problem. Data values are imported from an external data file called PackingSquares.dat, sideSize represents the side size of the square area where squares must be placed, squares is the quantity of squares (8) to place; PackingSquares.s is an assignment for the array of Square objects declared at line 5 of the model file. Here, a set of values is assigned to the third attribute (size) of each Square object of the array s. For instance, the value 3 is assigned to the attribute size of the first object of the array. The value 2 is assigned to the attribute size of the second, third and fourth object of the array. The value 1 is assigned to the attribute size of remaining objects of the array. We use standard modeling notation (_) to omit assignments. Let us remark that we can perform direct value assignment for attributes of an object in the data file (as we did for the attribute `size`), this particularity gives us some benefits:

- Allow us to avoid the definition of constructors[1] for each class.
- We do not need to call a constructor each time an object is stated. If we need to perform an assignment we done it directly in the data file.
- We believe the omission of these statements give us a cleaner class declaration.

At line 3 in the model file, the definition of the class begins, `PackingSquares` is the name given to this class. Then, an array containing objects from the class `Square` is defined. This class (declared at line 30) is used to model the set of squares. Attributes `x` and `y` represent respectively the x and y coordinates where the squares must be placed. So, s[2].x=1 and s[2].y=1 means that the second of the eight squares must be placed in row 1 and column 1, indeed in the upper left corner of the square area. Both variables (x,y) are constrained, they must have values into the domain `[1,sideSize]`. The last attribute called `size` represent the size of the square.

At line 7, a constraint zone called `inside` is declared. In this zone a `forall` loop contains two constraints to ensure that each square is placed inside the area, one constraint about rows and the other about columns. The constraint zone `noOverlap` declared at line 14 ensures that two squares do not overlap. The last constraint zone called `fitArea` ensures that the set of squares fits perfectly in the square area.

3.2 Extension Mechanism

Extensibility is an important feature of s-COMMA. Let us show this feature using the same problem. Consider that a programmer adds to the Gecode/J solver two new built-in functionalities: a constraint called `inside` and a function called `pow`. The constraint `inside` ensures that a square is placed inside a given area, and `pow(x,y)` calculates the value of x to the power of y. In order to use these functionalities we can extend the syntax of the constraint language by defining an extension file where the rules of the translation are described. This file is composed by one or more main blocks (see Figure 2). A main block defines the solver where the new functionalities will be defined. Inside a main block two new blocks are defined: a `Function` block and a `Relation` block. In the `Function` block we define the new functions to add. The grammar of the rule is as follows:

⟨*s-COMMA-code*⟩ (⟨*input-parameters*⟩) -> " ⟨*solver-code*⟩ ";

In the example the s-COMMA code is `pow(x,y)`. This is the code which will be used to call the new function from s-COMMA. The input parameters of the new s-COMMA function are x and y. Finally, the corresponding Gecode/J code is given to define the translation. The new function will be translated to `pow(x,y)`. This code calls the new built-in method from the solver file. The translator

---

[1] A constructor is a special function used to set up the class variables with values. It is used in most of object-oriented programing languages.

must recognize the correspondence between input parameters in s-COMMA and input parameters in the solver code. Therefore, variables must be tagged with \$ symbols. In the example, the first parameter and the second parameter of the s-COMMA function will be translated as the first parameter and the second parameter in the Gecode/J function, respectively.

In the **Relation** block we define the new constraints to add. We use the same grammar as for functions. In the example, a new constraint called **inside** is defined, it receives four parameters. The translation to Gecode/J is given. Once the extension file is completed, it can be called by means of an import statement. Let us note that we can define extensions for more than one solver in a same extension file. The resultant s-COMMA model using extensions is shown in Figure 2.

```
//Extension File                          4. class PackingSquares {
1.  GecodeJ {                             5.    Square s[squares];
2.     Function {                         6.
3.        pow(x,y) -> "pow($x$,$y$)";     7.    constraint placeSquares {
4.     }                                  8.      forall(i in 1..squares) {
5.     Relation {                         9.        inside(s[i].x,s[i].y,s[i].size,sideSize);
6.        inside(a,b,c,d) ->             10.        forall(j in i+1..squares) {
7.           "inside($a$,$b$,$c$,$d$);"; 11.          s[i].x + s[i].size <= s[j].x or
8.     }                                 12.          s[j].x + s[j].size <= s[i].x or
9.  }                                    13.          s[i].y + s[i].size <= s[j].y or
10.                                      14.          s[j].y + s[j].size <= s[i].y;
11. ECLiPSe {                            15.        }
12.    Function {                        16.      }
13.    ...                              17.    }
                                        18.
                                        19.    constraint fitArea {
//Model File                            20.      (sum(i in 1..squares)
1. import PackingSquares.dat;                        (pow(s[i].size,2))) = pow(sideSize,2);
2. import PackingSquares.ext;           21.    }
3.                                      22.}
```

Fig. 2. Extension for Gecode/J and the resultant model

3.3 The Engine Problem

Complex structures as mixers, engines, computer graphics, molecules are in general entities composed by many pieces. These pieces have often their own composition rules and constraints between other pieces and/or between its attributes.

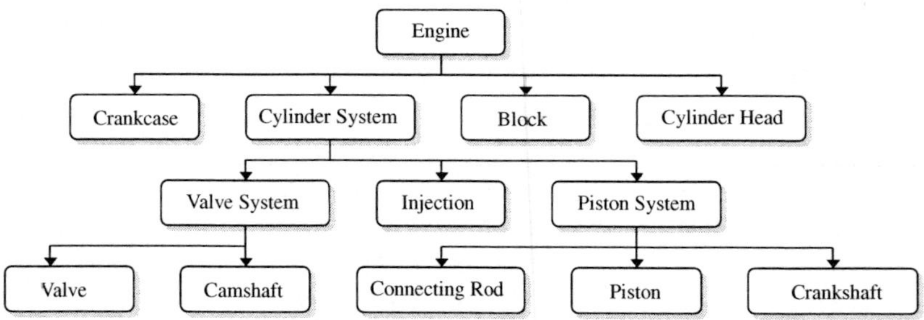

Fig. 3. The Engine Problem

```
//Data file                          21. class CylSystem {
1. enum size := {small,medium,large};  22.    int          quantity in [2,12];
2. enum flow := {direct,indirect};      23.    int          distBetCyl in [3,18];
                                        24.    Injection    inj;
//Model file                          25.    ValveSystem  vSyst;
1.  import Engine.dat;                 26.    PistonSystem pSyst;
2.                                     27.    constraint determinePressure {
3.  class Engine {                     28.       if (quantity = 6)
4.     CrankCase cCase;                29.         distBetCyl > 6;
5.     CylSystem cSyst;               30.       else
6.     Block     block;              31.         distBetCyl > 3;
7.     CylHead   cHead;              32.    }
8.     int       volume;            33. }
9.     constraint dim {             34.
10.       volume > cCase.volume;     35. class Injection {
11.    }                             36.    flow     gasFlow;
12. }                                37.    size     admValve;
13.                                  38.    int      pressure;
14. class CrankCase {                39.    constraint compValues {
15.    size base;                    40.       compatibility(gasFlow,admValve,pressure) {
16.    int  oilVesselVol;            41.          ("direct",   "small",  80);
17.    int  bombePower;              42.          ("direct",   "medium", 90);
18.    int  volume;                  43.          ("indirect", "medium", 100);
19. }                                44.          ("indirect", "large",  130);
20.                                  45.       }
                                     46.    }
                                     47. }
```

Fig. 4. An s-COMMA model for the Engine Problem

Modeling these structures is not quite natural using either a pure logic or a pure declarative CP language. It seems more appropriate to state the model as a hierarchic CSP where pieces of the system are represented by objects under constraints.

Let us now show an s-COMMA model for a hierarchic CSP from the configuration and design area. We consider the task of configuring a car engine using a compositional approach (see Figure 3). The engine at the first level is built from a crankcase, a cylinder system, a block and a cylinder head at the second level. The cylinder system is a subsystem made of a valve system, an injection and a piston system. Both valve and piston systems have their own composition rules.

Figure 4 shows an s-COMMA model for the Engine problem. The class Engine is at the top of the hierarchy. The attributes cCase, cSyst, block and cHead represent the subsystems of the engine. Those four objects are instances of classes to be declared. The last attribute volume defines the volume of the engine. Then, a constraint between volume and the volume attribute of cCase is posted.

The class CrankCase has four attributes, three integer decision variables and a decision variable with a type size. The type size is an enumeration defined in the data file, this means that the decision variable base adopts as domain the set of values of the enumeration size. Thus, base adopts as domain {small,medium,large}.

The class CylSystem has a more complex declaration. The first attribute called quantity represents the quantity of cylinders; and the second represents the distance between cylinders. The cylinder system has three subsystems denoted by `inj`, `vSyst` and `pSyst`. Then, a constraint zone called `determinePressure` is declared to state a conditional constraint. This conditional constraint represents that 6-cylinder-engines have a distance between cylinders bigger that 6. In others kinds of engines the distance must be bigger than 3. In order to represent this constraint, an `if-else` statement is stated. If the condition is true, the first constraint is activated. Otherwise, the second constraint is activated.

The subsystem injection is composed of three attributes called `gasFlow`, `admValve`, and `pressure`. The class `Injection` has also a compatibility constraint [12] between its components. A compatibility constraint limits the combination of allowed values for the decision variables to a given set. For example, for variables `gasFlow`, `admValve` and `pressure` just four combination of values are allowed. The possible values are described inside the `compatibility` built-in constraint.

Let us notice that remaining elements of the engine are not presented because they can be modeled with the elements already shown in previous subsections.

4   Implementation

The s-COMMA system is supported by a three-layered architecture: Modeling, Mapping and Solving (see Fig. 5). On the first layer, models are stated by the user, extension and data files can be given optionally. Models are syntactically and semantically checked by two ANTLR [1] tree-walkers. If the checking process succeeds, an intermediate model called Flat s-COMMA is generated (see Section 4.1). Finally, the Flat s-COMMA file is taken by the selected translator which generates the executable solver file.

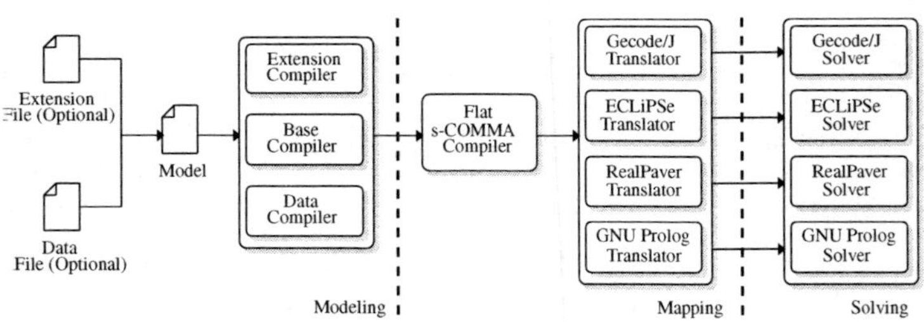

Fig. 5. The s-COMMA compiling system

4.1   From s-COMMA to Flat s-COMMA

A direct translation from s-COMMA to executable solver code is feasible (in fact, we have studied this in [16]). However, many statements provided by s-COMMA are not supported by solvers. Thus, for performing this direct mapping, many model-transformations must be carried out at the level of translators. This makes translators bigger (in terms of code lines) and, as consequence, difficult to develop. A well-known technique to simplify code generation is to include an intermediate phase where the non-supported features are transformed to simpler (or supported) features. We state this transformation on an intermediate model called Flat s-COMMA. The set of performed transformations from s-COMMA to Flat s-COMMA are described below.

Flattening composition. The hierarchy generated by composition is flattened. This process is done by expanding each object declared in the main class adding its attributes and constraints in the Flat s-COMMA file. The name of each attribute has a prefix corresponding to the concatenation of the names of objects of origin in order to avoid name redundancy. The expansion of objects cCase and cSyst is shown below

```
size cCase_base;                       int cSyst_distBetCyl in [3, 18];
int  cCase_oilVesselVol;               flow cSyst_inj_gasFlow;
int  cCase_bombePower;                 ...
int  cCase_volume;                     volume > cCase_volume;
int  cSyst_quantity in [2,12];
```

Loop unrolling. Loops are not widely supported by solvers, hence we generate an unrolled version of loops.

```
//s-COMMA                  //flat s-COMMA
forall(i in 1..3) {        a[1] < a[2];
  a[i] < a[i+1];           a[2] < a[3];
}                          a[3] < a[4];
```

Enumeration substitution. In general solvers do not support non-numeric types. So, enumerations are replaced by integer values, for example size:={small, medium,large} is replaced by size:={1,2,3}; original values are stored to give the results.

Data substitution. Data variables are replaced by its value defined in the data file.

Conditional removal. Conditional statements are transformed to logical formulas. For instance, if a then b else c is replaced by $(a \Rightarrow b) \cap (a \cup c)$.

```
//s-COMMA            //flat s-COMMA
if(x=y) w>=z;        (x=y) -> (w>=z) and (x=y) or (w<z);
else    w<z;
```

Compatibility removal. Compatibility constraints are also translated to a logical formula. We create a conjunctive boolean expression for each n-tuple of allowed values. Then, each constraint of the n-tuple is stated in a disjunctive constraint. The transformed compatibility constraint of the Engine problem is shown below. Non-numeric values were replaced by the corresponding integer values in the enumeration substitution step.

```
((gasFlow=1) and (admValve=1) and (pressure=80)) or
    ((gasFlow=1) and (admValve=2) and (pressure=90)) or ...
```

Array decomposition. Array containing objects are decomposed into single objects which are then expanded by the flattening composition described above. For instance, in the Packing squares problem, the array of objects called s is decomposed as follows:

```
int s_1_x in [1,5];      int s_2_x in [1,5];      int s_3_x in [1,5];
int s_1_y in [1,5];      int s_2_y in [1,5];      ...
```

The name of each variable is composed by the name of the array (s), the position of the object in the array, and the name of the attribute of the object. Only, variables x and y are considered decision variables, the attribute size is considered constant due to has an instantiation (in the data file at line 3) for the whole set of objects contained in the array. The value 5 in the variables' domain comes from the data substitution of the data variable sideSize.

Logic formulas transformation. Some logic operators are not supported by solvers. For example logical equivalence ($a \Leftrightarrow b$) and reverse implication ($a \Leftarrow b$). We transform logical equivalence expressing it in terms of logical implication ($(a \Rightarrow b) \cap (b \Rightarrow a)$). Reverse implication is simply inverted ($b \Rightarrow a$).

Finally, the generated Flat s-COMMA code is taken by the selected translator which generates the executable solver file.

5   Conclusion and Future Work

In this work we have attempted to make the following two contributions:

- A modeling language including an object-oriented framework, solver-independence and extension mechanisms.
- A concrete step towards the definition of a standard modeling language.

To reach this final purpose, several aspects could be developed, for instance: more work on benchmarks, solver cooperation, new global constraints and translation to new solvers. The development of a tool for modeling CSPs from a graphical standpoint will be useful too.

Again, we believe s-COMMA is a concrete step towards the pursuit of a standard modeling language. Or probably, in the near future s-COMMA may be a suitable starting point to define a standard language for object-oriented constraint modeling.

The s-COMMA system is available as open source under BSD license at www.inf.ucv.cl/~rsoto/s-comma. The distribution includes Javadoc, a user's manual, and a set of benchmarks.

References

1. ANTLR Reference Manual, http://www.antlr.org
2. Gecode System, http://www.gecode.org
3. s-COMMA System, http://www.inf.ucv.cl/~rsoto/s-comma
4. Bisschop, J.J., Meeraus, A.: On the Development of a General Algebraic Modeling System in a Strategic Planning Environment. Mathematical Programming Studies 20, 1–29 (1982)
5. Diaz, D., Codognet, P.: The gnu prolog system and its implementation. In: SAC (2), pp. 728–732 (2000)
6. Frisch, A.M., et al.: The design of essence: A constraint language for specifying combinatorial problems. In: IJCAI, pp. 80–87 (2007)
7. Wallace, M., et al.: Eclipse: A platform for constraint logic programming (1997)

8. Nethercote, N., et al.: Minizinc: Towards a standard cp modelling language. In: Bessière, C. (ed.) CP 2007. LNCS, vol. 4741, pp. 529–543. Springer, Heidelberg (2007)

9. Rafeh, R., et al.: From zinc to design model. In: Hanus, M. (ed.) PADL 2007. LNCS, vol. 4354, pp. 215–229. Springer, Heidelberg (2006)

10. Benhamou, F., et al.: Trends in Constraint Programming. ISTE (2007)

11. Fourer, R., Gay, D.M., Kernighan, B.W.: A Modeling Language for Mathematical Programming. Management Science 36, 519–554 (1990)

12. Gelle, E., Faltings, B.: Solving mixed and conditional constraint satisfaction problems. Constraints 8(2), 107–141 (2003)

13. Granvilliers, L., Benhamou, F.: Algorithm 852: Realpaver: an interval solver using constraint satisfaction techniques. ACM Trans. Math. Softw. 32(1), 138–156 (2006)

14. Puget, J.F.: A C++ implementation of CLP. In: SCIS, Singapore (1994)

15. Puget, J.F.: Constraint programming next challenge: Simplicity of use. In: Wallace, M. (ed.) CP 2004. LNCS, vol. 3258, pp. 5–8. Springer, Heidelberg (2004)

16. Soto, R., Granvilliers, L.: An extensible framework for mapping constrained objects to native solver models. In: IEEE ICTAI (1) 2007, pp. 243–250 (2007)

17. Van Hentenryck, P.: The OPL Language. The MIT Press, Cambridge (1999)

A Layered Ontology-Based Architecture for Integrating Geographic Information

Agustina Buccella[1], Domenico Gendarmi[2], Filippo Lanubile[2],
Giovanni Semeraro[2], Alejandra Cechich[1], and Attilio Colagrossi[3]

[1] GIISCO Research Group,
   Departamento de Ciencias de la Computación,
   Universidad Nacional del Comahue, Neuquen, Argentina
   {abuccel,acechich}@uncoma.edu.ar
[2] Dipartimento di Informatica,
   University of Bari,
   Via E. Orabona, 4 - 70125, Bari, Italy
   {gendarmi,lanubile,semeraro}@di.uniba.it
[3] Dipartimento Tutela delle Acque Interne e Marine,
   APAT, Via Curtatone, 3 - 00185, Rome, Italy
   attilio.colagrossi@apat.it

Abstract. Architectural solutions to information integration have extensively appeared during the last years, mostly from the federated system research field. Some of these solutions were created to deal with geographic information, whose inherent features make the integration process particularly complex. Among others, the use of ontologies has been proposed as a way of supporting an automated integration. However, how to specify and use a geographic ontology is not so clear in this context. In this paper, we introduce an ontology-based architectural solution as an extension of a federated system (Information Broker) built by the Italian Agency for Environmental Protection and Technical Services (APAT). Our extension is aimed at improving integration by adding semantic features through the use of ontologies and the ISO 19100 standards.

Keywords: Geographic Information Systems, Federated Systems, Ontology, ISO 19100 Standards.

1 Introduction

Currently, the information integration research is focusing on Geographic Information Systems (GIS). Most of the proposals in this area come from the federated system research field and use conventional information sources. However, the use of ontologies [1], thesaurus, metadata and other types of semantic resources for integrating Geographic Information Systems is increasingly common. For example, ontologies have become the main tools to solve heterogeneity problems. Therefore, proposals based on conventional systems might be extended to better analyze and compare geographic information. Extending a system by using ontologies implies two main tasks: *semantic enrichment* and *mapping*

N.T. Nguyen, R. Katarzyniak (Eds.): New Chall. in Appl. Intel. Tech., SCI 134, pp. 135–144, 2008.
springerlink.com

discovery. The former is aimed to reconcile semantic heterogeneity, so it involves adding more semantic information about the data. Many approaches add extra semantic information through the use of metadata or ontologies. For example, proposals extending common data models such as Entity-Relationship diagrams [2] and object-oriented ones [3] have been presented in order to add geographic features.

For the latter, mapping discovery, several surveys [4, 5] have emerged describing and analyzing methodologies, frameworks, and systems proposed for semantic matching, i.e. ontology matching. The proposal of Euzenat & Shvaiko [5] is one of the more recent works describing and analyzing a wide set of ontology matching proposals[1]. However, all these surveys are focused on conventional systems, and they do not analyze systems based on geographic information. Considering the hypothesis tested by [6] in which *"geographic and non-geographic entities are ontologically distinct in a number of ways"*, a different analysis must be performed when geographic elements are included. Therefore, the semantic enrichment task becomes crucial to reach a successful mapping discovery.

In our first works [7, 8] we have developed an Information Broker System together with a schema integration process focusing specially on syntactic interoperability. This integration process uses information mostly represented by XML data models, without modeling semantics. Therefore, the process is made manually, increasing the chance of introducing errors and inconsistency.

In this paper, we propose an extension of the integration process by adding semantic information through the use of ontologies. Thus, the Information Broker System will be implemented as an ontology-driven system in order to share the real common vocabulary contained in the sources. Our extension is based on previous work on integration of geographic information [9, 10], which focuses on ontology matching. In addition, we incorporate a modeling process (within the semantic enrichment task) in which ontologies are created towards integration by using a family of the ISO 19100 standards (prepared by ISO Technical Committee 211 (TC211)[2]). Specially, ISO 19109 [11], ISO 19110 [12], and ISO 19107 [13] are used in this work. Our main concern is improving our integration method by taking advantage of the semantic of ontologies and their specific representation.

This paper is organized as follows: next Section describes the current Information Broker System. Then, section 3 presents the extension describing its architecture. In Section 4 we discuss some related work. Future work and conclusions are discussed afterwards.

2   The Information Broker System

APAT was established in 2002 to carry out scientific and technical activities in the national interest of Italy to protect the environment, water resources and soil. Although the huge amount of information is owned by the same organization, it is

---

[1] http://www.ontologymatching.org
[2] http://www.isotc211.org/

managed by different departments and units. Besides, given the large diversity in syntax and semantic of data, measures are stored into several independent systems, which are based on the most appropriate technology for their data type. All these characteristics have made very hard to share information among the different systems. Thus, the main need of APAT is to develop a system to provide a fully and user-transparent integration of the heterogeneous data sources, ensuring at the same time, the existing legacy applications that operates on them will continue operating autonomously, without undergoing any sort of modification.

In previous works [7, 8], we have introduced and implemented an Information Broker System based on a layered-based architecture. In this paper we are interested in the federated layer, and more specifically, in the main processes to build its components. One main component is the *Federated Schema*, which is designed to provide a shared vocabulary of the information sources. Based on this vocabulary, we implemented the user interface and the query processor components in order to give a global view of the whole system. In this way, the federated schema constitutes the core of the Information Broker system. We have applied a bottom-up process consisting of four steps and taking into account syntactic interoperability. Figure 1 shows graphically the components created within each step.

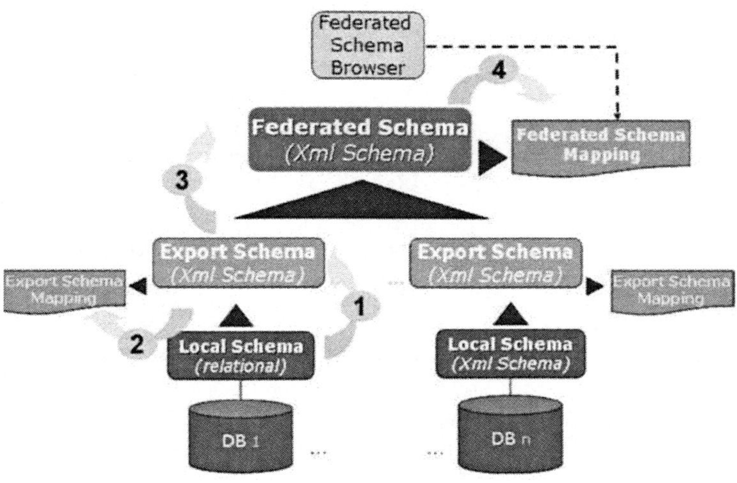

Fig. 1. Schema Integration Process

The first step transforms the local schemas into so-called export schemas, which are expressed in a common data model (CDM) and represented by XML data models. Thus, local schemas of the different databases of the federation converge on a common structure of data. Then, the second step creates the export-schema mappings, which are XML files manually generated at design time from each export schema. Such files contain the mappings between the local and export schemas. Finally, the third step builds the federated schema, which represents the logical model of the virtual database containing all data available within the federation. The federated schema is the result of merging all the export schemas. During this merging, all possible conflicts must be identified and solved manually. Once the federated schema has been generated, the last step in the process manually generates the federated-schema mapping file (in XML).

3   An Ontology-Based Extension for Generating a Federated Schema

Although the current Information Broker architecture is well suited for manipulating standard information through XML formatting rules, integration completely depends on users' interpretations and background. The task of building the federated schema is completely manual and in the case of large information sources (as we have to face in this project) it becomes tedious and error-prone. Aspects as adaptability and scalability were not taken into account because re-executing the integration process only for some changes on data can take several days.

In such conditions, the process of building the federated schema becomes difficult to standardize and evolve. Taking into account these two points we propose changes on the general process of building the federated schema in order to facilitate the use of more suitable processes. Figure 2 shows the main changes made on the original schema integration process. As in the original process (Figure 1), four bottom-up steps are necessary to build the federated schema. However, these steps are different this time. The first and second steps, which were in charge of transforming local schemas into export schemas and generating export schema mappings, are now responsible for standardizing the geographic information of sources through the definition of formal ontologies. The third and fourth steps, which were in charge of creating the federated schema and its mappings, are now responsible for applying the method for merging ontologies.

In this way, the four steps are combined into two main processes, *enriching local ontologies* and *the merging process* itself. The first process defines the steps to create formal ontologies by applying the ISO 19100 standard for geographic information. Then, the *merging process* implements our merging method by using the *Logic* and *Analysis* components showed in Figure 2. Next two sub-sections provide a brief description of these processes.

3.1   Enriching Local Ontologies

The use of the ISO 19100 standard gives a new perspective to face integration problems for the interoperability of geographic systems. New ontology modeling techniques of this type of systems should be based on this standard in order to allow integration methods for taking advantage of the benefits provided by standardization.

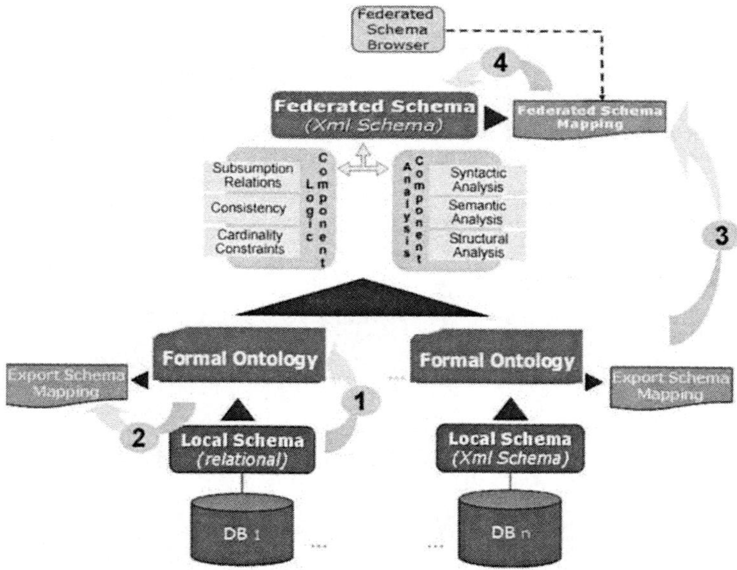

Fig. 2. Changes on the Schema Integration Process

Figure 3 shows the relation between the 4-layer architecture for modeling geographic information of the ISO 19109 std. and the ontologies used in our methodology. As we can see, the ontologies are specifically based on some levels of the standard in order to add interoperability aspects.

In our extension, a *top-level ontology* and a *domain ontology* are built based on the information provided by the models of the standard (ISO 19109 and 19107 std.). Gray arrows in the Figure show how the information flows among the models. Thus, the domain ontology is built considering the General Feature Model (GFM) and the Application Schema [11]. The GFM is a meta-model of feature types. It defines the structure for classifying features used then to build the application schema. In this way, the domain ontology is located between two levels of abstraction (application and meta level) because it will be based on the GFM and will have features as well as associations defined by the application schema. The point is that the information stored by ontologies is different from the information stored in the application schema. An ontology is defined by how a community perceives a specific concept, while an application schema is determined by how an application regards the same concept. A classical example is the concept *Car*. The application schema will store information about its model, color, function, etc., only if they are important for the application. On the other hand, the ontology should store all information about it because these features are necessary for being a car. Similarly, the top-level ontology is also in the middle of two layers, meta and meta-meta level layers. The information represented in this ontology will be based on both, the structure of the GFM and the general features of the model being built. In the case of the top-level

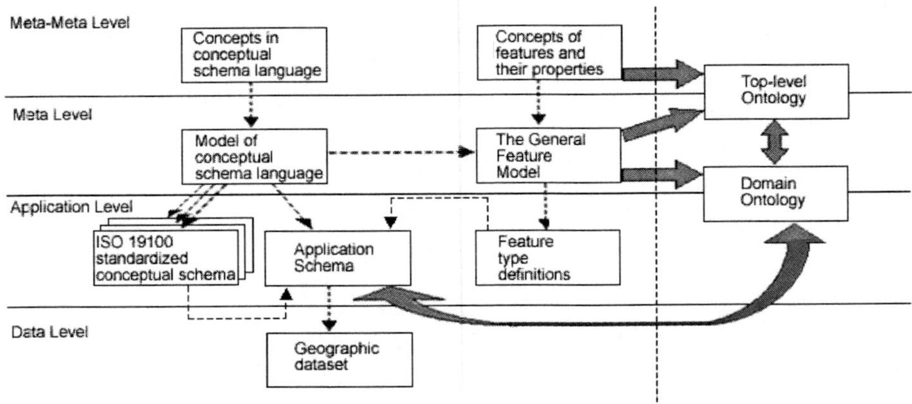

Fig. 3. The 4-layer Architecture for modelling geographic information

ontology, it is based on the structure of the GFM and the general features of the model being built.

Currently, there are new methodologies proposing the creation of ontologies such as [14, 15], including *semantic enrichment* as one of the most important steps. The main goal of this step is to reconcile semantic heterogeneity by adding more semantic information about data. In our work, as both ontologies – top-level and domain – have to be based on the standard before being created, we add a new step in the process named *the enrichment step*. In this step, the components of the ontologies are enriched in their descriptions, through the metaclasses (from GFM) which they are instance of and the schemas on which they are based. In this way, all metaclasses extracted from the GFM and representing information by the application schema are created as abstract classes in the local ontology. Creating an ontology with these characteristics is not a complex task because the information needed with respect to the GFM can be extracted from the Feature Catalogue. Besides, by using an ontology editor as Protégé[3] to model OWL ontologies [16], a list of existing OWL Ontologies based on ISO Standards[4] can be imported.

Thus, all the ontologies will have the same structure because all components are subclassifying the same model. The GFM acts as a top-level ontology classifying the elements of the ontology and making the integration easier. Figure 4 shows an example of one feature together with spatial and non-spatial attributes as subclasses of the GFM. The *Station* class is a subclass of the *GF_FeatureType* metaclass and both spatial and non-spatial attributes are subclasses of *GF_SpatialAtttributeType* and *GF_AttributeType* respectively. The type of the *centerPoint* spatial attribute is part of the ISO 19107 std.

---

[3] http://protege.stanford.edu/

[4] http://loki.cae.drexel.edu/~wbs/ontology/list.htm

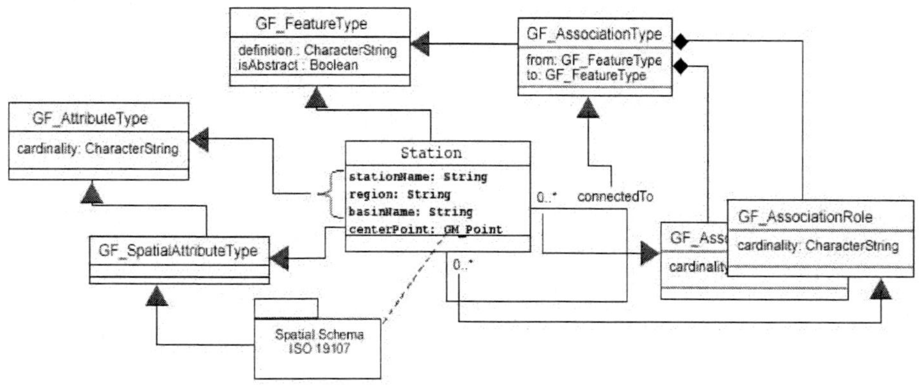

Fig. 4. An example of one enriched class

3.2   The Merging Process

The merging process involves the task of merging the geographic sources in order to create a global vocabulary (federated schema) by defining two main components (Figure 2), *logic* and *analysis*. Both processes are used in different parts of the merging process.

This process is composed of three main phases: *unit, integration* and *system*. In the *unit phase* each system is analyzed separately. The top-level and domain ontologies can be seen as a unique ontology in which generalization / specialization relations are the connectors between them. This ontology will be formally represented by using OWL.

Then, once the ontologies are correctly created, a reasoning system (such as RACER [17]) is applied in order to discover inferences not detected by users. We take advantage of the capability of inferring subsumption relations between classes and properties in the schema (TBox). As result for each system, a normalized ontology (that can be divided into a top-level and a domain ontology) is returned. This ontology will be based on the geographic standards containing metaclasses descriptions (GFM) and the geographic schemas on which they are based.

In the *integration phase* three processes are responsible for matching two normalized ontologies in order to create the global ontology. It contains the general concepts users will use to query the integrated system. In addition, a set of mappings are returned in order to represent the matching among the ontologies. *Merge, General Analysis,* and *Specialized Analysis* are the processes of this phase. In order to accomplish the first process, both ontologies of each system are joined by using generalization/specialization relations. In this way, the ontologies are taken as they are returned from the unit phase. Then, the two ontologies belonging to two different systems are merged. The merge process is performed by matching the classes that are part of the standard (metaclasses). As both ontologies have the same superclasses, merging is an easy task.

Once the merge process is finished, the *General Analysis* starts. During this step two types of analysis are performed: syntactic and semantic [9]. Within the syntactic analysis, three syntactic functions are used in order to compare the names of the concepts in a different way. Then, in the semantic analysis, a thesaurus as WordNet[5], is used to extract synonym relationships between the concepts of the ontologies. The *Specialized Analysis* performs a structural comparison by applying the similarity function described in [10, 18].

Finally, in case of the processes executed before had generated inconsistencies within this final ontology, the *system phase* re-normalizes the global ontology created in the last phase by using again a reasoning system.

Taking into account the enriched class of the Figure 4 and considering another ontology with an enriched class named *Facility*, our merging process starts analyzing the consistency of the two ontologies separately. If any inconsistency is found, an expert user is responsible of solving it. Then, as *Station* and *Facility* are subclasses of GF_FeatureType class, the process analyzes their components syntactically and semantically. In the first step attributes are compared considering names, data types, and synonyms relations (by using a thesaurus). In the second step, associations are also analyzed together with domain and range restrictions. Finally, in the third step an structural analysis is performed taking the information obtained in the two last steps. *Station* and *Facility* are compared considering how many attributes and associations they have in common. In addition, the method applies a syntactic and semantic analysis over these classes in order to analyze similarities at these levels.

4 Related Work

Mapping discovery by using ontologies has being extensively investigated during the last years. Various approaches have emerged proposing processes and techniques to find similarities between elements of different but related ontologies. In this work, we are interested in integration methods of geographic sources [19, 20, 18, 21, 22, 23, 24]. One particularity of all these proposals is the use of ontologies to represent either top-level information or domain information or both of them. In the case of ODGIS [21] several ontologies are built (top-level, domain, and application ontologies) in order to provide more information about the domain and thus facilitate the integration process. But the activity of creating these ontologies is not an easy task and demands a lot of effort. Other proposals as GeoNis [22], Aerts et al. [23] and Hakimpour et al. [20] use a top-level ontology together with the advantages of a formal language (to make inferences) as tools to find more suitable mappings. The use of similarity functions, in proposals as SIM-DL [24] and MDSM [18] involves a set of functions that analyze the concepts and properties syntactically and semantically. In MDSM functions comparing similar structures are applied. In particular the use of these types of functions is useful when the ontologies are not complete (that is, there is absent information about the domain) and/or as starting point of an integration process when a top-level ontology is not involved. Proposals performing some manual step within the integration process require the assistance of an expert

---

[5] http://wordnet.princeton.edu/

user to do so. For example, BUSTER [19] needs of an expert user although it uses inferences during the query process.

Our merging method applies three mechanisms to integrate ontologies. Firstly, top-level ontologies are created by using the information provided by the geographic standard. Secondly, logic capabilities and finally matching functions are combined in order to find more suitable mappings. The use of these three options makes our approach take advantage of the inherent benefits of using the standard in geographic information, the logic of data, and the semantic information from ontologies.

5   Conclusion and Future Work

In this work, we have presented an extension of a system for integrating information, the Information Broker System, in order to add capabilities which improve the generation of the federated schema. Particularly, our proposal aims at improving interoperability and consistency through the use of ontologies. However, there are still many issues that need further research. For example information sources in APAT Information Broker are not currently standardized, which may hinder consistency. The use of the ISO 19100 Stds. is a starting point for improving that.

With respect to the merging process, an implementation of the analysis component has been built as a plug-in for Protégé (it will be soon available to be downloaded from Internet). As this component was initially created for non-geographic ontologies, several proofs had been done using this type of ontologies. However, we have detected some weaknesses of this part of the process. For example, ontologies with concepts containing several attributes (for structural comparison) and without many hierarchical relationships returned the best results. The addition of reasoning capabilities (logic component) have improved the process by generating a complementary comparison to find more suitable mappings. In addition, further validation of the ontology merging process would be absolutely necessary for large ontologies – although our experiences [9] have shown good results when using small ones.

References

1. Gruber, T.: A translation approach to portable ontology specifications. Knowledge Acquisition 5(2), 199–220 (1993)
2. Parent, C., Spaccapietra, S., Zimányi, E.: Spatio-temporal conceptual models: data structures + space + time. In: Proceedings of the GIS 1999: 7th ACM international symposium on Advances in geographic information systems, pp. 26–33. ACM Press, New York, NY, USA (1999)
3. Borges, K., Davis, C., Laender, A.: Omt-g: An object-oriented data model for geographic applications. Geoinformatica 5, 221–260 (2001)
4. Kalfoglou, Y., Schorlemmer, M.: Ontology mapping: the state of the art. The Knowledge Engineering Review 18(1), 1–31 (2003)

5. Euzenat, J., Shvaiko, P.: Ontology matching. Springer, Heidelberg (2007)
6. Mark, D.M., Skupin, A., Smith, B.: Features, objects, and other things: Ontological distinctions in the geographic domain. In: Montello, D.R. (ed.) COSIT 2001. LNCS, vol. 2205, pp. 488–502. Springer, Heidelberg (2001)
7. Calefato, F., Colagrossi, A., Gendarmi, D., Lanubile, F., Semeraro, G.: An information broker for integrating heterogeneous hydrologic data sources: A web services approach. In: Xu, A.M., Chaudhry, L., Guarino, S.N. (eds.) Research and Practical Issues of Enterprise Inf. Sys. IFIP Series, vol. 205, Springer, Heidelberg (2006)
8. Gendarmi, D., Lanubile, F., Lichelli, O., Semeraro, G., Colagrossi, A.: Water protection information management by syntactic and semantic interoperability of heterogeneous repositories. In: Proceedings of the ISESS 2007 (2007)
9. Buccella, A., Cechich, A.: Towards integration of geographic information systems. Electronic Notes in Theoretical Computer Science 168, 45–59 (2007)
10. Buccella, A., Cechich, A., Brisaboa, N.R.: A three-level approach to ontology merging. In: Gelbukh, A., de Albornoz, Á., Terashima-Marín, H. (eds.) MICAI 2005. LNCS (LNAI), vol. 3789, pp. 80–89. Springer, Heidelberg (2005)
11. Geographic information. Rules for Application Schema. Draft Int. Std. 19109, ISO/IEC (2005)
12. Geographic information. Geographic Information and Methodology for Feature Cataloguing. Draft International standard 19110, ISO/IEC (2005)
13. Geographic information. Spatial Schema. Int. Std. 19107, ISO/IEC (2003)
14. Belussi, A., Negri, M., Pelagatti, G.: An iso tc 211 conformant approach to model spatial integrity constraints in the conceptual design of geographical databases. In: ER (Workshops), pp. 100–109 (2006)
15. Jang, S., Kim, T.J.: Modeling an interoperable multimodal travel guide system using the iso 19100 series of international standards. In: Proceedings of the GIS 2006, pp. 115–122. ACM Press, New York (2006)
16. Horrocks, I., Hayes, P., Patel-Schneider, P.: OWL web ontology language semantics and abstract syntax. Technical report (February (2004)
17. Haarslev, V., Moller, R.: Racer system description. In: Lambrix, P., Borgida, A., Lenzerini, M., Moller, R., Patel-Schneider, P. (eds.) Proceedings of the CEUR-WS International Workshop on DL, vol. 22 (1999)
18. Rodríguez, M.A., Egenhofer, M.J.: Comparing geospatial entity classes: An asymmetric and context-dependent similarity measure. International Journal of Geographical Information Science 18(3), 229–256 (2004)
19. Visser, U. (ed.): Intelligent Information Integration for the Semantic Web. LNCS (LNAI), vol. 3159. Springer, Heidelberg (2004)
20. Hakimpour, F.: Using Ontologies to Resolve Semantic Heterogeneity for Integrating Spatial Database Schemata. PhD thesis, Zurich University (2003)
21. Fonseca, F.: Ontology-driven Geographic Information Systems. PhD thesis, University of Maine (2001)
22. Stoimenov, L., Stanimirovic, A., Djordjevic-Kajan, S.: Discovering mappings between ontologies in semantic integration process. In: Proceedings of the AGILE 2006, Visegrád, Hungary, pp. 213–219 (2006)
23. Aerts, K., Maesen, K., van Rompaey, A.: A practical example of semantic interoperability of large-scale topographic databases using semantic web technologies. In: Proceedings of the AGILE 2006, Visegrád, Hungary, pp. 35–42 (2006)
24. Janowicz, K.: Sim-dl: Towards a semantic similarity measurement theory for the description logic cnr in geographic information retrieval. In: OTM Workshops (2), pp. 1681–1692 (2006)

Generalizing the QSQR Evaluation Method
for Horn Knowledge Bases

Ewa Madalińska-Bugaj and Linh Anh Nguyen*

Institute of Informatics, University of Warsaw
Banacha 2, 02-097 Warsaw, Poland
{ewama,nguyen}@mimuw.edu.pl

Abstract. We generalize the QSQR evaluation method to give a set-oriented depth-first evaluation method for Horn knowledge bases. The resulting procedure closely simulates SLD-resolution (to take advantages of the goal-directed approach) and highly exploits set-at-a-time tabling. Our generalized QSQR evaluation procedure is sound, complete, and tight. It does not use adornments and annotations. To deal with function symbols, our procedure uses iterative deepening search which iteratively increases term depth bound for atoms occurring in the computation. When the term depth bound is fixed, our evaluation procedure runs in polynomial time in the size of extensional relations.

1 Introduction

Horn knowledge bases are definite logic programs, which are usually so big that either they cannot be totally loaded into memory or evaluations for them cannot be done totally in memory. Thus, in contrast to logic programming, for Horn knowledge bases efficient access to secondary storage is an important aspect. Horn knowledge bases can be treated as extensions of Datalog deductive databases without the range-restrictedness and function-free conditions. Developing efficient evaluation methods for Horn knowledge bases is worth not only for practical applications but also for the theory of knowledge bases.

To develop evaluation procedures for Horn knowledge bases one can either adapt tabled SLD-resolution systems of logic programming to reduce the number of accesses to secondary storage or generalize top-down or bottom-up evaluation methods of Datalog queries to deal with non-range-restricted definite logic programs and goals that may contain function symbols.

Tabled SLD-resolution systems like OLDT [10], SLD-AL [12, 13], linear tabulated resolution [9, 14] are efficient computational procedures for logic programming without redundant recomputations, but they are not directly applicable to Horn knowledge bases to obtain efficient evaluation engines because they are not set-oriented (set-at-a-time). In particular, the suspension-resumption mechanism and the stack-wise representation as well as the "global optimizations of

---

* Nguyen's research on this work has been supported by Polish Ministry of Science and Higher Education under grant N N206 3982 33.

N.T. Nguyen, R. Katarzyniak (Eds.): New Chall. in Appl. Intel. Tech., SCI 134, pp. 145–154, 2008.
springerlink.com

SLD-AL" are all tuple-oriented (tuple-at-a-time). Data structures for them are too complex so that they must be dropped if one wants to convert the methods to efficient set-oriented ones. Of course, one can use, e.g., XSB [7, 8] (a state-of-the-art implementation of OLDT) as a Horn knowledge base engine, but as pointed out in [3], it is tuple-oriented and not suitable for efficient access to secondary storage. The try of converting XSB to a set-oriented engine [3] removes essential features of XSB and is not natural.[1]

In [13], Vieille adapted SLD-AL resolution for Datalog deductive databases to obtain the top-down QoSaQ procedure by representing goals by means of tuples and translating the operations of SLD-AL on goals onto operations on tuples. This evaluation method of Datalog queries is also tuple-oriented due to the stack-wise representation and the optimizations originated from SLD-AL. Besides, the specific techniques of QoSaQ like "instantiation pattern", "rule compilation", "projection" are heavily based on the range-restrictedness and function-free conditions. Thus, it is not appropriate to extend the QoSaQ evaluation method for Horn knowledge bases.

In [11], Vieille gave the query-subquery recursive (QSQR) evaluation method for Datalog deductive databases, which is a top-down method based on SLD-resolution and the set-at-a-time technique. Vieille's QSQR algorithm is incomplete [13], but in this paper we refer to the revised version of QSQR [1] (and treat it as "original" w.r.t. our generalization), which is complete due to the added outer loop. The QSQR method uses adornments to simulate SLD-resolution in pushing constant symbols from goals to subgoals. The annotated version of QSQR also uses annotations to simulate SLD-resolution in pushing repeats of variables from goals to subgoals [1].

The magic-set technique [2, 6] is another formulation of tabling for Datalog deductive databases. It simulates the top-down QSQR evaluation by rewriting a given query to another equivalent one that when evaluated using a bottom-up technique (e.g. the seminaive evaluation) produces only facts produced by the QSQR evaluation. Adornments are used as in the QSQR evaluation. To simulate annotations, the magic-set transformation is augmented with subgoal rectification (see, e.g., [1]). Some authors have extended the magic-set technique for Horn knowledge bases [5, 3]. To deal with non-range-restrictedness and function symbols, "magic predicates" are used without adornments.

As seen from the above discussion, there are tuple-oriented depth-first evaluation methods (e.g. [8]) and (set-oriented) breadth-first evaluation methods [5, 3] (based on the magic-set transformation and the bottom-up seminaive evaluation) for Horn knowledge bases. However, as far as we know, no set-oriented depth-first evaluation method was developed for Horn knowledge bases.

In this paper, we generalize the QSQR evaluation method to give a set-oriented depth-first evaluation method for Horn knowledge bases. The resulting procedure closely simulates SLD-resolution (to take advantages of the goal-directed approach) and highly exploits set-at-a-time tabling. Our generalized QSQR evaluation procedure is sound, complete, and tight. It does not use adornments and

---

[1] The original XSB uses depth-first search, while Breadth-First XSB [3] does not.

annotations. To deal with function symbols, our procedure uses iterative deepening search which iteratively increases term depth bound for atoms occurring in the computation. When the term depth bound is fixed, our evaluation procedure runs in polynomial time in the size of extensional relations.

2 Preliminaries

A signature for first-order logic consists of constant symbols, function symbols, and predicate symbols. Terms and formulae over a fixed signature are defined using the symbols of the signature and variables in the usual way. An *atom* is a formula of the form $p(t_1, \ldots, t_n)$, where p is an n-ary predicate and $t_1, \ldots, t_n$ are terms. The *term depth* of a term or a formula is the maximal nesting depth of function symbols occurring in that term or formula. By $Var(\Gamma)$ we denote the set of variables occurring in Γ.

If θ and δ are substitutions such that $\theta\delta = \delta\theta = \varepsilon$, where ε is the empty substitution, then we call them *renaming substitutions*. If E and E' are terms, non-quantified formulae or tuples of terms such that $E\theta = E'$ for some renaming substitution θ, then E is called a *variant* of E', and vice versa.

The restriction of a substitution θ to a set X of variables is denoted by $\theta_{|X}$.

If φ is a formula, then by $\forall(\varphi)$ we denote the *universal closure* of φ, which is the formula obtained by adding a universal quantifier for every variable having a free occurrence in φ.

2.1 Positive Logic Programs and SLD-Resolution

A (positive) *program clause* is a formula of the form $\forall(A \vee \neg B_1 \vee \ldots \vee \neg B_k)$, written as $A \leftarrow B_1, \ldots, B_k$, where $A, B_1, \ldots, B_k$ are atoms. A is called the *head*, and $(B_1, \ldots, B_k)$ the *body* of the program clause. If p is the predicate of A then the program clause is called a program clause defining p.

A *positive* (or *definite*) *logic program* is a finite set of program clauses.

A *goal* (also called a *negative clause*) is a formula of the form $\forall(\neg B_1 \vee \ldots \vee \neg B_k)$, written as $\leftarrow B_1, \ldots, B_k$, where $B_1, \ldots, B_k$ are atoms. If $k = 1$ then the goal is called a *unary* goal.

If P is a positive logic program and $G = \leftarrow B_1, \ldots, B_k$ a goal, then θ is called a *correct answer* for $P \cup \{G\}$ if $P \models \forall((B_1 \wedge \ldots \wedge B_k)\theta)$.

A goal G' is *derived* from a goal $G = \leftarrow A_1, \ldots, A_i, \ldots, A_k$ and a program clause $\varphi = (A \leftarrow B_1, \ldots, B_h)$ using A_i as the *selected atom* and θ as the most general unifier (mgu) if θ is an mgu for A_i and A, and $G' = \leftarrow (A_1, \ldots, A_{i-1}, B_1, \ldots, B_h, A_{i+1}, \ldots, A_k)\theta$. We call G' a *resolvent* of G and φ. If $i = 1$ then we say that G' is derived from G and φ using *the leftmost selection function*.

Let P be a positive logic program and G a goal.

An *SLD-derivation* from $P \cup \{G\}$ consists of a (finite or infinite) sequence $G_0 = G, G_1, G_2, \ldots$ of goals, a sequence $\varphi_1, \varphi_2, \ldots$ of variants of program clauses of P and a sequence $\theta_1, \theta_2, \ldots$ of mgu's s.t. each G_{i+1} is derived from G_i and φ_{i+1}

using θ_{i+1}. Each φ_i is a suitable variant of the corresponding program clause such that φ_i does not have any variables which already appear in the derivation up to G_{i-1}. Each program clause variant φ_i is called an *input program clause*.

An *SLD-refutation* of $P \cup \{G\}$ is a finite SLD-derivation of $P \cup \{G\}$ which has the empty clause as the last goal in the derivation.

A *computed answer* θ for $P \cup \{G\}$ is the substitution obtained by restricting the composition $\theta_1 \ldots \theta_n$ to the variables of G, where $\theta_1, \ldots, \theta_n$ is the sequence of mgu's used in an SLD-refutation of $P \cup \{G\}$.

2.2   Definitions for Horn Knowledge Bases

Similarly as for deductive databases, we classify each predicate either as *intensional* or *extensional*. A *generalized tuple* is a tuple of terms, which may contain function symbols and variables. A *generalized relation* is a set of generalized tuples of the same arity. A *Horn knowledge base* is defined to be a pair consisting of a positive logic program for defining intensional predicates and a *generalized extensional instance*, which is a function mapping each extensional n-ary predicate to an n-ary generalized relation. Note that intensional predicates are defined by a positive logic program which may contain function symbols and not be range-restricted. From now on, we use the term "relation" to mean a generalized relation, and the term "extensional instance" to mean a generalized extensional instance.

Note: We will treat a tuple $\bar{t}$ from a relation of a predicate p as the atom $p(\bar{t})$. Thus, a relation (of tuples) of a predicate p is a set of atoms of p, and an extensional instance is a set of atoms of extensional predicates. Conversely, a set of atoms of p can be treated as a relation (of tuples) of the predicate p.

Given a Horn knowledge base specified by a positive logic program P and an extensional instance I, a query to the knowledge base is a positive formula $\varphi(\bar{x})$ without quantifiers, where $\bar{x}$ is a tuple of all the variables of φ. A (correct) answer for the query is a tuple $\bar{t}$ of terms of the same length as $\bar{x}$ such that $P \cup I \vDash \forall(\varphi(\bar{t}))$. When measuring "data complexity", we assume that P and φ are fixed, while I varies. Thus, the pair $(P, \varphi(\bar{x}))$ is treated as a *query* to the extensional instance I. We will use the term "query" in that meaning.

It can be shown that, every query $(P, \varphi(\bar{x}))$ can be transformed in polynomial time into an equivalent query of the form $(P', q(\bar{x}))$ over a signature extended with new intensional predicates, including q. The equivalence means that, for every extensional instance I and every tuple $\bar{t}$ of terms of the same length as $\bar{x}$, $P \cup I \vDash \forall(\varphi(\bar{t}))$ iff $P' \cup I \vDash \forall(q(\bar{t}))$. The transformation is based on introducing new predicates for defining complex subformulae occurring in the query. For example, if $\varphi = p(x) \wedge r(x, y)$, then $P' = P \cup \{q(x, y) \leftarrow p(x), r(x, y)\}$, where q is a new intensional predicate.

Without loss of generality, we will consider only queries of the form $(P, q(\bar{x}))$, where q is an intensional predicate. Answering such a query on an extensional instance I is equivalent to finding (correct) answers for $P \cup I \cup \{\leftarrow q(\bar{x})\}$.

3   Generalizing the QSQR Evaluation Algorithm

3.1   Informal Description

We first adapt SLD-resolution to find all answers effectively. We set up the problem as follows: given a positive logic program P, an extensional instance I and an atom A of an intensional predicate p, construct an answer relation ans_p such that for every SLD-refutation of $P \cup I \cup \{\leftarrow A\}$ with computed answer θ, $A\theta$ is an instance of a variant of some atom from ans_p, i.e. ans_p contains a more general answer than θ. The mentioned property is called completeness (of the evaluation). We expect also two other properties: soundness and tightness. Soundness states that for every atom A' of ans_p, $P \cup I \vDash \forall(A')$, and tightness informally states that all atoms of ans_p closely relate to the query (that is, no irrelevant atoms are derived). The relation ans_p contains tuples (as for the predicate p) that are treated as atoms of p.

For each intensional predicate q, we use a global variable ans_q to keep an answer relation for q. Tuples of ans_q are treated as atoms of the predicate q. At the beginning, we set all of such variables to empty relations. Consider an SLD-refutation of $P \cup I \cup \{\leftarrow A\}$. Let the first input program clause applied to $\leftarrow A$ be $\varphi = (A' \leftarrow B_1, \ldots, B_n)$ and the used mgu (for A and A') be θ. The next goal is thus $\leftarrow (B_1, \ldots, B_n)\theta$.

Let $\delta_0 = \theta$. For each $1 \leq i \leq n$, we process $\leftarrow B_i\delta_{i-1}$ as follows, where δ_{i-1} is the substitution containing the bindings of variables after processing $\leftarrow B_{i-1}\delta_{i-2}$. Let p_i be the predicate of B_i.

1. Case p_i is an extensional predicate: If γ_i is an mgu for $B_i\delta_{i-1}$ and a fresh variant of some atom from $I(p_i)$ then let $\delta_i := \delta_{i-1}\gamma_i$ and continue to process the next goal atom.
2. Case p_i is an intensional predicate:
   a) Recursively process $\leftarrow B_i\delta_{i-1}$ in the same way as for $\leftarrow A$. This task does not pass bindings of variables directly outside but it updates the answer relations held by global variables.
   b) If γ_i is an mgu for $B_i\delta_{i-1}$ and a fresh variant of some atom from ans_p_i then let $\delta_i := \delta_{i-1}\gamma_i$ and continue to process the next goal atom.

Then δ_n holds a correct answer for $P \cup I \cup \{\leftarrow A\}$. Thus, if $A\delta_n$ is not an instance of a *fresh* variant of any atom from the answer relation ans_p, where p is the predicate of A, then we can add $A\delta_n$ to ans_p. (A *fresh* variant is a variant whose variables are not used anywhere else.)

To obtain all answers for the goal $\leftarrow A$, all the choices are systematically tried, and the process is repeated until no changes are made to the global variables during the last iteration. To guarantee the stop property, each goal like $\leftarrow A$ is processed only once. Furthermore, to avoid redundant recomputation we check that $\leftarrow A$ is not an instance of a fresh variant of any goal that has been processed before. To do this we record A in a relation held by a global variable $input\_p$, where p is the predicate of A. Such a relation is called an input/goal relation.

It can be represented as a generalized relation and we treat tuples of *input_p* as atoms of the predicate p.

Note that, for the adaptation, we concentrate on unary goals.

3.2 A Formal Tuple-at-a-Time Version of the Method

We now formally present the algorithm of the evaluation method described in the previous subsection. Let l be a fixed natural number, which we will use as the bound imposed on term depth of atoms occurring in SLD-derivations.

Algorithm 1

Evaluate a query $(P, q(\overline{x}))$ on an extensional instance I.

1. Initialize global variables *ans_p* and *input_p* to empty relations for every intensional predicate p of P.
2. Call Procedure 2 to process the goal $\leftarrow q(\overline{x})$.
3. Return *ans_q*.

Procedure 2

Process a goal $\leftarrow A$.

1. Let p be the predicate of A. If A is an instance of a fresh variant of some atom from *input_p* then exit, else add A to *input_p*.
2. Repeat until no new tuples are added to any global *ans_* variable: For each program clause φ defining p in P, call Procedure 3 to process the goal $\leftarrow A$ on a fresh variant of φ.

Procedure 3

Process a goal $\leftarrow A$ on a program clause $\varphi = (A' \leftarrow B_1, \ldots, B_n)$, where A has the same predicate as A'.

1. If A and A' cannot be unified then exit, else let δ_0 be an mgu of A and A'.
2. $sup_0 := \{\delta_0\}$. (*sup_i* denotes the so called ith "supplementary" relation.)
3. For each i from 1 to n do:
   a) Let p_i be the predicate of B_i.
   b) $sup_i := \emptyset$.
   c) Case p_i is an extensional predicate: For every $\delta_{i-1} \in sup_{i-1}$ and every atom $B_i' \in I(p_i)$, if $B_i\delta_{i-1}$ is unifiable with a fresh variant of B_i' using an mgu γ_i then add $\delta_{i-1}\gamma_i$ to sup_i.
   d) Case p_i is an intensional predicate: For every $\delta_{i-1} \in sup_{i-1}$ do:
      i. If the term depth of $B_i\delta_{i-1}$ is not greater than l then call Procedure 2 to process the goal $\leftarrow B_i\delta_{i-1}$.
      ii. For every atom $B_i' \in ans\_p_i$, if $B_i\delta_{i-1}$ is unifiable with a fresh variant of B_i' using an mgu γ_i then add $\delta_{i-1}\gamma_i$ to sup_i.
4. For each $\delta_n \in sup_n$: If $A\delta_n$ is not an instance of a fresh variant of any atom from *ans_p*, where p is the predicate of A, then: delete from *ans_p* every atom whose fresh variant is an instance of $A\delta_n$, and add $A\delta_n$ to *ans_p*.

3.3   Soundness, Completeness, and Tightness

In this subsection, we give our results on soundness, completeness, and tightness of the top-down evaluation method specified by Algorithm 1. Due to the lack of space, the proofs are presented only in the long version of this paper [4].

Theorem 1 (Soundness). *Let $(P, q(\overline{x}))$ be a query and I an extensional instance. Consider the execution of Algorithm 1 for that query on I. Then, for every atom A'' added to ans_p at Step 4 of Procedure 3, $P \cup I \vDash \forall(A'')$.*

Theorem 2 (Completeness). *Let $(P, q(\overline{x}))$ be a query, I an extensional instance, and θ the computed answer of an SLD-refutation of $P \cup I \cup \{\leftarrow q(\overline{x})\}$ that does not contain any goal with term depth greater than l. Then the execution of Algorithm 1 (with parameter l) for the query $(P, q(\overline{x}))$ on I returns ans_q that contains a tuple $\overline{t}$ such that $\overline{x}\theta$ is an instance of a variant of $\overline{t}$.*

Note that in the above theorem $\overline{x}\theta$ is an instance of a variant of $\overline{t}$ but is not $\overline{t}$ nor a variant of $\overline{t}$ because of the optimization made at Step 4 of Procedure 3. For knowledge bases, it is inessential to require $\overline{x}\theta$ to be $\overline{t}$ or a variant of $\overline{t}$.

For queries and extensional instances without function symbols, we take term depth bound $l = 0$ and obtain the following strong completeness result, which immediately follows from the above theorem.

Corollary 1. *Let $(P, q(\overline{x}))$ be a query and I an extensional instance over a signature without function symbols. Let θ be a computed answer for $P \cup I \cup \{\leftarrow q(\overline{x})\}$. Then the execution of Algorithm 1 with $l = 0$ for the query $(P, q(\overline{x}))$ on I returns ans_q that contains a tuple $\overline{t}$ s.t. $\overline{x}\theta$ is an instance of a variant of $\overline{t}$.*

Definition 1. *An* unrestricted SLD-derivation *is an SLD-derivation, except that we drop the requirement that the used substitutions θ_i are most general unifiers. They are only required to be unifiers.*

The following theorem states that Algorithm 1 derives no irrelevant atoms. All input_ and ans_ atoms produced by the algorithm closely relate to the query.

Theorem 3 (Tightness). *Let $(P, q(\overline{x}))$ be a query and I an extensional instance. Consider the result of the execution of Algorithm 1 for that query on I. Then:*

1. *For every intensional predicate p of P and every atom $A' \in$ input_p, there is a variant A of A' that appears in an unrestricted SLD-derivation from $P \cup I \cup \{\leftarrow q(\overline{x})\}$.*
2. *For every intensional predicate p of P and every atom $A'' \in$ ans_p, there exists $A \in$ input_p such that A'' is an instance of A and $P \cup I \cup \{\leftarrow A''\}$ has an SLD-refutation with the empty substitution as the computed answer.*

The first assertion states that every input_ atom closely relates to the given query, while the second assertion states that every ans_ atom closely relates to some input_ atom, and therefore closely relates to the given query. That is, Algorithm 1 derives no irrelevant atoms.

Note that one cannot make the theorem stronger by deleting the word "unrestricted" because a goal $\leftarrow r(\bar{t})$ may trigger a subgoal $\leftarrow r(\bar{t}')$, which in turn may produce an answer $\overline{t''}$, which in turn may be used to answer $\leftarrow r(\bar{t})$ and restrict subqueries. This is a common problem of approaches with tabling (e.g. Theorem 13.4.1 of [1] considers only tightness w.r.t. ground facts).

3.4 Doing It Set-at-a-Time

Operations for databases and knowledge bases are often done set-at-a-time instead of tuple-at-a-time in order to reduce the number of accesses to secondary storage. This approach allows various optimizations like sorting, indexing, and clustering. In this subsection, we reformulate Algorithm 1 using the set-at-a-time technique. For the new algorithm, we use the following relational operators:

- $eliminate\_subsumed\_tuples(J, J')$
  - where J and J' are generalized relations of the same arity,
  - returns the set of tuples of J that are not instances of a fresh variant of any tuple from J'.
- $merge(J, J')$
  - where J and J' are generalized relations of the same arity,
  - is $eliminate\_subsumed\_tuples(J, J') \cup eliminate\_subsumed\_tuples(J', J)$.
- $resolve\_with\_head\_atom(J, A')$
  - where J is a goal relation of the same predicate as A',
  - returns the set of tuples (A, δ_0) for each atom $A \in J$ such that δ_0 is an mgu of A and A'.
- $resolve\_with\_body\_atom(K, B_i, R, X)$
  - where K has the format as $resolve\_with\_head\_atom(J, A')$ for some J and A', B_i is an atom, R is a generalized relation of the predicate of B_i, and X is a set of variables,
  - returns the set of tuples $(A, \delta_{i\,|Var(A)\cup X})$ for each $(A, \delta_{i-1}) \in K$ and each $B_i' \in R$ such that: γ_i is an mgu for $B_i\delta_{i-1}$ and a fresh variant of B_i', and $\delta_i = \delta_{i-1}\gamma_i$.

Here is our reformulation of Algorithm 1:

Algorithm 4

Evaluate a query $(P, q(\overline{x}))$ on an extensional instance I.

1. Initialize global variables ans_p and $input\_p$ to empty relations for every intensional predicate p of P.
2. Call Procedure 5 to process the goal relation $\{\overline{x}\}$ of predicate q.
3. Return ans_q.

Procedure 5

Process a goal relation J of a predicate p.

1. $J := eliminate\_subsumed\_tuples(J, input\_p)$.
2. Exit if J is empty.

3. $input\_p := input\_p \cup J$.
4. Repeat until no new L-tuples are added to any global $ans\_$ variable:
   - For each program clause φ of P that defines p, let $A' \leftarrow B_1, \ldots, B_n$ be a fresh variant of φ and do:
     a) $K := resolve\_with\_head\_atom(J, A')$.
     b) $i := 0$.
     c) While $i < n$ and K is not empty do:
        i. $i := i + 1$.
        ii. $X := Var(\{B_{i+1}, \ldots, B_n\})$.
        iii. If the predicate p_i of B_i is an extensional predicate then:
           - $K := resolve\_with\_body\_atom(K, B_i, I(p_i), X)$,
        iv. else:
           - Recursively call Procedure 5 for $\{B_i\delta_{i-1} \mid (A, \delta_{i-1}) \in K$ and the term depth of $B_i\delta_{i-1}$ is not greater than $l\}$.
           - $K := resolve\_with\_body\_atom(K, B_i, ans\_p_i, X)$.
     d) $ans\_p := merge(ans\_p, \{A\delta_n \mid (A, \delta_n) \in K\})$.

Theorem 4. *The evaluation by Algorithm 4 is sound, complete, and tight (in the sense stated for Algorithm 1).*

Theorem 5. *For a fixed query and a fixed bound l on term depth, Algorithm 4 runs in polynomial time in the size of the extensional instance.*

The *data complexity* of an evaluation algorithm is the computational complexity of that algorithm measured w.r.t. the size of the extensional instance when the query is fixed.

Corollary 2. *Algorithm 4 with parameter $l = 0$ is a complete evaluation algorithm with PTIME data complexity for the class of queries over a signature without function symbols.*

3.5   Iterative Deepening Search

Suppose that we want to compute as many as possible but no more than k correct answers for a query $(P, q(\overline{x}))$ on an extensional instance I within time limit T. Then we can use iterative deepening search as follows:

1. Initialize term depth bound l by 1 (or another small integer value).
2. Run Algorithm 4 for evaluating $(P, q(\overline{x}))$ on I within the time limit.
3. While ans_q contains less than k tuples and the time limit was not reached yet, do:
   a) Increase term depth bound l by 1.
   b) Run Algorithm 4 without resetting the *input_* and *ans_* global variables (i.e. without Step 1).
4. Return ans_q.

4 Conclusions

We have generalized the QSQR evaluation method to give the first set-oriented depth-first evaluation method for Horn knowledge bases. The resulting procedure closely simulates SLD-resolution (to take advantages of the goal-directed approach) and highly exploits set-at-a-time tabling. Its operations on relations can be implemented to really reduce the number of accesses to secondary storage. In the case when the used Horn knowledge base is specified in the top-down manner as in logic programming, where the order of clauses is essential, and the user is interested only in finding some but not all answers for a query, our method would be more efficient than the breadth-first evaluation method based on the magic-set transformation and the bottom-up seminaive evaluation.

References

1. Abiteboul, S., Hull, R., Vianu, V.: Foundations of Databases. Addison-Wesley, Reading (1995)
2. Bancilhon, F., Maier, D., Sagiv, Y., Ullman, J.D.: Magic sets and other strange ways to implement logic programs. In: Proceedings of PODS 1986, pp. 1–15. ACM Press, New York (1986)
3. Freire, J., Swift, T., Warren, D.S.: Taking I/O seriously: Resolution reconsidered for disk. In: Naish, L. (ed.) Proc. of ICLP 1997, pp. 198–212. MIT Press, Cambridge (1997)
4. Madalińska-Bugaj, E., Nguyen, L.A.: The long version of this paper (2008), http://www.mimuw.edu.pl/~nguyen/papers.html
5. Ramakrishnan, R., Srivastava, D., Sudarshan, S.: Efficient bottom-up evaluation of logic programs. In: Vandewalle, J. (ed.) The State of the Art in Computer Systems and Software Engineering, Kluwer Academic Publishers, Dordrecht (1992)
6. Rohmer, J., Lescouer, R., Kerisit, J.-M.: The Alexander method – a technique for the processing of recursive axioms in deductive databases. New Generation Computing 4(3), 273–285 (1986)
7. Sagonas, K.F., Swift, T.: An abstract machine for tabled execution of fixed-order stratified logic programs. ACM Trans. Program. Lang. Syst. 20(3), 586–634 (1998)
8. Sagonas, K.F., Swift, T., Warren, D.S.: XSB as an efficient deductive database engine. In: Snodgrass, R.T., Winslett, M. (eds.) Proceedings of the 1994 ACM SIGMOD Conference on Management of Data, pp. 442–453. ACM Press, New York (1994)
9. Shen, Y.-D., Yuan, L.-Y., You, J.-H., Zhou, N.-F.: Linear tabulated resolution based on Prolog control strategy. TPLP 1(1), 71–103 (2001)
10. Tamaki, H., Sato, T.: OLD resolution with tabulation. In: Shapiro, E. (ed.) ICLP 1986. LNCS, vol. 225, pp. 84–98. Springer, Heidelberg (1986)
11. Vieille, L.: Recursive axioms in deductive databases: The query/subquery approach. In: Proceedings of Expert Database Conf., pp. 253–267 (1986)
12. Vieille, L.: A database-complete proof procedure based on SLD-resolution. In: Proceedings of ICLP, pp. 74–103 (1987)
13. Vieille, L.: Recursive query processing: The power of logic. Theor. Comput. Sci. 69(1), 1–53 (1989)
14. Zhou, N.-F., Sato, T.: Efficient fixpoint computation in linear tabling. In: Proceedings of PPDP 2003, pp. 275–283. ACM Press, New York (2003)

On Vowels Segmentation and Identification Using Formant Transitions in Continuous Recitation of Quranic Arabic

Hafiz Rizwan Iqbal, Mian Muhammad Awais,
Shahid Masud, and Shafay Shamail

Department of Computer Science,
Lahore University of Management Sciences,
DHA 54792, Lahore, Pakistan
{rizwani,awais,smasud,sshamail}@lums.edu.pk

Abstract. This paper provides an analysis of cues to identify Arabic vowels. A new algorithm for vowel identification has been developed that uses formant frequencies. The algorithm extracts the formants of already segmented recitation audio files and recognizes the vowels on the basis of these extracted formants. The investigation has been done in context of recitation principles of Holy Quran which are commonly known as Tajweed rules. Primary objective of this work is to be able to identify zabar /a/, zair /e/ and pesh /u/ mistakes of the recitor during the recitation. Acoustic Analysis was performed on 150 samples of different recitors and a corpus comprising recitation of five experts was used to validate the results. The vowel identification system developed here has shown up to 90% average accuracy on continuous speech files comprising around 1000 vowels.

Keywords: Tajweed, Formant transition track(s),Wavelet transforms, Location, Trend, Gradient, Vowels, Zabar, Zair, Pesh, Laam, Meem, Noon, Continuous Arabic speech.

1 Introduction

Keeping in view the emerging demands of speech recognition, a prototype application to understand Arabic recitation has been developed. This system acts as a language tutor to correct mistakes in pronunciation and recitation. The application is developed around an automated speech recognition system for which speech segmentation and identification are essential components. High segmentation accuracies are required for such systems to work. A brief description of the phoneme segmentation in continuous Arabic speech has been discussed in our prior work [1]. This paper is focused on identification of vowels from segmented speech.

The standard Arabic language has 34 phonemes out of which there are 6 vowels and 28 consonants [2, 3]. Vowels are the fundamental speech units present in every spoken language. The Arabic vocalic system is composed of three short vowels /a/, /e/, /u/ and three vowels of the same quality but of longer duration /aa/, /ee/, /uu/.

Several features and techniques of vowels identification in Arabic language have been discussed in the literature [4 - 9]. Most of the existing schemes are based on standard set of features such as spectral densities, intensities or formant frequencies. These techniques are known to result in Recognition Error Rate (RER) of around 10% [10].

N.T. Nguyen, R. Katarzyniak (Eds.): New Chall. in Appl. Intel. Tech., SCI 134, pp. 155–162, 2008.

In this research we have used formant transition tracks along with the phoneme duration cues for vowels identification in Quranic recitation. The identification algorithm can be divided into different stages. In the first phase, the output of the segmentation algorithm is provided to Praat speech processing tool [11] which returns all formants existing in a particular time slot. In the second phase, the segmented vowels are separated from the nasals (/l/ laam, /m/ meem, /n/ noon) and also identified as /a/, /e/, /u/ on the basis of formants obtained in the first step.

This work aims at establishing the relationship between values of formants corresponding to different vowels in Arabic recitation. Previous research in this area has been focused on the first three formants i.e. F1, F2 and F3. In comparison, this work uses only two formants F1 and F2 for vowel identification. Experimental results obtained using these two formants show 90% accuracy for the vowels identification.

The rest of the paper is organized as follows: Section 2 gives an outline of the cues used during different experiments in the vowel identification system. Section 3 explains the methodology that has been adopted for vowel identification. Section 4 describes the results and analysis of the proposed algorithm. Section 5 gives the conclusion and proposed future work while section 6 gives the references.

2 Features Analysis

Properties related to phoneme are embedded in different type of signals which can act as cues for vowels identification. Different combinations of these cues can generate different results with varying accuracy levels. These features include formants transition tracks. The formant transition track(s) have very useful information hidden in the formant frequency trends of all the phonemes [4]. Each group of formant transition tracks in F1 and F2 possess features which could be used for unique identification of the standard Arabic phonemes. This is described below.

2.1 Formant Analysis

The vowels identification algorithm developed here is independent of the speaker. The speaker independence has been achieved through a preprocessing step that relies on Location, Trend and Gradient (LTG) from the graphical analysis of formant tracks [4].

Typical formant transitions for zabar /a/, zair /e/ and pesh /u/ are shown in figures 1, 2 and 3 respectively. It can be observed from these figures that the Trend and gradient sub-cues show almost the same characteristics for all the three vowels. A prominent difference in the location of formants F1 and F2 was observed. When a speaker recites /a/ (zabar) vowel the distance between formant F1 and F2 is about 800-900 Hz as shown in the Figure 1. When /u/ (pesh) is recited, this difference between F1 and F2 decreased to about 400-500 Hz (half of the zabar) as shown in Figure 3. Difference between the locations of F1 and F2 for the vowel /e/ (zair) is about 1700-1900 Hz (twice as of zabar) as depicted in Figure 2. Formant transitions of F1 and F2 are also observed for the nasals (/l/ laam, /m/ meem, /n/ noon) sounds. It has been analyzed that F1 ranges from 300-500 Hz while F2 lies between 1250-1650 Hz, for all of the three nasals.

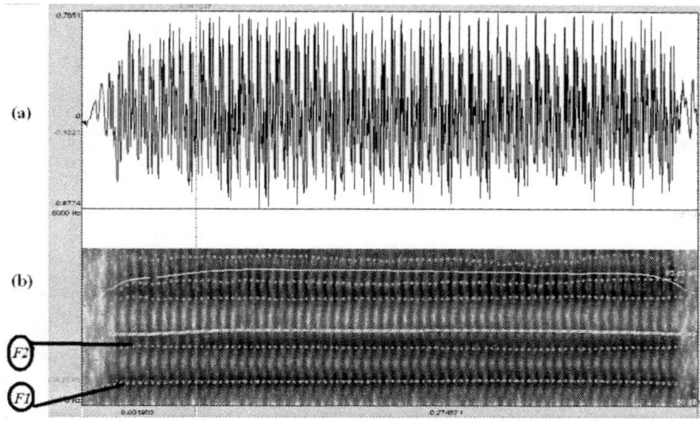

Fig. 1. Representation of /a/ **(a)** Waveform of signal **(b)** Formant transitions

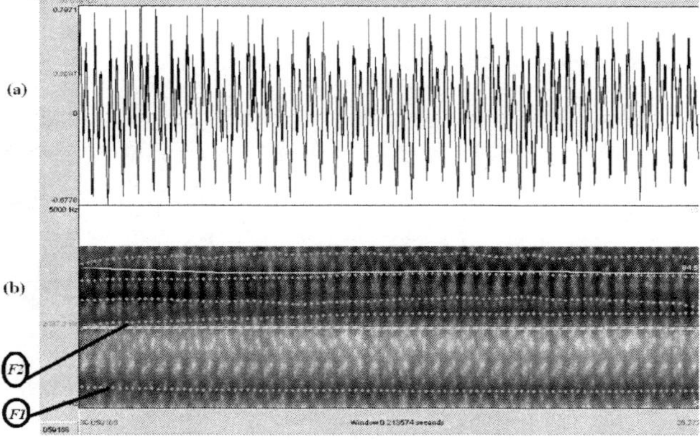

Fig. 2. Representation of /e/ **(a)** Waveform of signal **(b)** Formant transitions

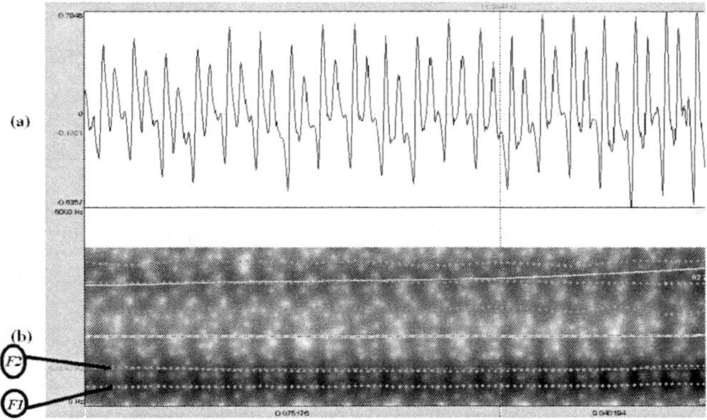

Fig. 3. Representation of /u/ **(a)** Waveform of signal **(b)** Formant transitions

3   Methodology of Vowel Identification

This section outlines in detail the settings, constraints, algorithm and calculations conducted with reference to the research presented here.

3.1   Experimental Setup

The recordings were conducted in a noise-free environment using 8 kHz sampling rate. At first, 12 speakers from an age group of 15 years to 30 years were selected for the recordings. All recitors in this set belonged to the same region and were experts in the recitation of the Holy Quran according to Tajweed rules. The segmentation and identification algorithms are developed in C++. After segmentation, the samples of vowels (/a/, /e/, /u/) and consonants uttered by a particular speaker were obtained as wave files. Praat tool was used in the analysis of segmented data. Over 150 samples of vowels (/a/, /e/, /u/) were used in this analysis.

3.2   Algorithm Implementation

Segmentation algorithm generates the time boundaries of the vowels. In some cases (e.g. vowel and nasal come together during recitation), these boundaries consist of a vowel part and a nasal (/l/ laam, /m/ meem, /n/ noon) part. Formants location against each point are calculated for each of the given time slot. Through these formants, the vowels are separated from nasals and also classified automatically using the application developed in C++. Figure 4 shows an abstract level diagram of the vowels segmentation and identification system.

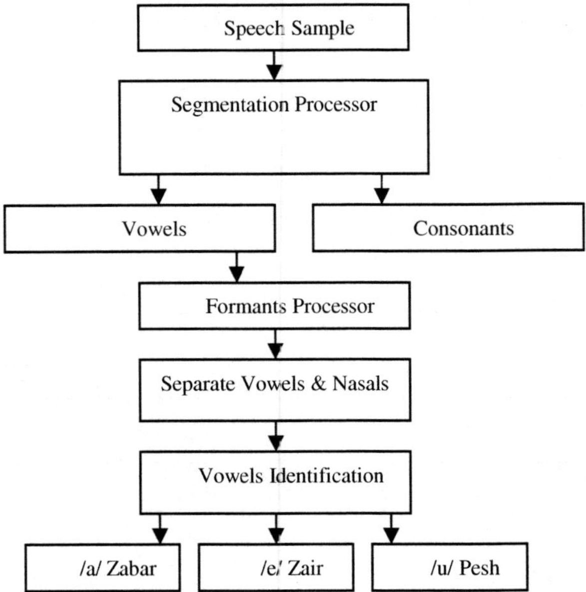

Fig. 4. Architecture of Vowels Identification System in Arabic Recitation

An audio file has been used for processing in the system. This audio file is stored in a numerical format in an array. Input speech is sampled at 8 KHz and the windows of 128 samples are taken for further processing. This data passes to the segmentation processor which generates the classified phonemes vowels and consonants. These time boundaries of vowels are sent to the Formant processor, where Praat tool calculates all the formants (F1, F2, F3 and F4) of each time slot. Now, the identification module uses these formant values to separate vowels from nasals and also classify the vowel part as /a/, /e/, /u/ (zabar, zair and pesh respectively). Figure 5 shows the algorithm that has been developed for vowels separation from nasals and also identification these vowels. It has been concluded from the experiments that the Lower Formant (LF) and Upper Formant (UF) limits consistently correspond to the following vowel ranges: These are used for different formants (F1 or F2 for vowel identification).

LF for /a/= 550   Hz, UF for /a/= 900   Hz
LF for /e/= 1800 Hz, UF for /e/= 2550 Hz
LF for /u/= 750   Hz. UF for /u/= 1100 Hz

---

1. *For 'i' from 1 to 4, get Formants F_i from the Praat tool and read each F_i against a particular time slot.*
2. *Check; If the formant F_1 lies between the "LF for /a/" and "UF for /a/" then it is vowel /a/.*
3. *If F_1 goes down to "LF for /a/" then Check F_2*
3.a. If F_2 lies between "LF for /e/" and "UF for /e/" then it is vowel /e/.
3.b. If F_2 lies between "LF for /u/" and "UF for /u/" then it is vowel /u/.
3.c. If F_2 lies between "UF for /u/" and "LF for /e/" then it is a nasalized vowel.
4. *Repeat until all the time slots are finished.*

Fig. 5. Algorithm to Identify Vowels in Arabic Recitation

4 Results and Analysis

Speech signal was divided into different time slots and for each slot the location of the formants has been probed to find the number of consecutive slots specifying a certain vowel or a non-vowel. For each detected phoneme (vowel or non-vowel), the starting time, ending time and proposed classification was evaluated. As an example, summarized results for each vowel, generated from the method described in section 3 for five different speakers are shown in Tables 1, 2 and 3. The results for non-vowels (/l/ laam, /m/ meem, /n/ noon) classification are also shown in Table 4. Each table shows the total number of vowels manually identified, total number of vowels identified by the proposed algorithm, (V_{asO}), actual vowels which were termed as other (vowels or non-vowels), (O_{asV}), actual other (vowels or non-vowels) termed as a particular vowel. Mathematical relationship for calculating Recall and Precision values are as follows:

$$VowelRecall = \frac{VowelsIdentifiedCorrectly}{VowelsIdentifiedCorrectly + V_{asO}} \tag{1}$$

$$VowelPrecision = \frac{VowelsIdentifiedCorrectly}{VowelsIdentifiedCorrectly + O_{asV}} \tag{2}$$

Precision defines the proportion of the classified phonemes which are actually correct whereas recall depicts the sensitivity, or the proportion of the correct results obtained. The overall accuracy of the system for/a/, /e/, /u/ is 96%, 92.5% and 84% respectively. For nasals the accuracy level is 87%. Average recall for all of the vowels and nasalized sounds is 93% and average precision for both types of sounds is 86%. Accuracy for the whole system is about 90%.

Table 1. Vowel V1 (/a/) Recall & Precision

Files	V1 $_{(Manual)}$	V1 $_{(Algo)}$	V1 $_{as\,O}$	O $_{as\,V1}$	V1 $_{Recall}$	V1 $_{Precision}$
1	37	37	0	1	100%	97%
2	35	35	0	3	100%	92%
3	35	34	1	0	94%	97%
4	40	39	1	1	95%	95%
5	56	55	1	6	96%	89%

Table 2. Vowel V2 (/e/) Recall & Precision

Files	V2 $_{(Manual)}$	V2 $_{(Algo)}$	V2 $_{as\,O}$	O $_{as\,V2}$	V2 $_{Recall}$	V2 $_{Precision}$
1	15	14	1	1	87%	87%
2	15	15	0	3	100%	83%
3	20	20	0	0	100%	100%
4	22	22	0	0	100%	100%
5	10	10	0	5	100%	67%

Table 3. Vowel V3 (/u/) Recall & Precision

Files	V3 $_{(Manual)}$	V3 $_{(Algo)}$	V3 $_{as\,O}$	O $_{as\,V3}$	V3 $_{Recall}$	V3 $_{Precision}$
1	7	7	0	1	100%	87%
2	13	12	1	3	86%	75%
3	8	8	0	3	100%	73%
4	9	9	0	2	100%	82%
5	13	11	2	5	73%	61%

Table 4. Non-Vowel V4 (/'n'/) Recall & Precision

Files	V4 $_{(Manual)}$	V4 $_{(Algo)}$	V4 $_{as O}$	O $_{as V4}$	V4 $_{Recall}$	V4 $_{Precision}$
1	63	60	3	1	91%	94%
2	62	53	4	1	80%	84%
3	51	48	3	1	89%	92%
4	63	60	3	1	91%	94%
5	71	57	0	1	80%	79%

5 Conclusion and Future Work

Formant transition track(s) are the cues which play a major role in the identification of vowels in Arabic recitation. A new algorithm using formant transitions was developed for vowels identification. The algorithm has been shown to provide over 90% accuracy for vowels identification in continuous speech samples. The approach developed here can be used in speech recognition solutions operating in the environment of recitation of religious scriptures, poetry or learning of a foreign language.

The scheme proposed here can be extended to use other features like phoneme duration, wavelet transforms, MFCC and cochleagram along with formant transitions to further increase the accuracy of the system. Further investigation on the effects of vowel lengthening, Qalqalah vowels and other tajweed rules for Quranic Arabic recitation is also being carried out. The objective of further work is to be able to identify the recitation mistakes in real-time for use in an interactive learning environment.

Acknowledgments. The authors acknowledge the funding provided by Lahore University of Management Sciences (LUMS), Lahore for the research conducted.

References

1. Ahmad, W., Awais, M.M., Shamail, S., Masud, S.: Continuous Arbic Speech Segmentation Using FFT Spectrogram. In: Innovations in Information Technology Conference, pp. 1–6. IEEE Press, Dubai (2006)
2. Maryati, M.: Man-Machine Communication and Arabic Language. Technical report, Scientific Studies and Research Center, Syria (1987)
3. Abady, Z.A.: Arabic Speech Processing. In: International Conference on Electronics Circuits and Systems, Jordan, pp. 647–650 (1995)
4. Shoaib, M., Rasheed, F., Akhtar, J., Awais, M., Masud, S., Shamail, S.: A Novel Approach to Increase the Robustness of Speaker Independent Arabic Speech Recogniton. In: INMIC Conference, pp. 371–376. IEEE Press, Pakistan (2003)
5. Selouani, S.-A., Caelen, J.: Recognition of Arabic Phonetic Features Using Neural Networks and Knowledge-Based System: a Comparative Study. In: Intelligence and Systems Conference, pp. 404–411. IEEE Press, USA (1998)

6. Gendrot, C., Adda-Decker, M.: Impact of duration on F1/F2 formant values of oral vowels: an automatic analysis of large broadcast news corpora in French and German: a Comparative Study. In: INTERSPEECH, Portugal, pp. 2453–2456 (2005)
7. Al-Tamimi, J., Carre, R., Marsico, E.: The status of vowels in Jordanian and Moroccan Arabic: Insights from production and perception. In: 48th meeting of the Acoustical Society of America, USA, p. 2629 (2004)
8. Al-Anani, M.: Arabic Vowel Formant Frequencies. In: International Congress of Phonetic Sciences, USA, pp. 2117–2119 (1999)
9. Gendrot, C., Adda-Decker, M.: Impact of duration and vowel inventory size on Formant values of earal vowels: an automated formant analysis from eight languages. In: International Congress of Phonetic Sciences, pp. 1417-1420. Germany (2007)
10. Quran Phonetic Search Engine, `http://www.islamicity.com/ps/default.asp`
11. Speech Analysis and Processing Tool, `http://www.fon.hum.uva.nl/praat/`
12. S.: IGNATIUS HIGH SCHOOL, `http://www2.ignatius.edu/faculty/turner/arabicspanish.htm`

Adaptation of FCANN Method to Extract and Represent Comprehensible Knowledge from Neural Networks

Sérgio M. Dias[1], Bruno M. Nogueira[2], and Luis E. Zárate[1]

[1] Department of Computer Science
   LICAP - Applied Computational Intelligence Laboratory*
   Pontifical Catholic University of Minas Gerais
   Av. Dom José Gaspar, 500 - ZIP: 30.535-610
   Belo Horizonte, Minas Gerais, Brazil
[2] Department of Computer Science
   Institute of Mathematics and Computer Science
   LABIC - Laboratory of Computational Intelligence
   University of São Paulo
   sergiomariano@gmail.com, brunomn@icmc.usp.br, zarate@pucminas.br
   *http://www.inf.pucminas.br/projetos/licap

Abstract. Nowadays, Artificial Neural Networks are being widely used in the representation of physical processes. Once trained, the nets are capable to solve unprecedented situations, keeping tolerable errors in their outputs. However, humans cannot assimilate the knowledge kept by these networks, since such knowledge is implicitly represented by their structure and connection weights. Recently, the FCANN method, based in Formal Concept Analysis, has been proposed as a new approach in order to extract, represent and understand the behavior of the process through rules. In this work, it is presented an adaptation of the FCANN method to extract more comprehensible variables relationships, obtaining a reduced and more interesting set of rules related to a predefined domain parameters subset, which provides a better analysis of the knowledge extracted from the neural networks without the necessity of a posteriori implications mining. As case study the approach FCANN will be applied in solar energy system.

Keywords: Extract knowledge, Artificial Neural Networks, Formal Concept Analysis.

1 Introduction

The real world processes understanding is not an easy task. Real problems, especially those related to industrial processes, can be difficult to model and understand. They can involve a high complexity level in their domains making the knowledge about their behavior very hard. In this context, Artificial Intelligence (AI) methods have been proposed as an alternative to represent knowledge in computer systems. AI can be subdivided in some fields of knowledge representation, for example, the connectionist field, where Artificial Neural Networks (ANN) are inserted, dealing with implicit, numerical or sub-symbolic knowledge.

N.T. Nguyen, R. Katarzyniak (Eds.): New Chall. in Appl. Intel. Tech., SCI 134, pp. 163–172, 2008.
springerlink.com

ANN have the capacity to obtain the relationship between the input and output parameters of the considered system, where the input parameters are mapped into outputs of the net by an implicit function. This behavior turns ANN a "black-box" [1] and no information which could be helpful to decision taken processes can be explicitly obtained from its internal structure. Due to this reason, new methods to extract knowledge from ANN are necessary. This knowledge can be represented by problem domain rules, extracted from previously trained neural networks, promoting the knowledge about the process in its real context.

Many researchers have been discussing the knowledge extraction from ANN [1, 2, 3]. Different methodologies have been presented for rules extraction. Some methods extract rules through the analysis of the net structure (i.e. weights, topology etc), while others extract rules analyzing the data set used to train the net. There are also methods that use both, the net structure and the data set to extract the rules. For other methods, when the refining of the rules is necessary, a net retraining has to be applied. This is a limitation of the proposed methods that does not exist in the FCANN method used in this work.

Recently, Formal Concept Analysis (FCA) has been proposed as a powerful technique in knowledge representation and extraction. In [4] and [5], FCA has been used to extract rules from previously trained neural networks. The method presented in [4] consists in a visual analysis of a line diagram [6] in order to extract the rules. In [7] the FCANN method was compared with conventional algorithms used in knowledge extraction, such as TREPAN and C4.5. Those comparisons showed the method relevance in extracting processes qualitative knowledge through implication rules in "If... Then..." format. The FCANN method extracts qualitative relations learned by the network, independently of its input-output structure, while conventional techniques look for it to extract the knowledge, working simply as a classifier. Moreover, in [7] the FCANN method revealed capable to represent processes of different classes learned by ANN, such as Cold Rolling Process, Climatic Behavior, and Evaluation of Urban Real Estate, among others. The main idea of that approach is to join Formal Concept Analysis [8, 9] and the Next Closure algorithm [10] to extract knowledge from ANN. The Next Closure algorithm permits to obtain implication rules of the type "If... Then..." without redundancies.

In this paper, the FCANN method will be adjusted in order to extract more comprehensible relationships, providing better analysis of the extracted knowledge from the neural networks. Thus, making possible to work with predefined interest parameters subsets, obtaining a representative, intuitive and concise rules set about the process behavior without the necessity to do a posterior implications mining in the sense of rules selection. This paper is organized in five sections. In the second one, a short review of the formal concept analysis is showed. In the third section, the FCANN method is presented and the adjustments of the FCANN method are discussed. In the fourth one, a case study is realized. Finally, the contributions and the conclusions of this work are presented.

2 Formal Concept Analysis - Short Review

Formal Concept Analysis (FCA) is a field of mathematics presented in the early eighties [8, 9]. Its main goal involves the knowledge representation by means of specific diagrams called line diagrams. In FCA, formal contexts are a primordial definition. They are represented by cross tables and have the notation $K(G, M, I)$, where G is a set of objects (rows headers), M is a set of attributes (columns headers) and I, an incidence relation ($I \subseteq G \times M$). If an object $g \in G$ and an attribute $m \in M$ are in the relation I, this is represented by gIm or $(g,m) \in I$ and is read as *"the object g has the attribute m"*.

A sort of algorithms can be applied to formal contexts (cross tables) in order to determine its formal concepts and its line diagrams. Formal concepts are pairs *(A, B)*, where $A \in G$ (called extent) and $B \in M$ (called intent). When the set of all formal concepts of a formal context $K:=(G, M, I)$ is ordered hierarchically according to the Complete Reticulate Theory [6], it is called conceptual reticulate with the notation $\underline{\beta}(G, M, I)$. The formal concepts are related as $(A_1, B_1) \leq (A_2, B_2)$ when $A_1 \subseteq A_2$ and $B_2 \subseteq B_1$.

2.1 The Next Closure Algorithm

The Next Closure algorithm was proposed in 1984 by Bernhard Ganter [10] as an algorithm with capacity to find closure systems. The main characteristic of this algorithm is to extract the minimum set of implication rules (*Steam-Base* or *Duquenne-Guigues base*) on a formal context [8, 10]. The minimal implication base, which is non-redundant, provides a complete implication set, so that any valid implication on formal context can be obtained through the combination of rules of the minimal base. The removal of any rule of the minimal base makes it an incomplete base.

The number of possible implication rules that can be extracted from formal context $K=(G,M,I)$ can be exponential. If M has n elements, then 2^{2n} possible implications will exist. Nevertheless, many implications generated are unnecessary or irrelevant due to the fact that they can be deducted from other implications. After the *Next Closure* algorithm is applied, it is possible to obtain, an implication base £, with the following characteristics:

1. Sound: each implication in £ is valid in $K=(G,M,I)$
2. Complete: each implication pertinent of $K=(G,M,I)$ is in £
3. Non redundant: no implication in £ is originated from other implications of £

Through these rules, it is possible to identify the parameters and their intervals that determine the operating point of the dependent variable. These rules can help the identification of the parameters that can change the operational point, such as controllable variables, in a control system.

3   FCANN for Knowledge Extraction from ANN

In this section, the steps to extract knowledge from previously trained neural networks, discussed in [7], will be presented in a summarized way:

1. Select a process representative data set in order to train the neural network. It is defined as:

$$X = [x_{ij}]m \times n \tag{1}$$

   where n is the number of parameters; X_{ij} to $i= 1, ...,m$ and $j=1, ..., n$-1 are the input parameters and x_{in} to $i=1, ..., m$ is the output parameter. This output parameter should have a known probability distribution, for example a normal distribution $N_1(\overline{x}, S(X))$.

2. Define the structure of the multi-layers neural network (with N input parameters; H hidden layers and M outputs), and train it. In this work $M = 1$ for any situation.

3. Build a synthetic database to operate the net considering the domain range of the input parameters. This synthetic database is defined as:

$$Y = [y_{ij}]_{p \times n-1} \tag{2}$$

   where: Y has only elements generated with the purpose of knowledge extraction. Thus, no one of such elements has been collected from the physics process. Each input parameter has minimal and maximal values, which defines the domain range of each parameter. The minimum and maximum values are vectors that are respectively defined as:

$$inf = \{u_1, u_2, ..., u_{n-1}\} \ sup = \{v_1, v_2, ..., v_{n-1}\} \tag{3}$$

The vector W defines the number of data that will be generated for each parameter, between the minimum and maximum values, Eq. (3):

$$W = [W_j]; j = 1, ..., n - 1 \tag{4}$$

Hence, the variation of each parameter, that will compose the synthetic database, is represented by the following expression:

$$Int = \{I_1, I_2, ..., I_{n-1}\} \ where \ I_j = \frac{|v_j - u_j|}{w_i} \ for \ j = 1, ..., n - 1 \tag{5}$$

The values of each parameter used to generate the synthetic database can be represented by:

$$S = \begin{bmatrix} S_{1,1} & S_{1,2} & \cdots & S_{1,n-1} \\ S_{2,1} & S_{2,2} & \cdots & S_{2,n-1} \\ \vdots & \vdots & \ddots & \vdots \\ S_{w_1,1} & S_{w_2,2} & \cdots & S_{w_{n-1},n-1} \end{bmatrix}$$

Or expressed implicitly:

$S_{1j} = u_j + \frac{I_j}{2}$ para $j= 1, ...,n\text{-}1$

$S_{kj} = S_{k-1,j} + \frac{I_j}{2}$ para $j= 1, ...,n\text{-}1;\ k= 1, ...,w_j$

It could be observed that the number of sets p that will be generated depends on the number of data of each parameter. So p can be defined as:

$$p = W_1 \times W_2 \times, ..., \times W_{n-1} = \prod_{i=1}^{n-1} W_i \tag{6}$$

4. Present the synthetic database Y to the net in order to obtain the output parameter $Z = [z_{ij}]_{p \times 1}$ which should have the same probability distribution. For example, to a normal distribution $N_2(\overline{z}, S(Z))$, it is possible to verify the net generalization through the comparison of the probability distributions $N_1(\overline{x}, S(X))$ and $N_2(\overline{z}, S(Z))$. If $e_{x,z} = |\overline{x}_1 - \overline{z}_2|$ and $e_{S(x,z)} = |S(x)_1 - S(z)_2|$ represent high errors, then, back to step 1.

5. Classify the parameters (columns) of the matrix $U = [Y, Z]_{p \times n}$ into discrete intervals. As some datasets considered by this method has continuous values, the better context to represent this data is a many-valued context. So data should be classified into ranges.

6. Build a *formal context cross table* to classify in intervals (discretization) the n objects variables, establishing a binary relationship between objects and attributes (incidence) where an object has or not an attribute.

7. Obtain the formal concepts, which are the ordered pairs *(Object, Attribute)*, from the formal context.

8. Apply the Next Closure algorithm in order to obtain the implication rules of the type: "If... Then...".

3.1   Analysis of a Parameters Subset through FCANN Method

While studying a physical process, it may be necessary the analysis of domain parameters subset. In such a case, it could be assigned a constant value to the not analyzed parameters and vary the values of the other parameters in order to determine the qualitative behavior of the process. This study can be done by FCANN method changing the steps 3 and 6 that compose the method. Following, these changes are shown:

Let C be the set of all parameters indices that will be fixed. So, the steps 3 and 6 may be redefined as:

– **Step 3.** On the third FCANN step, a synthetic dataset $[S]$ should be built. To assign a constant value to a parameters subset C, do:

$$\textit{If } j \in\ C \textit{ Then } S_{kj} = q;\ \textit{For } k = 1..., W_j \textit{ where } U_j \leq q \leq V_j \tag{7}$$

– **Step 6.** The formal context $K{:=}(G,M,I)$ is built on step 6. In this case, consider $M = M'$, where:

$$M' = \{m_i | m_i \in M;\ \textit{where } j \in C \textit{ and } i \neq j\} \tag{8}$$

4 Case Study - Solar Energy System

The main objective of this section is to show the method application, to extract knowledge from previously trained ANN and to demonstrate that it could be possible to extract more comprehensible variables relationships by adopting the modifications shown at section 3.1. As case study the approach FCANN will be applied in solar energy system.

The solar energy systems, specifically solar water heaters, have considerable importance as substitutes of traditional electrical systems. An example of water heating system is called thermosiphon, the most widely used of all solar energy thermal convention devices. Thermosiphon [11, 12] systems are cost competitive compared with the conventional energy systems available in the whole world.

4.1 Neural Representation

Input water temperature (Tin), solar irradiance (G), ambient temperature ($Tamb$), flow rate ($\overline{m}$), inclination of the solar collector (I) and height of the water storage tank (H) are variables used as inputs to the ANN. The output water temperature ($Tout$) is the desired output from the net. The thermosiphon system is represented through Eq. (9).

$$f(Tin, G, Tamb, \overline{m}, I, H) \xrightarrow{ANN} (Tout) \tag{9}$$

The neural network used corresponds to a multi-layer perceptron, feedforward and totally connected one with $N=6$ inputs and $M=1$ output, where the number of neurons in the hidden layer was chosen as 13 (2N+1), as proposed in [13]. As axon transfer function, the sigmoid non-linear function was chosen and the network was trained through "back-propagation" algorithm.

For the training and test process, 117 data have been collected (see Table 1) directly from the system and 90% of them used in the training stage. Since that the objective is to obtain a correctly trained network to extract trustful knowledge a larger amount of data for training were considered. The following procedure has been adopted to build training sets:

1. For each variable of the net, the maximum and the minimum values found in the set have been selected.
2. The remaining elements have been randomly selected, with the purpose of reaching the size (90%) of the training set.

In order to improve convergence of the ANN training process, the normalization interval [0, 1] was reduced to [0.2, 0.8], because in the sigmoid function the values [0, 1] aren't reached: $f \rightarrow 0$ *for net* $\rightarrow -\infty$ *and* $f \rightarrow 1$ *for net* $\rightarrow +\infty$. The data was normalized through the following formula:

$$Ln = \frac{(Lo - Lmin)}{(Lmax - Lmin)} \quad and \quad Lo = Ln * Lmax + (1 - Ln) * Lmin \tag{10}$$

Table 1. Minimum and maximum parameter values

	H (m)	I (cm)	Tin (°C)	$\overline{m}$	G (W/m^2)	$Tamb$ (°C)	$Tout$ (°C)
Minimum	70.00	30.00	27.38	0.0082	993.47	29.07	79.89
Maximum	20.00	20.00	19.41	0.0006	297.87	18.21	40.14
Average	43.25	24.91	23.35	0.0038	745.12	23.58	62.84

Table 2. Training and test process

Error (°C)	Training Process	Test Process
Minimum	0.01	0.16
Maximum	1	3.38
Average	0.32	0.99
Standard Deviation	0.18	1.15

For the thermosiphon systems the average error in *(Tout)* recommended to analyze is 1°C. After the training process the average error reached for output temperature was of 0.32°C with the minimum and maximum errors of 0.01°C and 1°C respectively. While the test process presented the average error of 0.99°C with the minimum and maximum errors of 0.16°C and 3.38°C respectively (Table 2). Those results show the optimal approaches of ANN.

4.2 Knowledge Extraction through FCANN Method

After a satisfactory training process the FCANN method can be applied to extract the parameters relationship learned by the neural network. In this case study, the SOPHIANN tool [14, 7] was used to apply the FCANN approach. Using the FCANN method with 2 data per parameter and 2 discretization intervals (see section 3) it was obtained a formal context $K:=(G,M,I)$ with 64 rows e 14 columns, i.e. objects and attributes. This formal context originated a line diagram with 1084 formal concepts and an implication base with 80 rules. Some examples of these rules are presented below:

- **If** $I{=}2$ **and** $G{=}1$ **and** $Tamb{=}1$ **Then** $Tout{=}1$
- **If** $H{=}2$ **and** $Tin{=}1$ **and** M$=2$ **and** $Tamb$ $=1$ **and** $Tout{=}2$ **Then** $I{=}1$
- **If** $I{=}1$ **and** $M{=}1$ **and** $G{=}2$ **and** $Tamb$ $=2$ **and** $Tout{=}2$ **Then** $H{=}1$

Even using only 2 data per parameter and 2 discretization intervals, which have resulted in the shortest formal context, the density of the line diagram has been very high and therefore the implications base £ have presented an also high cardinality. This occurred because the number of formal concepts and extracted rules depends on the number of objects G, attributes M and of the incidence relationship gIm[8]. A subsequent analysis of these results can be difficult because of the number of extracted relations from the neural networks could be great.

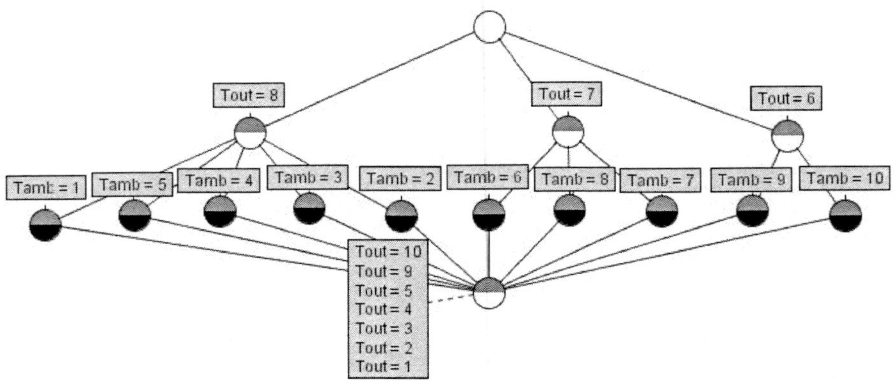

Fig. 1. Lines diagram with $Tamb \rightarrow Tout$ relations

However, the relations could be extracted from a parameters subset, which results in a reduced formal context, providing better results analysis. On the other hand, using 10 data per parameter, 10 discretization intervals and extracting the relations learned by the neural network for the ambient temperature *(Tamb)* and the output water temperature *(Tout)* and setting all other parameters in average values (Table 1), the line diagram (Figure 1) and the following implications were obtained:

- **If** $Tamb = 1$ **Then** $Tout = 8$ -**If** $Tamb = 6$ **Then** $Tout = 7$
- **If** $Tamb = 2$ **Then** $Tout = 8$ -**If** $Tamb = 7$ **Then** $Tout = 7$
- **If** $Tamb = 3$ **Then** $Tout = 8$ -**If** $Tamb = 8$ **Then** $Tout = 7$
- **If** $Tamb = 4$ **Then** $Tout = 8$ -**If** $Tamb = 9$ **Then** $Tout = 6$
- **If** $Tamb = 5$ **Then** $Tout = 8$ -**If** $Tamb = 10$ **Then** $Tout = 6$

Note that, all extracted relations are direct, i.e. "If a Then b". These relationships are interesting because they are more intuitive and therefore facilitate a subsequent analysis, especially for those people that seek for process knowledge acquirement through machine learning techniques. It is also possible to see that all *Tamb* intervals were mapped in a *Tout* interval. Moreover, *Tout* was mapped into the intervals number 6 to 8, so, if all other parameters are in average conditions, regardless of *Tamb* values, *Tout* values will always be in those intervals.

It could be also noted that, even increasing the number of discretization intervals (to build the formal context), which provides more process details, the line diagram density and the implication base cardinality were shorter. This has occurred due to the fact that the objects number and the new formal context density are much smaller.

When analyzing the lines diagram (Figure 1) it is possible to see, as on the implications base, three value intervals that *Tout* can assume and the respective *Tamb* intervals. It can also be noticed a set of attributes (*Tout* intervals 1 to 5, 9 and 10) which values have not suffered influence from *Tamb* values.

The lines diagram and the implications base represents the variables relations learned by the neural network to the analyzed parameters. In some cases, the analysis of the data structured on the lines diagram can be more intuitive than the study of the implications base, since that this diagram represents data hierarchically.

5   Conclusions

In this work, the approach (FCANN) for knowledge extraction and representation from previously trained neural networks was adapted in order to extract the relations among process variables learned by a neural network in a more comprehensible way.

FCANN produces qualitative behavior rules where the dependent parameters of the process can or cannot be the consequent variables of the rules extracted. This can be useful in industrial processes for automation, control and design of supervisory systems, where it is important to consider rules that have the dependent parameters as consequent. This can help to identify which parameters can change the operational point, like the controllable variables for a control system.

Using the FCANN method, it is possible to extract the variables relations using parameters subsets without the necessity of ANN retraining, what could demand large computational efforts. In this sense, it is possible to determine an interest parameters subset, which consists on the most important process variables to the apprentice, and obtain specific rules that can be helpful to understand the process domain. The proposed modifications presented the relations learned by the ANN, through lines diagram and implications base, in a more comprehensible way. Once the obtained relations set is relatively small, concise and related only to some predefined especial attributes, it is not necessary to do a posterior mining of the obtained rules to select the most representative ones (the adapted process itself do the selection). Those relations are much more intuitive, what can facilitate one posterior analysis, especially for those people who want to learn a process domain using machine learning.

Acknowledgments. The authors would like to thank the financial support of the National Research Council CNPq - Brazil. Project CT-INFO/ MCT/CNPq n° 031/2004, Process: 506512/2004-5.

References

1. Towell, G.G., Shavlik, J.W.: The extraction of refined rules from knowledge-based neural networks. In: Machine Learning Research Group Working, pp. 71–101 (1993)
2. Hilario, M.: An overview of strategies for neurosymbolic integration. In: Wermter, S., Sun, R. (eds.) Hybrid Neural Systems, Springer, Germany (2000)
3. Craven, M.W.: Extracting Comprehensible Models from Trained Neural Networks. PhD Thesis, department of Computer Sciences, University of Wisconsin-Madison (1996)

4. Vimieiro, R., Zárate, L.E., Silva, J.P.D., Pereira, E.M.D., Diniz, A.S.C.: Rule extraction from trained neural networks via formal concept analysis. In: del Pobil, A. (ed.) Proc. Eighth IASTED International Conference on Artificial Intelligence and Soft Computing. IASTED, pp. 334–339. ACTA Press (2004)
5. Vimieiro, R., Zárate, L.E., Pereira, E.M.D., Vieira, N.J.: Using the Next Closure Algorithm to Extract Rules from Trained Neural Networks Application in solar energy systems. In: IEEE SMcia, Finland, pp. 184–189 (2005)
6. Gratzer, G.: General lattice theory. ch. Applied lattice theory: Formal concept analysis, 2nd edn., pp. 591–605. Birkhäuser Verlag (1998)
7. Zárate, L.E., Dias, S.M., Song, M.A.J.: FCANN: A new approach for extraction and representation of knowledge from ANN trained via Formal Concept Analysis Neurocomputing, Amsterdam (to appear, 2007)
8. Ganter, B., Wille, R.: Formal concept analysis: Mathematical foundations. Springer, Germany (1996)
9. Carpineto, C., Romano, G.: Concept Data Analysis: Theory and Applications. John Wiley & Sons, Chichester (2004)
10. Ganter, B.: Formal concept analysis: algorithmic aspects. TU Dresden, Germany, Tech. Rep. (2002)
11. Zárate, L.E., Pereira, E.M.D., Soares, D.A., Silva, J.P.D., Vimieiro, R., Diniz, A.S.C.: Optimization of neural networks training sets via clustering: Application in solar collector representation. In: Seruca, I., Filipe, J., Hammoudi, S., Cordeiro, J. (eds.) Proceedings of 6th International Conference on Enterprise Information Systems, Porto, Portugal. ICEIS, vol. 2, pp. 147–152. INSTICC Press (April 2004)
12. Zárate, L.E., Pereira, E.M.D., Silva, J.P.D., Vimieiro, R., Diniz, A.S.C.: Neural representation of a solar collector with optimization of training sets in Lecture Notes in Artificial Intelligence. In: Orchard, B., Yang, C., Moonis, A. (eds.) ISAI, AAAI, ACM/SIGART, INNS, Springer, Heidelberg (2004)
13. Kovács, Z.L.: Redes Neurais Artificiais. Edição Acadêmica São Paulo, Cap. 5, 75–76 (1996)
14. Zárate, L.E., Song, M., Alvarez, A., Soares, B., Nogueira, B., Vimieiro, R., Diaes, S., Santos, T., Vieira, N.: An Approach to Knowledge Extraction From ANN Through Formal Concept Analysis - Computational Tool Proposal: SOPHIANN. In: Proc. ISIE, Montreal Canada (2006)

Video Similarity Measurement Based on Attributed Relational Graph Matching

Ines Karouia, Ezzeddine Zagrouba, and Walid Barhoumi

Equipe de Recherche en Systèmes Intelligents en Imagerie et Vision Artificielle,
Institut Supérieur d'Informatique,
2 Rue Abou Rayhane Bayrouni, 2080 Ariana, Tunisia
ines.karouia@ensi.rnu.tn, ezzeddine.zagrouba@fsm.rnu.tn,
walid.barhoumi@laposte.net

Abstract. In this paper, an original scheme for video similarity detection is proposed in order to establish correspondence between two video sequences. This scheme consists first to summarize the visual contents of a video sequence in a small set of images. Each image is then modeled, by an Attributed Relational Graph (ARG), as the composition of salient objects with specific spatial relationship. Matching two video sequences is thereby reduced to the ARG similarity problem. The proposed approach offers a principled way to define the ARG similarity that accounts for both the attribute and topological differences of the two considered ARGs. Indeed, we proposed herein a cost-efficient solution to find the best alignment between two ARGs. This consists to the minimization of a similarity measure between the two graphs using dynamic programming. This measure can be considered as a matching rate which can be very useful for Content Based Video Retrieval (CBVR) applications. The suggested scheme was preliminary tested on real-world databases and very promising results were observed.

Keywords: Computer Vision, Video Sequence, Matching, Attributed Relational Graph, Dynamic Programming, Video Similarity.

1 Introduction

Video sequences' matching is a long-standing challenging issue in computer vision. Along with the rapid development of computer networks, video acquisition devices and Internet, the amount of video data have grown immensely over past years. For this, automated video matching and recognition has emerged in many underlying applications in multimedia and database related areas [1] [2]. Therefore, many works on video similarity detection have been proposed. Interested readers on the state of art can refer to [3] and [4].

Most video matching schemes first reduce videos to a small set of images which will be then matched using image matching schemes [5]. This reduction can be done either by summarizing the visual content of the video sequence on one image representing the mosaic of the video [6] or by extracting a set of key-frames [1]. While most approaches use interest points and curve fragments for image matching, there is also a significant amount of work on region-based matching to address problems from stereo matching to Content Based Image Retrieval (CBIR) [7]. In this paper, we propose a novel image matching method based on the visual content and on

N.T. Nguyen, R. Katarzyniak (Eds.): New Chall. in Appl. Intel. Tech., SCI 134, pp. 173–182, 2008.
springerlink.com

objects spatial interrelationships. Indeed, the input images are modeled by relational attributed graphs, which will be then matched using a cost-efficient algorithm based on dynamic programming. To this end, we defined a measure of similarity between two graphs. This measure can be considered as a matching rate which can be very useful in Content Based Video Retrieval (CBVR) applications. Preliminary simulations show that the suggested scheme achieves better quality results than the similar conventional ones. The rest of this paper falls into four sections. In the next section, we describe the process permitting to abstract the visual content of a video sequence in terms of attributed relational graphs. Section 3 is devoted to the presentation of the cost-efficient graph matching solution. The experimental results and performances study is reported in section 4. A summary of the results of this research is presented at the end with some of the perspectives.

2   From Video Sequence to Attributed Relational Graph

To abstract the visual content of a video sequence in terms of attributed relational graphs, we begin by summarizing the video data in a small set Λ of stationary images. Then, each image ($\in \Lambda$) is modeled by a graph G illustrating the composition of regions with spatial/attribute relationship.

2.1   From Video Sequence to Image Set

Video data are first transformed from their sequential and redundant frame-based representation, in which the information about the scene is distributed over many frames, to an explicit and compact scene-based representation [6]. Many researchers are working on mechanisms for generating a short summary of a video, what is known as video abstraction [12]. On one hand, the abstraction can be done while presenting the panoramic spatio-temporal view of the entire scene in the form of a mosaic (Card(Λ)=1). Seen the small motion between two successive images of a video shot, several methods allow fast and reliable construction of a mosaic from a video sequence by alignment of the different images of the sequence. On the other hand, the content of the video can be summarized on an ordered set of key-frames (Card(Λ)$\geq$1). Many works have addressed the problem of automated extraction of key frames by frame difference, clustering, motion information, etc [8].

In our case and in order to guarantee the flexibility of the proposed scheme, the both solutions for summarizing the visual content of a video sequence are proposed (Fig. 1). For the mosaic building, we used an efficient method based on multi-feature matching [9]. On the other side, we adopted a solution based on frame difference for the key-frames extraction. In fact, key-frames correspond to those frames characterized by the presence of at least one new significant object.

2.2   From Image to Attributed Relational Graph

Attributed Relational Graph (ARG) is a very useful model for representing the visual appearance of an image. It is an extension of the ordinary graph by associating discrete or real-valued attributes to its nodes and edges. The use of attributes allows

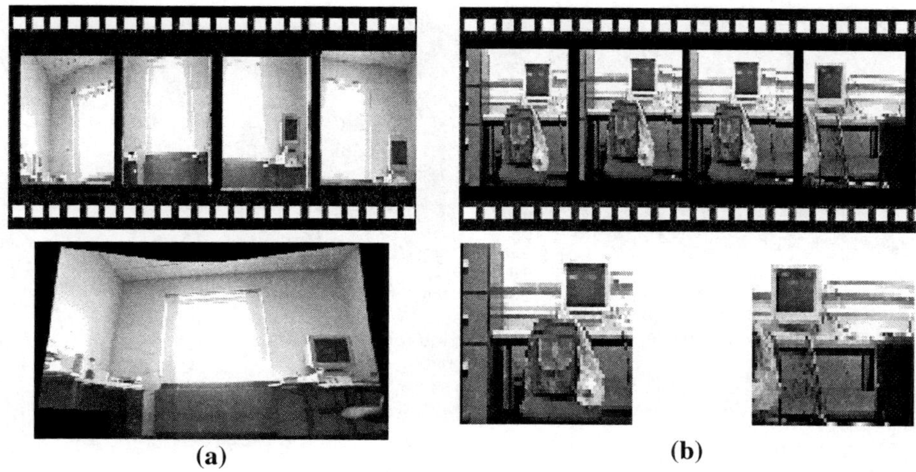

(a)   **(b)**

Fig. 1. From video sequence to image set. At the top: sub-sequence of the video data. At the bottom: summary of the input sub-sequence in: a- a video mosaic, b- an ordered set of two key-frames.

ARG not only be able to model the topological structure of an entity but also its non-structural properties, which can be represented as feature vectors [11]. Formally, an attributed relational graph is a quadruple $G=(V,E_S, v, \xi)$, where V is a finite set of nodes, E_S $(\subseteq V{\times}V)$ is a finite set of edges between nodes, $v: V \rightarrow A_V$ is the function generating unary node attributes and $\xi: E_S \rightarrow A_{ES}$ is the function generating binary edge attributes. In our case, each image is modelled using a complete non-oriented graph where each node represents a salient region. Indeed, many works demonstrate several advantages of using regions over interest points or edge fragments for image matching [7]. In fact, the higher dimensionality of regions presents them as the richest descriptors of the geometric and photometric objects appearances. Besides, the elevated dimensional character of regions makes them a stable primitive for matching under small illumination and viewpoint changes. ARG allows also the comparison of two images in terms of geometric and photometric properties of regions as well as region topology.

For regions detection, we used a fast fuzzy technique of unsupervised coarse segmentation. It is an ameliorated version of the Fuzzy C-means algorithm [13] which permits to automatically summarize the visual content of an image in a reduced number of intuitive and visually characteristic regions of interest (Fig. 2) while preserving essential information needed for the image understanding task. In fact, we tried to replicate the Human Visual System (HVS) which coarsely perceives the scene zones with their visual appearances and relative topological dispositions [16]. Besides, small regions, which have no significant impact on the image content, were merged to the including regions and/or to the ones maximizing the length of the common border with these undesirable regions. This post-treatment allows again the reduction of the graph size what reduces considerably the complexity of the ARG matching process, without significant loss in the matching accuracy.

(a)   (b)

Fig. 2. The produced regions maps relatively to the images used in Fig. 1

Each node ($v \in V$) in the graph G denotes an image region and an edge ($e_S \in E_S$) represents the spatial relationship between two regions not necessary adjacent. Node attributes A_V identify the visual content of the corresponding region in terms of colour (colour histograms), shape (invariant geometric moments) and texture (co-occurrence matrix) descriptors [10]. These descriptors compose the feature vector of each segmented region, which captures the distinctive feature inside one region finely. The spatial edge attributes A_{ES} indicate the relative spatial relationships between two nodes while specifying the fuzzy degrees of positioning and inclusion between the two corresponding regions. Each edge ES ($\in VxV$) connecting two nodes V_i et V_j (representing two regions R_i and R_j) is labelled by two 5-dimensional vectors describing the degrees of inclusion and positioning of R_i relatively to R_j, and vice versa. Indeed, the spatial disposition of a region R_i, relatively to another one R_j, is illustrated in terms of five measures reflecting successively the fuzzy position of R_i relatively to R_j at one of the following positions: on the left, in the bottom, on the right, in the top and inside (1).

$$\forall pos \in \{left, bottom, right, top, inside\}, \quad \mu_{i\%j}^{pos} = \frac{\left|\{p \in R_i / p \text{ is at 'pos' of } R_j\}\right|}{Size(R_i)} \quad (1)$$

where, | | denotes the set cardinality operator. For example, if $pos='left'$, the fuzzy disposition of R_i on the left of R_j (Fig. 3) is defined as follows (2):

$$\mu_{i\%j}^{left} = \frac{\left|\{p \in R_i / abs(p) \in [xBegin(R_i), xBegin(R_j)]\}\right|}{Size(R_i)} \quad (2)$$

In what follows, we illustrate an example of graph building for the mosaic image of Fig. 1.a. After extracting only seven dominant regions (Fig. 4) composing the studied image (of size $512x1024$), we compute for each couple (R_i, R_j) $(1 \leq i, j \leq 7$ and $i \neq j)$ two 5-dimensional vectors (Table. 1). The used segmentation technique provides a good trade-off between computation complexity and image segmentation efficiency. It is fast and detects coarse user-intuitive regions which can encompass different shades of the same hue with isolated spatial details. Note that the concept of coarse segmentation and fine region description is very close to the HVS, in which human first roughly identify few perceptual salient objects in the image and then recognize them with fine visual features. Besides, the proposed spatial relations descriptors reflect accurately the position of a region over another one even if they are not adjacent. It optimizes the reduction of the semantic gap between the studied image and the correspondent ARG [10]. Indeed, most of the classical used techniques describe the topological inter-regions relations by "Allen-relations" [15] or by a simple discretization of the geometric space into a set of zones [14]. However, these two solutions are deterministic: a region is either absolutely on the right, on the left, in top or in bottom of another region. The proposed description of region spatial disposition allows the definition of a certain probability for positioning a region compared to another one. These probabilities will be very determinants for the validation during the matching process. This minimizes the effects of the accumulated errors, generated by the segmentation and the region description steps, on the matching results.

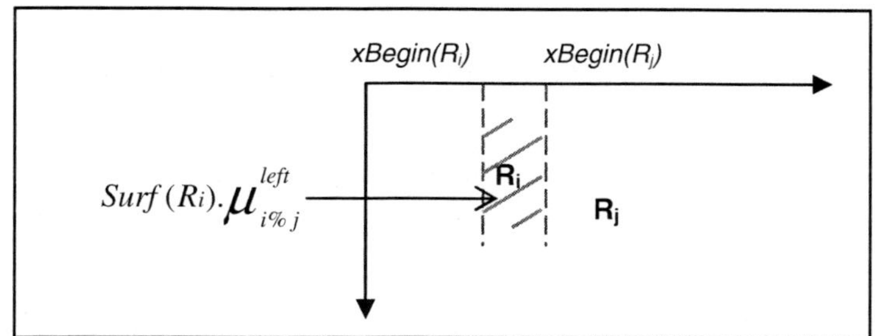

Fig. 3. The fuzzy disposition of the region R_i on the left of the region R_j

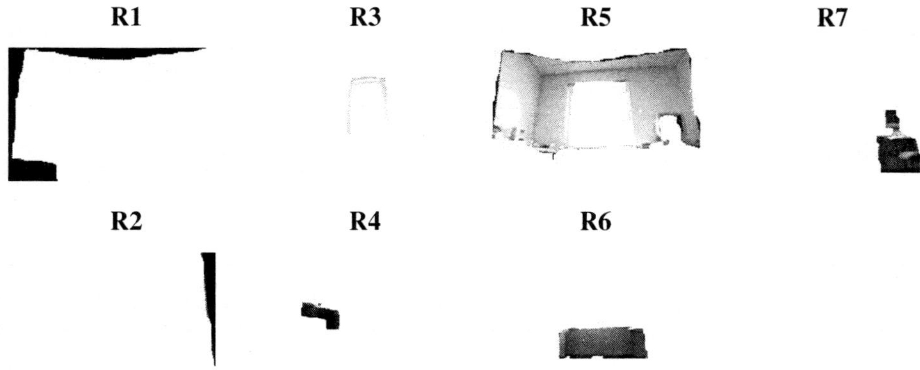

Fig. 4. Dominant regions extraction (*c.f.* Fig. 2.a.)

Table 1. Edges labelling. Note: the recorded values are computed with a precision close to 10^{-1}.

	R1	R2	R3	R4	R5	R6	R7
R1		(1, 0, 0, 0.4, 0)	(0.7, 0.5, 0.04, 0.3, 0)	(0.2, 0.7, 0.3, 0, 0.3)	(0.3, 0, 0, 0.3, 0.5)	(0.7, 0.7, 0, 0, 0)	(1, 0.6, 0, 0, 0)
R2	(0, 0, 1, 0, 0)		(0, 0.4, 1, 0, 0)	(0, 0.9, 1, 0, 0)	(0, 0, 0.5, 0, 0.5)	(0, 0.9, 1, 0, 0)	(0, 0.8, 0, 0, 0.2)
R3	(0, 0, 0, 0, 1)	(1, 0, 0, 0, 0,)		(0, 0.9, 1, 0, 0)	(0, 0, 0, 0, 1)	(0, 1, 0, 0, 0)	(1, 0.5, 0, 0, 0)
R4	(0, 0, 0, 0, 1)	(1, 0, 0, 0.3, 0)	(1, 0, 0, 0.9, 0)		(0, 0, 0, 0.5, 0.5)	(1, 0.1, 0, 0, 0)	(1, 0, 0, 0, 0)
R5	(0, 0, 0.2, 0, 0.9)	(1, 0, 0, 0, 0)	(0.5, 0.4, 0.4, 0, 0)	(0, 0.9, 0.6, 0, 0)		(0.4, 1, 0.2, 0, 0)	(0.7, 0.7, 0, 0, 0)
R6	(0, 0, 0, 0, 1)	(1, 0, 0, 0.6, 0)	(0.1, 0, 0.2, 1, 0)	(0, 0, 1, 0, 0)	(0, 0, 0, 0.7, 0.3)		(1, 0, 0, 0, 0)
R7	(0, 0, 0.7, 0, 0.4)	(0.8, 0, 0, 0.5, 0)	(0, 0, 1, 0.8, 0)	(0, 0.1, 1, 0, 0)	(0, 0, 0, 0.6, 0.4)	(0, 0.2, 1, 0, 0.)	

3   Graphs Similarity Measurement

Given two graphs, this stage outputs a score evaluating the similarity between them while displaying the couples of matched nodes. The major used approaches for ARG matching include energy minimization framework, spectral method, Markov Random Field labeling and Bayesian approach [11]. The key idea behind our matching algorithm, based on energy minimization, is to use of as much image information as possible, without being misled by image regions which are poorly matched. For this, we start by extracting the best matching based only on optimal node similarity. This matching is then iteratively improved by taking into account sub-optimal node matching and edge matching. In other words, once a set of initial correspondences has been found, a regression model, based on dynamic programming, estimates visual and topological correspondence across the entire image. This estimation allows further matches to be discovered and refined. We use probabilistic weights for each correspondence, allowing the algorithm to detect and discard mismatches. These weights are mainly defined according to likelihood of the nodes and of the edges connecting them.

3.1   Similarity Degree Measurement

To measure the dissimilarity degree D (3) between two matched ARGs, $G=(V, E_S, v, \xi)$ and $G'=(V', E'_S, v', \xi')$, we evaluate the similarity D_N of the matched nodes as well as the similarity D_E of the corresponding matched edges. Let $|V|<|V'|$, the visual appearance similarity of the couple $(R_i \in V, Homologous(R_i) \in V')$ is expressed in terms of local geometric and photometric properties (shape, colour and texture). Then, topological properties are involved by comparing the corresponding matched edges.

$$D(G, G') = \frac{\sum_{R_i \in V} d(R_i, Homologous(R_i))}{|V|} + \frac{2. \sum_{A_i \in E_S} d(A_i, Homologous(A_i))}{|V|.(|V|-1)} \quad (3)$$

To compare two n-dimensional vectors X and X', we used an adopted variant of the Minkowski distance (4). The normalized coefficients α_j aim to weigh the importance of the used attributes for the description of each region. In fact, the matching results depend strongly on the nodes matching quality which depends on the weighting of the colour, the shape and the texture descriptors (Fig. 5). For example, to match natural images, it seems to be very convincing to give more importance for texture descriptor. Thus, the values of these coefficients are interactively selected.

$$d(X, X') = \frac{\sum_{j=1}^{n} \alpha_j \|X_j - X_j'\|}{n} \tag{4}$$

3.2 Attributed Relational Graphs Matching

The goal of the ARG matching is to find the best mapping, of the smallest graph G on the higher-order graph G' ($m=|V|<|V'|=n$), that leads to the smallest dissimilarity measurement D (3). This minimization process is NP-complete if we look to generate all possible mappings and pick the best one. In order to reduce the combinatory complexity, we proposed a method to find near-optimal mappings between two ARGs using dynamic programming. It basically consists in solving an instance of a problem by taking advantage of already computed solutions for smaller instance of the same problem. To this end, we decompose the matching process into K iterations ($1 \leq K \leq |V|$). Each iteration t ($1 \leq t \leq K$) consists on defining the best global bijective mapping while considering for each node of G only the best t potential homologous in G' (minimizing the distance D_N). Thus, the minimization problem at the iteration t can be defined as a small instance (sub-problem) of a similar problem at the next iteration. At the end of every iteration t, the best bijective mapping among all the precedent iterations is retained. To characterize the correctness of correspondences, each bijective mapping (*resp.* non-bijective) is weighted by its similarity score (D^{-1}) (*resp.* by 0 and then discarded). A simple normalization of these weights present them as probabilities of correct mapping. On the other hand, an iteration t is composed of m steps, such that each step s ($1 \leq s \leq m$) tests the possible t sub-optimal mapping for the node V_s. The dynamic programming avoids the re-consideration of the old mapping since they could be treated in one of the precedent iterations. We note that the choice of the value of K is let to the user and the ratio K/m defines the degree of approximation between the extracted mapping and the exact optimal one. Indeed, when $K=m$, the proposed algorithm converge to the exact optimal mapping.

The efficiency of the proposed matching scheme is well illustrated with small values of K, where optimal or near-optimal solution is often obtained in a reduced calculation time. Indeed, while testing various values of K, sufficient results (given a ground truth) were recorded with $K=3$. The used graph matching algorithm reduces clearly the combinatory of graphs matching. In fact, the complexity of the algorithm in the worse case is $O(m^2 K^m)$. In comparison with the complexity of other similar well-known algorithms [17], such as the algorithm A* ($O(m^2 n^{m+1})$) and the one of Ullmann ($O(m^2 m^n)$), the proposed matching seems to be more optimal. The dynamic programming reduces considerably the number of the needed operations. In fact, a sub-graph of G' which appears multiple times is compared only once to G. Besides, matching solutions which are not bijective are not considered. Finally, the computing process of the matching cost stops evolving as soon as the dissimilarity score gotten during the previous iterations is reached, what reduces again the computational complexity.

4   Experimental Evaluation

For the case of video abstraction by key-frames, a complete oriented ARG is defined for each video where nodes are the selected key-frames, which are themselves modeled by ARGs (as it was explained in sub-section 2.2). The edges are labeled by the dissimilarity degrees between the corresponding key-frames (them ARGs) according to the equation (3). The dissimilarity between key-frames of two videos is computed by applying the suggested ARG matching approach. Finally, we look for the optimal key-frames matching between the two input video sequences in order to deduce the dissimilarity score between these two videos. This video matching method makes effective search of the best match for each key-frame (Fig. 5).

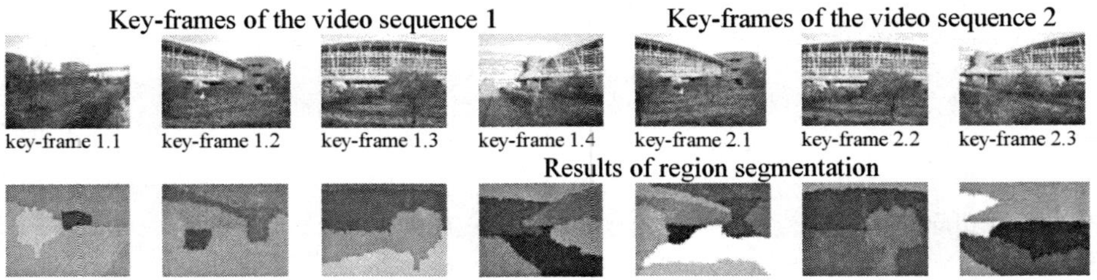

Fig. 5. Videos matching where each video is summarized with a small set of its sampled frames (similarity measurement of the two videos is 97.5%). The defined correspondence between the two video key-frames: 1.2 to 2.1 with a dissimilarity score of 0.036, 1.3 to 2.2 with a dissimilarity score of 0.013 and 1.4 to 2.3 with a dissimilarity score of 0.029.

To evaluate the dependence of the matching results on the regions visual appearance as well as on the relative spatial relationships, we used the standard recall assessment criterion. Recall is defined as the ratio of the number of relevant or perceptually similar matched regions (given a ground truth) by the total number of matched regions ($|V|$). Five curves are dressed to compare the recall score for 15 couples of indoor and outdoor images, in the case of one region-descriptor (colour, shape or texture) without using the relative spatial relationships, in the case of the all three descriptors without using the relative spatial relationships (*without RS*) and finally in the case of the all three descriptors while using the relative spatial relationships (*complete algo*) (Fig. 6.a). It is clearly seen that introducing spatial disposition information provides better matching effectiveness (a mean recall score of 95.6% was recorded for only 3 iterations). Besides, the colour descriptor appears more discriminative than the texture descriptor which is more discriminative than the shape one. In Fig. 6.b, we compared the efficiency of the two used methods of video abstraction (mosaic building *vs.* key-frames extraction) for matching 10 couples of

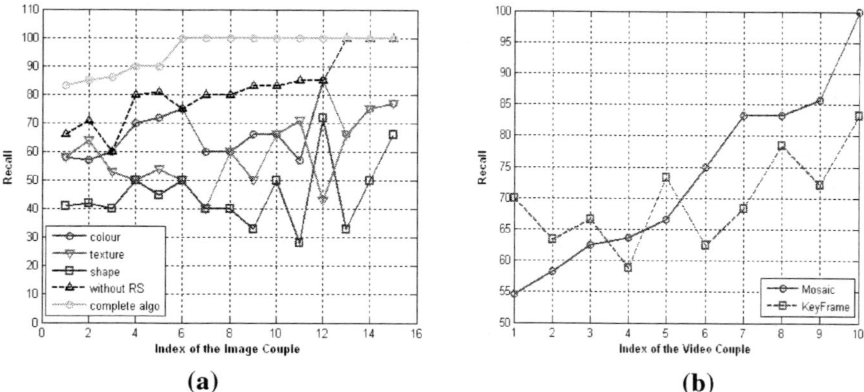

Fig. 6. Experimental evaluation: a- dependence of matching results on the regions visual appearances and on the relative spatial relationships, b- comparison of the two used techniques of video abstraction (Note: video sequences were indexed relatively to them sizes).

video sequences belonging to different categories ($K=3$). According to the produced curves, no one of the two abstraction techniques outperforms the other one. The mosaic building (*resp.* key-frames extraction) is well adopted for long (*resp.* short) video sequences. Besides, the video category (subject, environment, etc.) influences strongly the choice of the video abstraction technique.

5 Conclusion

In this paper, we presented a novel scheme for matching two video sequences. This scheme consists first to abstract the visual contents of each video on a small set of stationary images. Then, these images are modeled, by Attributed Relational Graphs, as the composition of the salient regions with specific topological relationships. Thus, matching two video sequences is defined as an ARG similarity problem. Computational complexity and the lack of good approximations methods make the use of graph matching techniques very difficult. For this, we proposed a cost-efficient solution to find the best mapping between two ARGs. This consists of the minimization of a similarity measure between the two graphs using dynamic programming. The contributions of our matching technique reside particularly on the use of the relative spatial relationships between regions (belonging to the same image or to different images) and on the reduction of the graph matching computational cost. Experimental simulations and objective assessment show the efficiency and the effectiveness of the proposed method for region-based image/video matching and retrieval. Indeed, although the proposed method uses an approximate matching algorithm, an optimal match (recall score greater than *80%*) between graphs is often obtained after a small number of iterations. The suggested scheme is developed to be fully extensible to any domain with a continuous media stream. Actually, we are validating our matching scheme for content-based video retrieval from a real-world dataset of Web video and MPEG-7 test sequences with artificially generated similar versions. Preliminary results are very promising. We are also working on the application of the proposed method to combine series of video frames in the framework of wide-area video surveillance. Besides, the objects to compare can have different granularity levels. Univocal matching of the components is not sufficient. It is necessary to be able to match a component to many others. A possible extension of our approach can make it multivalent (each region can be mapped to a set of regions). Furthermore, possible future directions would be made to attempt the use of moving images (video skims) for summarizing video visual content.

References

1. Yijun, L., Jin, J.S., Xiaofang, X.: Video Matching using Binary Signatures. In: IEEE ISPACS, Hong Kong, pp. 317–320 (2005)
2. Milrad, M., Rossmanith, P., Scholz, M.: A Web-Based Video Retrieval System of Educational Films: Design Architecture and Implementation Aspects. In: IEEE International Conference on Advanced Learning Technologies, Finland, pp. 111–115 (2004)
3. Adjeroh, D., Lee, M.C., King, I.: A Distance Measure for Video Sequences. Computer Vision and Image Understanding 75(1), 25–45 (1999)
4. Mohan, R.: Video Sequence Matching. In: IEEE International Conference on Acoustics, Speech and Signal Processing, Washington, pp. 3697–3700 (1998)
5. Irani, M., Anandan, P.: Video Indexing Based on Mosaic Representation. IEEE Trans. on PAMI. 86(5), 905–921 (1998)
6. Todorovic, S., Ahuja, N.: Region-Based Hierarchical Image Segmentation. International Journal of Computer Vision (2007)
7. Narasimha, R., Savakis, A., Rao, R.M., De Queiroz, R.: Key Frame Extraction using MPEG-7 Motion Descriptors. In: Asilomar Conference on Signals, Systems and Computers, pp. 1575–1579 (2003)
8. Zagrouba, E., Barhoumi, W., Amri, S.: An Efficient Image Mosaicing Method based on Multifeature Matching. In: Machine Vision and Applications, pp. 1–24. Springer, Heidelberg (2007)
9. Zagrouba, E., Ouni, S., Barhoumi, W.: A Reliable Image Retrieval System Based on Spatial Disposition Graph Matching. Review on Computers and Software 2(2), 108–117 (2007)
10. Zhang, D.Q., Chang, S.F.: Detecting Image Near-Duplicate by Stochastic Attributed Relational Graph Matching with Learning. In: ACM International Conference on Multimedia, New York, pp. 877–884 (2004)
11. Truong, B.T., Venkatesh, S.: Video Abstraction: A systematic Review and Classification. ACM Trans on Multimedia Computing, Communications and Applications 3(1) (2007)
12. Zagrouba, E., Barhoumi, W.: Semiautomatic Detection of Tumoral Zone. Image Analysis and Stereology 21(1), 13–18 (2002)
13. Hlaoui, A., Sun, H., Wang, S.: Image Retrieval using Fuzzy Segmentation and Graph Matching Technique. In: Conference on Machine Learning and Cybernetics, Beijing (2002)
14. Malki, J., Zahzah, E., Mascarilla, L.: Indexation et Recherche d'Images Fondées sur les Relations Spatiales entre Objets. Traitement du Signal 19(4), 235–250 (2002)
15. Philipp-Foliguet, S., Lekkat, M.: Recherche d'Images à partir d'une Requête Partielle utilisant la Disposition des Régions. In: Congrès Francophone de Reconnaissance des Formes et Intelligence Artificielle (RFIA), Toulouse, pp. 123–132 (2004)
16. Wong, A.C., You, M.: Entropy and Distance of Random Graphs with Application to Structural Pattern Recognition. IEEE PAMI 7(3), 509–609 (1985)

A Knowledge-Based System for CMM Evaluation

Javier Andrade, Juan Ares, Rafael García, Santiago Rodríguez,
María Seoane, and Sonia Suárez

Software Engineering Laboratory, University of A Coruña
Campus de Elviña s/n, 15071, A Coruña, Spain
{jag,juanar,rafael,santi,mariaseoane,ssuarez}@udc.es

Abstract. Even though the possession of a high CMM level undoubtedly implies prestige and competitive advantages for a software development organisation, its attainment may imply a considerable economic burden because of potentially necessary audits. It is therefore very interesting to minimise the costs by paying only for the truly indispensable audits. This article proposes a Knowledge-Based System that makes it possible to evaluate an organisation at a determined CMM level and as such limit the services of an auditor to those cases in which the system's response complies with the requested CMM level and the necessary associated skills. This clearly implies an important cost reduction for audits with a negative result. The design of this system is based on the CommonKADS methodology, and its implementation was carried out with the Clips tool.

Keywords: Audit, Capability Maturity Model, CommonKADS, Key Process Area, Knowledge-Based System, Clips.

1 Introduction

The purposes of the Capability Maturity Model (CMM) are twofold: on the one hand, to evaluate how organisations develop software (i.e., the quality of the processes they follow and the mechanisms they use), and on the other hand to serve as a guide towards continuous improvement [1]. This model contemplates five maturity levels that represent the quality of the software development process in the organisation. These levels are, ranged from minor to major, the following: initial, repeatable, defined, managed, and optimised, with a major CMM level implying more quality.

Each maturity level, with the exception of the first, defines a series of Key Process Areas (KPAs) with which the company must comply in order to be positioned in that level, as well as a series of skills, also known as preconditions set, that must exist in the organisation in order to implant the software processes appropriately. These skills typically involve resources, organisational structures, and trainings.

It is commonly accepted that the satisfaction of a high maturity level by an organisation implies prestige and competitive advantages for that organisation [2]. It may however be very expensive to evaluate whether or not an organisation disposes of the necessary preconditions or skills to obtain a given CMM level: this implies the repeated services of an auditor before and after the correction of any detected insufficiencies (lacking skills). Also, once the necessary skills are acquired, the

N.T. Nguyen, R. Katarzyniak (Eds.): New Chall. in Appl. Intel. Tech., SCI 134, pp. 183–191, 2008.

auditor must intervene once more to evaluate the organisation at the desired CMM level. The entire process may turn out to be very costly.

This article proposes a Knowledge-Based System (KBS) for the evaluation of an organisation at a specific CMM level. The application of this system will substantially reduce the need for expensive audits because these will only take place after the system has issued a positive report on compliance with skills or level—a report that has many possibilities to coincide with the auditor's review but is not necessarily identical, since the auditor may weigh certain aspects that are not considered by the KBS. These differences between expert and KBS can actually be exploited, since the inclusion of new knowledge will lead to the improvement of the system. There are other similar projects using ontology-based decision support agents, which can be found in [3, 4, 5]. These systems represent the domain knowledge in an ontology that will be adopted by the computational intelligent multi-agent. Their goal, like ours, is to get an intelligent based system for CMM assessment.

The design of the proposed KBS is detailed in Section 2 of this paper. Section 3 shortly describes the system implementation, and Section 4 sets out the conclusions.

2   Design of the Proposed System

The quality of KBS design depends on the knowledge engineer's programming skills, and on his ability to devise, remember, and dynamically update a design specification. This is a difficult task for all but the smallest KBSs.

Difficulties like these can be alleviated by producing representations of the expert's knowledge and of the design specification in the shape of text or diagrams. The best known approach towards the production of such documents is the CommonKADS methodology [6, 7, 8, 9]. It now is the European de facto standard for knowledge analysis and knowledge-intensive systems development, and it has been adopted as a whole or has been partly incorporated in existing methods by many major companies in Europe, as well as in the US and Japan [9]. By CommonKADS we elaborate a list of potential components of the model for the KBS, select the adequate template for the task, and construct the initial domain scheme. The last stage is a complete specification of the knowledge model. The following sections describe how each of these activities was carried out.

2.1   List of Potential Model Components

The task of the proposed KBS belongs to a highly specialised field (a concrete and classified theme within Software Quality Management). It is perhaps for this reason that we dispose of reliable and empirically proved information on how to carry out CMM audits [10]. Consequently, the knowledge of the domain can be said to be formal.

On the one hand, the above documentation shows evidence of the existence of a commonly accepted structure in the sphere of the CMM model—shown in Figure 1— that represents an initial candidate for the domain model. This structure reflects the existence of maturity forms related to each CMM level (except level 1). Also, a

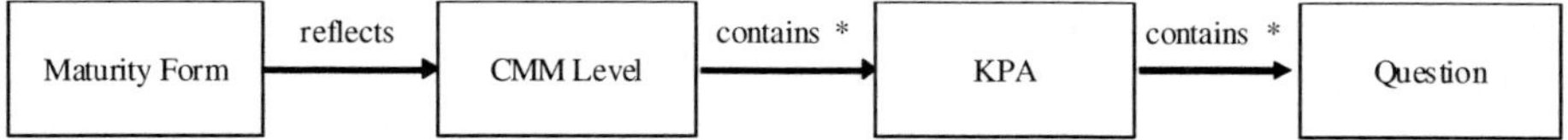

Fig. 1. Initial relationships structure

maturity level requires compliance with a series of KPAs [11], and each KPA contains in turn a series of questions; consequently, the compliance of all these questions implies the compliance of the KPA as a whole, and the compliance of all the KPAs associated to a level is equal to the compliance of that level.

On the other hand, it is fundamental to record the performed audits and their results in, for example, a database: when we consult the system with respect to the convenience of an audit of CMM level n, we must be able to check whether the organisation was successfully audited in CMM level $n-1$ and whether its skills were audited for CMM level n. If the answer is negative, there is no possibility whatsoever to compete for the desired level, because neither in the system, nor in reality, is it allowed to "skip" CMM levels. For example, in order to reach level 3 we must previously have obtained both level 2 and the skills for level 3.

2.2 Selection of the Task Template

The final purpose of the proposed KBS is to provide an organisation with the possibility to fill out a form for a given CMM level and consult the system regarding its viability: *"Given the data contained in this form, is it possible for the organisation to successfully pass an audit for CMM level x?"*.

In this context, and from the point of view of the task, this is an activity that fits into the category of *assessment*. These activities are provided with various *templates*, from which we have selected the one mentioned in [6].

The main motive for this choice is that the associated inferential structure matches the purpose of the application. A good technique to establish this adequacy to the problem consists in building an annotated inferential structure in which the dynamic roles are annotated or made to correspond with specific elements of the domain. This inferential structure is shown in Figure 2.

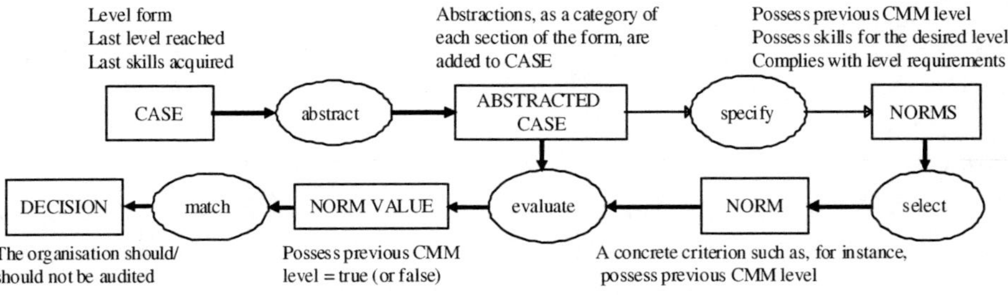

Fig. 2. Annotated inferential structure

2.3 Construction of the Initial Domain Scheme

As recommended in [6], this activity was carried out in parallel with the previous one. The result is a set of *domain-specific conceptualisations*—shown in Figure 3—and a set of *method-specific conceptualisations*—shown in Figure 4.

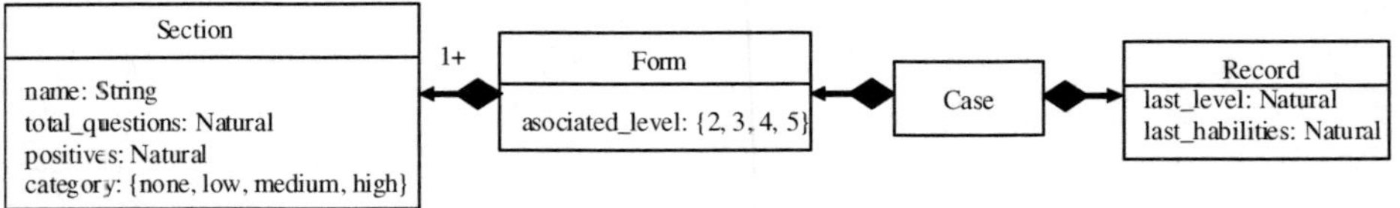

Fig. 3. Domain-specific conceptualisations

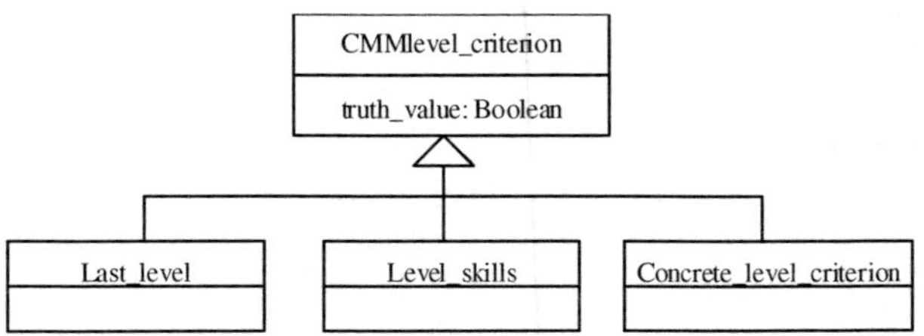

Fig. 4. Method-specific conceptualisations

We have detected two main concept types in the problem domain: *Form* and *Section*. We also need some historical information, such as the last CMM level that was reached and the last skills for which the organisation was successfully audited. To this effect, we model a concept *Record* with two attributes that represent these needs. The concepts Form and Record constitute the initial reasoning case. A form refers to a specific CMM level and consists of a series of sections that each refer to a KPA of the specified level. This fact is reflected by modelling an aggregation relationship between the concepts Form and Section. The concept Form has an "associated-level" attribute that indicates the CMM level to which it corresponds. The concept Section presents four attributes: "name", "total-questions", "positives" and "category". The first refers to the name of the section—e.g. Project Management—, the second indicates the total number of questions in the section, the third represents the total number of questions that were answered positively, and the last attribute refers to the organisation's level of compliance in the section.

The previously mentioned level of compliance is obtained in function of the attributes "total-questions" and "positives" in the following manner:

- If the positive answers represent less than 25% of the total, the level of compliance is considered "none". This means that the organisation does not comply with the KPA represented by the section.
- If the positive answers represent between 25-50% of the total, the level of compliance is considered "low".

- If the positive answers represent between 50-75% of the total, the level of compliance is considered "medium".
- If the positive answers represent between 75-100% of the total, the level of compliance is considered "high".

In addition to the above, the notion of criterion plays an important role in the problem domain. In this case we can distinguish three types of criteria, each with a "truth-value" attribute that indicates whether it is true or false:

- Last-level: Was the organisation successfully audited in the level previous to that at which it aspires? In other words, does it meet the requirement of having been successfully audited at a CMM level that precedes the desired level (except for level 2)?
- Level-skills: Was the organisation successfully audited with respect to the skills that are required to start an improvement process and reach the desired CMM level? In other words, in order to reach a certain CMM level, the organisation must not only have reached the previous level, it must also have passed the audit of the skills that are required for that level.
- Concrete-level-criterion: Does the organisation meet the specific requirements of the level for which it wants to be audited? If the organisation wishes to be successfully audited for a specific CMM level, it must meet its KPAs at certain rates or levels (many possibilities are accepted).

Finally, we wish to emphasise that the system only offers a positive response if all the criteria present the value *true*.

2.4 Complete Specification of the Knowledge Model

As explained before, the activity to be modelled is an instance of the task type *assessment*. Also, the selected template shows an adequate inferential structure for the purpose of this KBS, in which the inferences present sufficient detail. It is for this reason that the construction process was executed by means of a "middle-out" approach [6].

Figure 5 shows the template that was chosen for the modelling and in which the global task is subdivided into two subtasks. As we can observe, the *task method* for the general task structures the reasoning process into two steps:

- Abstraction: the purpose of this step is for the organisation to obtain the level of compliance for each section (KPA of the desired maturity level). As explained above, this level of compliance can be "none", "low", "medium" or "high". The motive for this abstraction is the fact that what matters in a decision is not so much the number of positive answers by the user, but rather the meaning of this number. In other words, the reasoning of an expert auditor will be as follows: "The organisation complies with all the sections at a medium level, but Project Management is indispensable (must have a high level of compliance) and I therefore consider that there must be made improvements in that area ...".
- Matching: the abstractions are matched in order to take the final decision on whether or not there is compliance with the established criteria.

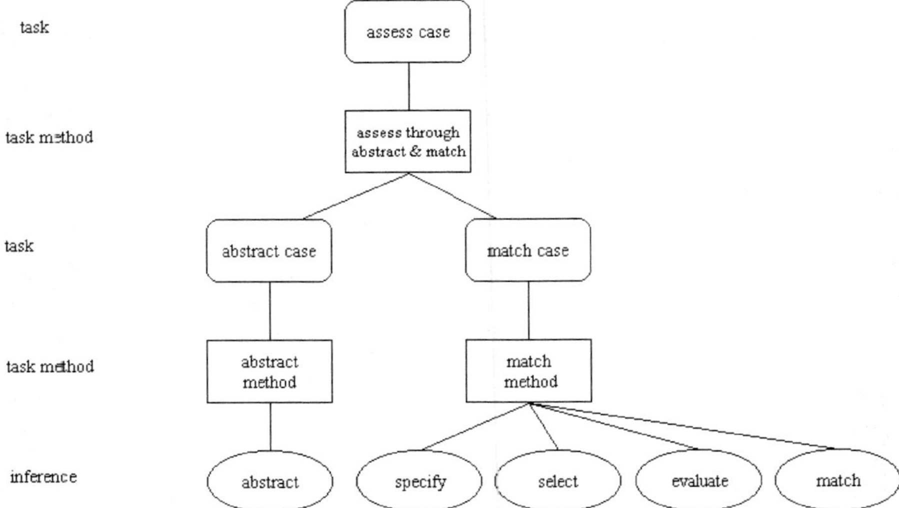

Fig. 5. Decomposition of the task

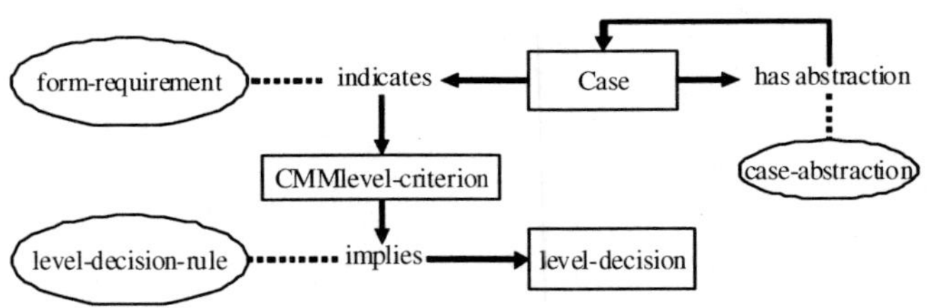

Fig. 6. Final knowledge scheme

On the other hand, and thanks to the "middle-out" reasoning process, we have obtained the final knowledge scheme that is shown in Figure 6.

Finally, we can observe that the final domain scheme incorporates three types of rules:

- "case-abstraction": the abstractions that are required for the application refer to the attainment of the sections' compliance level by using the "total-questions" and "positives" attributes as previously mentioned. Even though the abstraction really refers to the Section concept, it is modelled in the Case concept as it possesses an aggregation relationship with Section.

- "form-requirement": this type of rule aims at offering truth values to the norms "Last-level", "Level-skills" and "Concrete-level-criterion". Their instances therefore indicate the compliance with the previous CMM level, the required skills for the desired level, and the acceptable compliance levels of the maturity form's sections for a determined level.

- "level-decision-rule": we need some type of knowledge that refers to the final decision offered by the system to the user. This decision is represented by a "Level-decision" concept with a "value" attribute that indicates whether or not the organisation has real possibilities of successfully passing an audit for the desired CMM level. Also, the "level-decision-rule" type of rule expresses the relation between the different criteria and the final decision. In this case, all the criteria must have the truth value *true* for the system to respond that the audit is *possible*.

3   Implementation of the Proposed System

The system was implemented according to the design presented in Section 2 and by means of the Clips tool [12]. In order to provide the application with modularity and make the development and depuration processes easier, we have defined the following seven knowledge bases:

- General: Contains all the definitions of classes, objects, and properties.
- Abstract: Contains the abstraction rules that allow the organisation to obtain the compliance level of each KPA of the desired CMM level.
- Level2, Level3, Level4, and Level5: Contain the rules for the evaluation of the three criteria "Last-level", "Level-skills" and "Concrete-level-criterion".
- Decision: The rules of this database refer to the achievement of the final decision, i.e. the compliance with the three abovementioned criteria.

A special menu invites the user to specify the CMM level for which the organisation will be evaluated. The Clips inference engine is then started and the

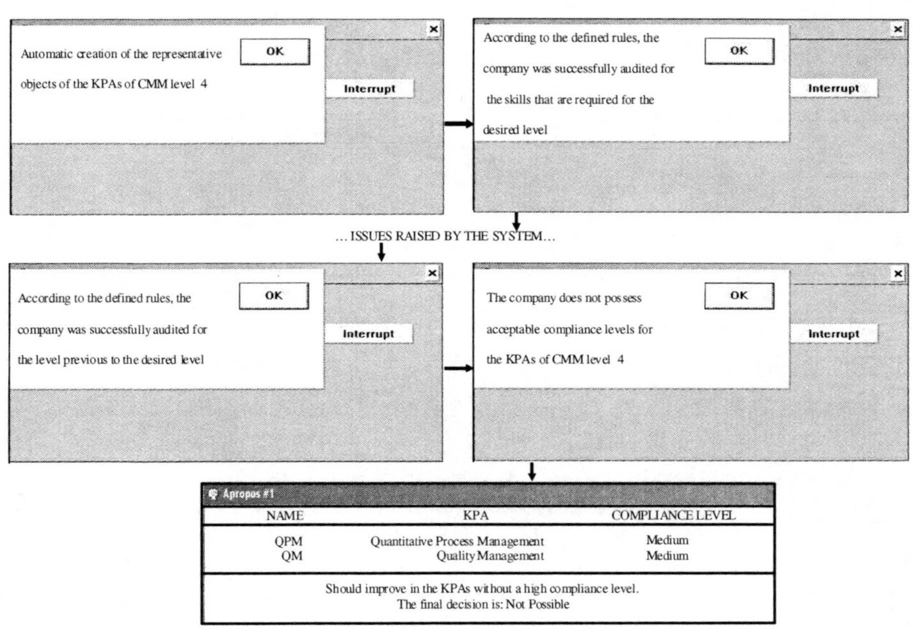

NAME	KPA	COMPLIANCE LEVEL
QPM	Quantitative Process Management	Medium
QM	Quality Management	Medium

Should improve in the KPAs without a high compliance level.
The final decision is: Not Possible

Fig. 7. An execution example

corresponding knowledge bases are loaded. Once the developed graphic interface is initiated, the inferential process begins. Figure 7 shows an execution example in which a company wishes to be evaluated at CMM level 4 after having been successfully audited at CMM level 3; the organisation has the necessary skills for the desired level, but it lacks an acceptable level of compliance with the KPAs.

4 Conclusions

In today's competitive context, any software development organisation aims to provide excellent products within the limits of a certain time schedule and financial planning. This is precisely the purpose of CMM and the reason why organisations dedicate numerous resources to the adaptation of their processes to the requirements of this model. However, until it reaches the desired CMM level, an organisation usually needs to pass a series of audits. This article proposes a KBS that considerably reduces the economic burden of such audits by limiting their number in function of the compliance with the desired CMM level or according to the required skills for that level.

Finally, the developed KBS is currently being installed and tested in various companies at A Coruña, Spain, with which the authors have collaborated in previous occasions.

References

1. McCollum, W.R.: Process Improvement in Quality Management Systems: A Case Study Analyzing Carnegie Mellon's Capability Maturity Model (CMM). Trafford Publishing (2006)
2. Persse, J.R.: Implementing the Capability Maturity Model. Wiley, Chichester (2001)
3. Lee, C.S., Wang, M.H.: Ontology-based Computational Intelligent Multi-Agent and Its Application to CMMI Assessment. Applied Intelligence (2007)
4. Wang, M.H., Lee, C.S.: An Intelligent Fuzzy Agent based on PPQA Ontology for Supporting CMMI Assessment. In: IEEE International Conference on Fuzzy Systems. London, UK (2007)
5. Lee, C.S., Wang, M.H., Chen, J.J., Hsu, C.Y.: Ontology-based Intelligent Decision Support Agent for CMMI Project Monitoring and Control. In: North American Fuzzy Information Processing Society (NAFIPS). Montreal, Quebec, Canada (2006)
6. Schreiber, G., de Hoog, R., Akkermans, H., Anjewierden, A., Shadbolt, N., de Velde, W.V.: Knowledge Engineering and Management: The CommonKADS Methodology. The MIT Press, Cambridge (2000)
7. Kingston, J.: Designing Knowledge Based Systems: The CommonKADS Design Model. Knowledge-Based Systems 11(5-6), 311–319 (1998)
8. Valente, A., Breuker, J., van de Velde, W.: The CommonKADS Library in Perspective. International Journal of Human-Computer Studies 49(4), 391–416 (1998)
9. CommonKADS home page (last access 18/02/2008),
   `http://www.commonkads.uva.nl/frameset-commonkads.html`
10. Software Engineering Institute home page (last access 31/10/2007), (2007),
   `http://www.sei.cmu.edu/`

11. Carnegie Mellon University. In: Software Engineering Institute: The Capability Maturity Model: Guidelines for Improving the Software Process, Addison-Wesley Professional, Reading (1995)
12. Riley, G.: Clips. A Tool for Building Expert Systems (last access 31/10/2007), http://www.ghg.net/clips/CLIPS.html

Gating Artificial Neural Network Based Soft Sensor

Petr Kadlec and Bogdan Gabrys

Computational Intelligence Research Group
Bournemouth University
Fern Barrow, Poole
BH12 5BB
United Kingdom

Abstract. This work proposes a novel approach to Soft Sensor modelling, where the Soft Sensor is built by a set of experts which are artificial neural networks with randomly generated topology. For each of the experts a meta neural network is trained, the gating Artificial Neural Network. The role of the gating network is to learn the performance of the experts in dependency on the input data samples. The final prediction of the Soft Sensor is a weighted sum of the individual experts predictions. The proposed meta-learning method is evaluated on two different process industry data sets.

1 Introduction

Modern production plants in the process industries are extensively instrumented with the data primarily recorded for process control purposes. But in recent years the data has found another form of application. Drawing upon techniques from statistics, pattern recognition, and machine learning, the data is being used to build predictive models which are within the process industry called *Soft Sensors*.

There are several reasons for the interest of the process industry in the development of data-driven Soft Sensors. One of the most important reasons is the difficulty in the development of model-driven Soft Sensors like First Principle Models (FPM). FPMs usually take a form of mathematical equations which make use of the knowledge of the physical and chemical laws for building models of the processes. Remarkably both static (energy or mass balance based) and dynamic simulators exist but the processes are usually too complex to be correctly and precisely described. There are also lots of external influences, e.g. the environmental temperature, the purity of the educts, the abrasion of different mechanical parts, which make the modelling of the exact process dynamics very difficult. For these reasons, the models have to be abstracted from the reality and focus on the important aspects of the process. An alternative way to make predictions about the state of the process or the product quality is to use the data, which is measured during the operation of the process and apply so called data-driven predictive methods. The advantage of using these methods, when compared to FPMs, is the ease of deployment. In contrast to FPM, extensive knowledge of the modelled process when developing the models is not a must

N.T. Nguyen, R. Katarzyniak (Eds.): New Chall. in Appl. Intel. Tech., SCI 134, pp. 193–202, 2008.
springerlink.com

although it can be of advantage if available. Data-driven techniques extract their process knowledge from the measured data automatically by the virtue of their nature and design.

A further development of Soft Sensors may bring numerous additional benefits to the process industry. The main goal of the Soft Sensors is to gain more information about the process. This information may be, for example, a description of the process state which may be extracted from observing a group of relevant measurements. This kind of process state monitoring could provide additional cue for the process operator and may help to predict and, thus, to prevent possible dangerous process states. Another benefit could be the additional information about the product quality. This has to be often evaluated by carrying out expensive laboratory-based analysis which may usually be performed only few times a day. In this case, the Soft Sensor may deliver continuous information stream about the product quality. In a more advanced scenario, Soft Sensors could also be involved in the automated process control loops which would help to increase the plant effectiveness and, thus, for example, reduce the energy consumption of the plant.

In terms of soft sensing, the most commonly applied techniques are Principle Component Regression (PCR) [1] from the statistical methods pool or Artificial Neural Networks (ANN) [2] from the computational intelligence field. Recently, hybrid techniques, which are combinations of the techniques discussed before, have become very popular. Especially neuro-fuzzy methods [3] posses a lot of potential for approaching the solutions of some of the challenges in Soft Sensors modelling. These methods can be easily modified into adapting or evolving methods, which are able to react to changes in the data and thus to change the learnt knowledge base if necessary. There is a large number of evolving neuro-fuzzy methods, for example evolving Takagi-Sugeno (eTS) system [4], [5], Dynamic Evolving Neuro-Fuzzy Inference System (DENFIS) [6] or General Fuzzy Min-Max (GFMM) system [7].

As it was already mentioned, most of the publications dealing with Soft Sensors are based on of either multivariate statistics (e.g. PCA), ANN or neuro-fuzzy approaches to solve process industry related problems. A typical application of Soft Sensors are process monitoring (see for example [8] or [9] for monitoring Soft Sensors based on PCA), prediction of values, which can not be measured on-line (e.g. neural networks based Soft Sensors [10]) or process fault detection Soft Sensors (e.g. [11], [12], [13]). Recently, adaptive Soft Sensors based on evolving neuro-fuzzy methods were published [14].

This work is motived by Jacobs and Jordan [15] [16], where a gating network is used to decide which of the models from a set of available local models, or local experts in the terminology of the cited work, is responsible for the prediction of the given input sample. The predictions of the particular local experts are weighted using weights, which are predicted by the gating networks. In the work of Jacobs and Jordan, there is a special algorithm for the training of the gating networks, which learns and stores the experts responsible for a significant improvement of the performance of the global model, defined. The algorithm is a kind of winner-takes-all approach which tends to assigns single local experts to partitions of the input space.

2   Gating Artificial Neural Network

In contrast to [15], in this work the responsibility of each of the experts is predicted based on their past performance on similar input samples. The final response of the model is a sum of the expert predictions weighted by the predicted performance of the experts in the current part of the input space. The performance is predicted by the gating Artificial Neural Network (gANN). The aim of the gANN is therefore to learn the performance of the experts in dependency on the position of the input sample in the input space. Thus the input to the gANN are the input samples. The target values of the gANNs has to indicate the performance of the particular experts. The most straight forward way to describe the performance of the experts is to use a measure, which is proportional to the inverted values of the prediction error, for example the Squared Error (SE). The most effective approach to train the gANN is to train one gating network for each of the experts. In this way it is guaranteed that the gANN becomes an *expert* for the performance prediction of the assigned model.

Once trained, the gating networks are able to predict the performance of the particular experts for the test samples x^{test}. Together with the particular predictions, the final response of the model is calculated as:

$$y^p_{final}(x^{test}) = \sum_{i=1}^{N} w_i(x^{test})y^p_i(x^{test}), \tag{1}$$

where $y^p_{final}(x^{test})$ is the final predicted output of the model given the input test samples x^{test}, y^p_i is the prediction of the ith expert, w_i is the weight of the expert i predicted by the gating network and N the number of available experts.

It is of advantage to apply a feature-selection or PCA/PLS algorithm to the usually high dimensional input data before feeding them to the gating networks. This will limit the input space of the gating networks to the most relevant features and thus allow them to put only the significant patterns of the input space into a relation with the expert's performance.

3   Soft Sensor Based on gANN

Based on the approach described in Sect. 2 a Soft Sensor, which is a model combination approach using the gating Artificial Neural Network (gANN), is presented. The structure of the Soft Sensor is shown in Fig. 1. The Soft Sensor consists of a set of experts, which are trained using the labelled training data set $< x^{train}; y^{train} >$. After the training of the experts, next step is the training of the gANN, for this purpose the performance of the experts on a validation data set $< x^{val}; y^{val} >$ has to be evaluated. The target values vector for the gANN

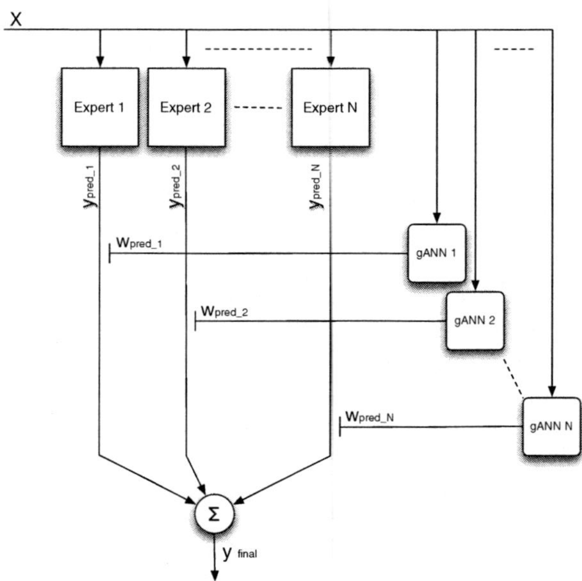

Fig. 1. The structure of the gating Artificial Neural Network based Soft Sensor

training w_i^{train} are calculated based on the prediction error e_i^p of the particular experts:

$$w_i^{train} = \frac{1}{1 + \alpha e_i^p} \quad \text{with e.g. } e_i^p = \left(y_i^p - y^{val}\right)^2, \tag{2}$$

where α is a scaling constant which helps to make an efficient use of the range $[0, 1]$, typical values of this constant are in the range $[1, 100]$, y_i^p is the ith expert prediction and e_i^p is the vector of squared prediction errors. The advantage of this performance measure is that, in combination with the scaling constant, it scales the weights to the range $[0, 1]$ automatically. After the calculation of the training weights, the ith gating network can be trained using the labelled data: $< x^{val}, w_i^{train} >$, where x^{val} are the validation data samples.

The experts themselves as well as the gANN are Multi-Layer Perceptrons (MLP) with randomly generated number of hidden units. One has only to specify the range, within which the number of hidden units has to be generated. The advantage of this approach is that one can skip the issue of the a-priori selection of the network topology, because the networks with well-performing topology will automatically get higher weights and be thus prioritised in comparison to experts with less appropriate topology.

Another common issue of ANN and other non-deterministic models solved by this approach is the problem of local minima. Neural Network models are prone to get stuck in local minima during the training and thus achieve a sub-optimal performance on the test data. This is not the case for the proposed Soft Sensor, because again the sub-optimally performing models will get lower weights assigned.

4   Experiments

This section demonstrates the performance of the proposed Soft Sensor by applying it to the prediction of continuous target values of two industrial data sets.

4.1   Methodology

For the training of the experts and of the gANNs, two-fold cross-validation was used. After running some preliminary experiments, two folds gave the best results. Further on, the term *expert* is used for the set of two networks resulting from the cross-validation, and each of the experts consist of two *partial-experts*. The partial-experts are trained using labelled training data set $Z^{train} :=< x^{train}, y^{train} >$. After the training the performance of the partial-experts is evaluated using the exclusive validation data Z^{val}. The validation results of the partial-experts represent the target values of the training data for the gating networks $Z^{gateTrain}$. For the evaluation of the gANN, there is another exclusive validation data partition $Z^{gateVal}$ necessary. The gANN validation data is the same for both of the gating networks from the cross-validation, which guaranties that both gating networks are assessed using the same independent data. The last partition of the data is the test data Z^{test}, which is being used for the performance evaluation of the whole Soft Sensor. The partitioning of the data is presented graphically in Fig. 2.

The parameters of the data partitioning for the experiments are the following, for the cross-validation $(Z^{train}+Z^{val})$ 50% of the data samples has been used, which means that 25% of the samples are used for the actuall training Z^{train} and the other 25% of the total number of samples for the validation Z^{val} of the particular CV-folds. The gate validation set $Z^{gateVal}$ are another 20% of the data and the remaining 30% were alocated for the test purposes. Because of the application of the cross-validation, there is two degrees of freedom for combining the results of the partial-experts. Firstly, one can combine the partial-experts to obtain the experts in different ways. The traditional approach is averaging the partial-experts predictions. Additionally, the presented approach allows to build a weighted sum of the partial-experts by using the weights predicted by the gANN. The second degree of freedom for building the combinations is at the level of the experts.

Partial-expert 1:	Z_{train}	Z_{val}		Z_{test}
gANN 1:		$Z_{gateTrain}$	$Z_{gateVal}$	

Partial-expert 2:	Z_{val}	Z_{train}		Z_{test}
gANN 2:	$Z_{gateTrain}$		$Z_{gateVal}$	

Fig. 2. Partitioning of the data to the training, gate validation and test data

Table 1. Considered model combinations approaches

	partial-experts combination	experts combination
Type 1	Mean	Mean
Type 2	Random selection	Weighted
Type 3	Weighted	Best performance
Type 4	Weighted	Weighted

The aim of the proposed approach is to build a set of experts and combine them to a final prediction. For the experiments the following combination types were considered:

Type 1: This is the traditional approach, where to obtain the prediction of the cross-validation ensemble, the mean value of the individual partial-expert predictions is built. For the experts combination the same is done, namely an average over the responses of the experts is built. As it is the simplest way of combining the models without involving the weights from the gating networks, this method represents the performance base-line for the comparison with the other methods.

Type 2: In this case, the combination at the level of the partial-experts is done by randomly selecting one of both partial-experts. The experts combination is a weighted sum of the experts, where the weights are obtained from the gating networks.

Type 3: Here, the experts are built as weighted sums of the partial-expert predictions. At the expert level, the selected expert is the one with the best performance on the gate validation data set $Z^{gateVal}$, which corresponds to the winner-takes-all approach.

Type 4: This is the approach discussed in Section 2 and Section 3. At the cross-validation level as well as at the expert level the output is a weighted sum of the individual predictions.

4.2   Results of the Drier Soft Sensor[1]

The drier Soft Sensor was developed using the methodology described in Section 4.1. Because of the high dimensionality of the input data it turned out to be of advantage to limit the input space of the gating networks. This has been achieved by applying the PCA algorithm [1] and taking the first five PCA features for further use.

There were 100 experts simulated. The number of hidden units of the experts was generated randomly within the range $[1, 10]$ by sampling from an uniform distribution. For each partial-experts a set of five different gANN with random number of hidden units (within the range $[6, 14]$) was trained. The performance of the gANN was assessed using the $Z^{gateVal}$ data partition. The gating network

---

[1] Data set provided by Evonik Degussa AG.

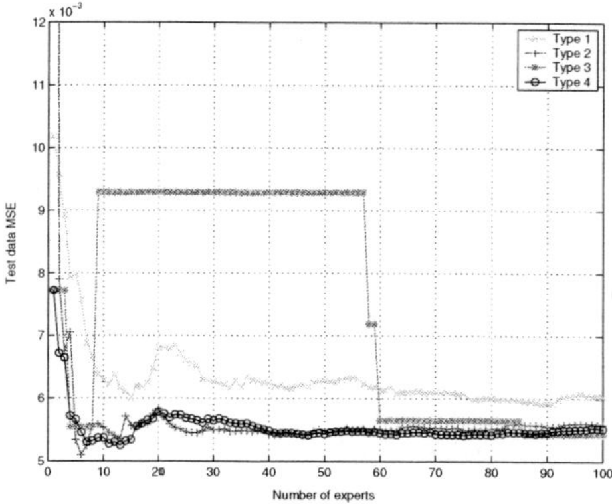

Fig. 3. The MSE performance of the drier Soft Sensor as a function of the number of experts N

with the best performance on that data was selected and stored for modelling the weights.

The following results evaluation focus on the MSE performance as a function of increasing number of involved experts. Fig. 3 compares Mean Squared Errors of the Type1 to Type4 models, as a function of the number of involved experts. One can see that, with an exception the winner-takes-all method, each of the methods converges with increasing number of involved experts to a stable performance level. The convergence value of the approaches using the weights of the gANN (Type2,3,4) are in general lower than the convergence of the base-line approach. In case of the winner-takes-all approach, there cannot be any convergence guaranteed, because in this case the output corresponds to the expert with the best performance on the validation data set, but in general this does not necessarily correspond to the best performance on the test data set. Fig. 4 presents the MSE statistics of the particular combination methods and the individual partial-experts in form of boxplots. From Fig. 4a it is obvious that there is large variance in the performance of the partial-experts. This shows one of the advantages of model combination approaches, namely the *stabilisation* of the results. In Fig. 4b one can see that besides of few outliers the performance of the weighted combinations (Type2 and Type4) is better than the performance of the best individual partial-expert. Additionally, Fig. 4b shows that in terms of convergence speed (median value of the boxplots) and stability (size of the boxplots) the Type2 and Type4 combination methods achieve superior performance compared to the base-line combination approach (Type1).

4.3   Results of the Debutanizer Column Soft Sensor[2]

Again, the methodology described in Section 4.1 was applied to develop this Soft Sensor. For the gating networks partial correlation based feature selection (see e.g.

---

[2] Data set available at: www.springer.com/1-84628-479-1

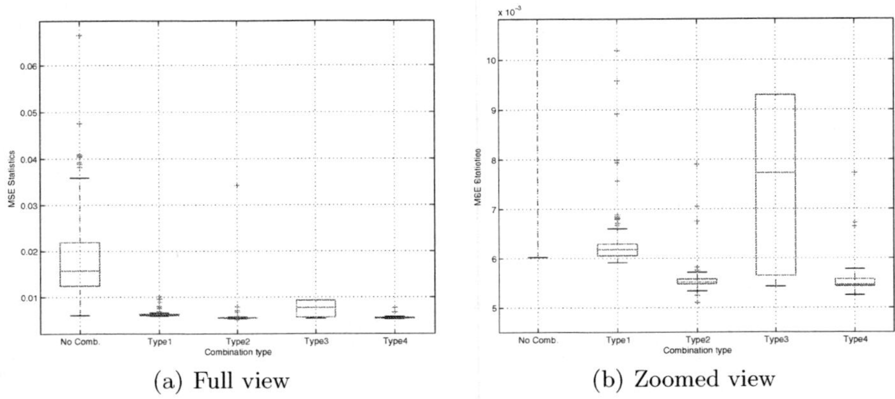

(a) Full view   (b) Zoomed view

Fig. 4. Statistics of the combination approaches together with the performance of the single partial-experts

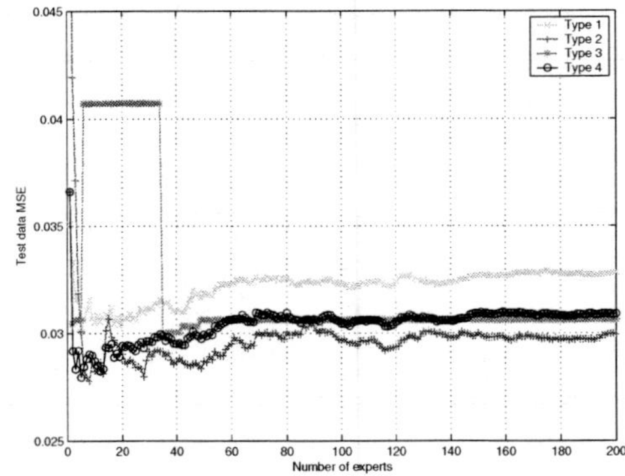

Fig. 5. The MSE performance of the debutan Soft Sensor as a function of the number of experts N

[10]) was applied. There were 200 experts built and combined for the Soft Sensor. The number of hidden units of the experts was generated randomly within the range $[1, 10]$. For each partial-expert a set of five different gANN with random number of hidden units (within the range $[6, 14]$) was trained, from this set the best gANN in terms of the gate-validation data was selected for further processing.

The dynamics of the MSE shows again convergence of the performance towards a stable level, as can be observed in Fig. 5. If assuming the distributions of the particular curves from Fig. 5 as normal and having the same standard deviation, then compared to the base-line approach (Type1) the approaches Type2 and Type4 achieve a significant performance gain.

From the boxplot statistics presented in Fig. 6, one can again observe high variance of the performances of the individual partial-experts. Also in this case

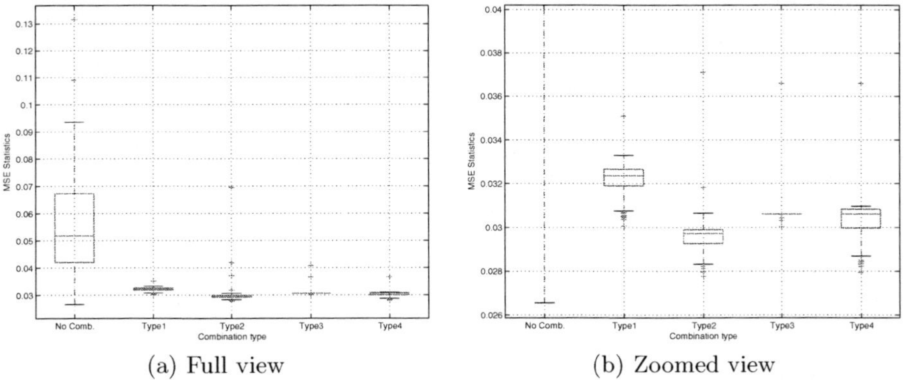

(a) Full view   (b) Zoomed view

Fig. 6. Statistics of the combination approaches together with the performance of the single partial-experts

study, the approaches involving the gating networks outperform both the individual partial-experts and the base-line combination method.

5   Summary

Training a set of models, or experts, and combing their predictions has in context of process industry data proven as a powerful approach to handle two issues of the traditional modelling practice, namely the a-priori selection of best model parameters and the handling of local minima problem. The a-priori model parameter, like the number of hidden units in the case of an ANN model, selection is handled by using so called gating networks, which are trained to predict the performance of the experts. These networks will predict lower weights for experts with lower performance and thus decrease their influence on the final prediction. The problem of local minima is solved in the same way. When a model gets stuck in a local minimum during the training, it will achieve sub-optimal performance on the validation and test data and the gating network will automatically assign lower weights to such an expert.

Although there are single models in the expert pool, which achieved better performance, because of the problems discussed before one cannot rely on the fact, that these optimally performing models will be identified during the training phase. The proposed approach performed better than both, the average expert in the pool and the base-line approach to model combination, namely the mean building of the prediction of the experts.

References

1. Jolliffe, I.T.: Principal Component Analysis. Springer, Heidelberg (2002)
2. Bishop, C.M.: Neural Networks for Pattern Recognition. Oxford University Press, Oxford (1995)

3. Jang, J.S.R., Sun, C.T., Mizutani, E.: Neuro-fuzzy and soft computing. Prentice Hall Upper Saddle River, Englewood Cliffs (1997)
4. Angelov, P.P., Filev, D.P.: Flexible models with evolving structure. International Journal of Intelligent Systems 19(4), 327–340 (2004)
5. Angelov, P.P., Filev, D.P.: An approach to online identification of takagi-sugeno fuzzy models. Systems, Man and Cybernetics, Part B, IEEE Transactions on 34(1), 484–498 (2004)
6. Kasabov, N.K., Song, Q.: Denfis: dynamic evolving neural-fuzzy inference system and itsapplication for time-series prediction. Fuzzy Systems, IEEE Transactions on 10(2), 144–154 (2002)
7. Gabrys, B., Bargiela, A.: Neural networks based decision support in presence of uncertainties. Journal of Water Resources Planning and Management 125(5), 272–280 (1999)
8. Champagne, M., Dudzic, M., Inc, T., Temiscaming, Q.: Industrial use of multivariate statistical analysis for process monitoring and control. In: American Control Conference, 2002 Proceedings of the 2002, vol. 1 (2002)
9. Li, W., Yue, H.H., Valle-Cervantes, S., Qin, S.J.: Recursive pca for adaptive process monitoring. Journal of Process Control 10(5), 471–486 (2000)
10. Fortuna, L.: Soft Sensors for Monitoring and Control of Industrial Processes. Springer, Heidelberg (2007)
11. Dunia, R., Qin, J., Edgar, T.F., McAvoy, T.J.: Sensor fault identification and reconstruction using principal component analysis. In: Proceedings of the 13th Triennial World Congress, pp. 259–264 (1996)
12. Dunia, R., Qin, S.J.: Joint diagnosis of process and sensor faults using principal component analysis. Control Engineering Practice 6(4), 457–469 (1998)
13. Amazouz, M., Pantea, R.: Use of multivariate data analysis for lumber drying process monitoring and fault detection. In: Crone, S.F., Stahlbock, S.L., R. (eds.) International Conference on Data Mining, pp. 329–332 (2006)
14. Macias, J.J., Zhou, P.X.: A method for predicting quality of the crude oil distillation. In: Evolving Fuzzy Systems, 2006 International Symposium on, pp. 214–220 (2006)
15. Jordan, M.I., Barto, A.G.: Task decomposition through competition in a modular connectionist architecture: The what and where vision tasks. Cognitive Science 15, 219–250 (1991)
16. Jacobs, R.: Adaptive mixtures of local experts. Neural Computation 3(1), 79–87 (1991)

A Classification Method of Users Opinions Using Category-Based Dictionary Generated from Answers in Open-Ended Questionnaire System

Keisuke Negoro[1], Hiroaki Oiso[2], Masanori Akiyoshi[1], and Norihisa Komoda[1]

[1] Osaka University, 2-1 Yamadaoka Suita Osaka, Japan
[2] Codetoys Ltd., 2-6-8 Nishitenma Kita-ku Osaka, Japan

Abstract. This paper addresses a classification method on users opinions for a content provider to grasp useful opinions from large amount of answers in open-ended questionnaires. Users opinions are categorized into known and unknown opinions by identifying known ones with typical words pattern-based extraction technique, however, the unknown opinions database still includes known opinions. Therefore our proposed method here introduces the category-based dictionary generated from the known opinions database and judges an opinion from two criteria; category typical words involvement and category words coincidence. We also discuss applied results of this method to mobile game content users opinions.

Keywords: open-ended questionnaires, category-based dictionary, typical words involvement, words coincidence.

1 Introduction

The mobile game market has been expanding rapidly since the late 1990s. The users pay their monthly fee in addition to packet charge. From game providers income viewpoints, they have to attract more users and prolong the subscription period of users by improving their service contents.

To improve the contents, answers to open-ended questions supplied by mobile game users when they unsubscribe the services are quite important.The questionnaire consists of multiple-choice and open-ended questions that appear on the screen of their cellular phones. Using multiple-choice questions, the questionnaire asks users to choose from some possible answers and these responses can rapidly be statistically analyzed[1][2]. Answers to the multiple-choice questions, however, are known from the contents provider's perspective. To tap into unexpected ideas, answers to open-ended questions must be analyzed.

There are no restrictions on open-ended questions and they are generally answered with natural language, so it is hard for contents provider to read all opinions. Therefore, a supporting system for analyzing answers to open-ended questions has been proposed[3]. The system are intended to extract useful opinions that include unexpected profitable ideas by categorizing such opinions

N.T. Nguyen, R. Katarzyniak (Eds.): New Chall. in Appl. Intel. Tech., SCI 134, pp. 203–211, 2008.

into known and unknown opinions. However, many known opinions can not be correctly discriminated, because the system simply uses typical words pattern-based extraction technique, which causes that an analyst still needs hard work to find useful opinions.

This paper addresses a classification method to remove these unextracted opinions as known ones in that system. Extracted known opinions by the proposed system are usually categorized by their meaning, and there could be key features that each known opinion that cannot be removed by the system has some similarity to a certain category in some sense. In this paper, new judgment criteria whether an opinion should be extracted or not is done by existence of similar category. The similarity between an opinion and a category is judged by using the category-based dictionary. Such a category-based dictionary may involve specific words and some relationships among them. The relationships are, in a sense, ontology to measure the conceptual similarity that are expected to construct automatically by using a method[4]. However, our approach is slightly biased to identify specific words to measure data-oriented similarity in target opinions. Therefore, we construct a category-based dictionary using index to decide words typical degrees in opinions.

2   Questionnaire Analysis

2.1   Questionnaire on Unsubscription

In order to improve the contents, the contents provider executes the questionnaire when consumers unsubscribe, and consider ideas from answers of the questionnaire. There are two kinds of questionnaires. In the multiple-choice questions, consumers just choose from a limited number of prelisted answers. On the other hand, in the open-ended questions, consumers can answer without restriction. Due to freely written style, there can be unexpected ideas in answers of open-ended questionnaire.

2.2   Analysis of Open-Ended Questions

The answers from consumers include important remarks related to dissatisfaction that cannot be grasped through the multiple-choice questions. The answers also include honest impressions or practical needs that the provider may not imagine. If similar opinions frequently happen, they are included in the multiple-choice questions. Additionally, frequent dissatisfaction and requests opinions motivate improvements to the contents.

However, useful opinions comprise only about 5% of all the opinions. Most answers reflect opinions already known by the provider or duplicate the meaning of other answers.

To grasp unexpected opinions, we usually classify the opinions of open-ended questionnaire data into known and unknown opinions.

The definition of known and unknown opinions is as follows:

- Known opinions:
  1. Opinions that reflect the meaning of items included in the multiple-choice questions. (e.g., The packet charge is too expensive.)
  2. Frequently claimed opinions that the provider already knows. (e.g., My knowledge increased.)
  3. Irrelevant opinions. (e.g., I had a baby!)
- Unknown opinions:
  Opinions that are not known ones. (e.g., Quizzes for kids are needed.)

However, the boundary between the known and the unknown is quite ambiguous and the distinction differs according to the analyst's background knowledge. An unknown opinion might change to a known opinion when the provider reads many opinions.

The proposed system classifies opinions in open-ended questions into known or unknown ones according to the above definition, and supports their analysis[3].

2.3   Outline of the Support System

Our developed system aims to support to analyze opinions in open-ended questions that are answered by users when they unsubscribe the service, and provides the following supportive functions.

- Analysis on known opinions is done by showing the transition of the number of grouped opinions, and that on unknown opinions is done by showing intuitively graphical map to be read.
- It should enable the boundary between known and unknown opinions to be flexibly changed by the analyst.

Fig. 1 shows the overview of our system, which consists of data input part, and opinion DB and typical words pattern DB management part, and data output part.

When new questionnaire data is inputted, opinions from the open-ended questions extracted from questionnaire data are divided into word lists by morphological analysis with "ChaSen"[5]. From the word lists, we extract nouns, independent adjectives, and independent verbs as the minimum words required for characterizing a sentence.

Applying typical words pattern-based extraction method to the word lists inputted from data input part, known opinions and unknown opinions are classified. The known opinions are classified into opinion statistics DB, and unknown opinions are stored to opinion display DB. In opinion statistics DB, opinions are registered to categories that were set according to the meanings. Contents of the categories are, for example, "Quezzes are very funny." or "The packet charge is too expensive." and so on.

Data of opinion statistics DB is shown as a graph which has axes of category numbers and time and the numbers of opinions, so that transition of opinions can be grasped as a whole. Data of opinion display DB is shown as opinion cards on the 2-dimensional plane, so intuitively tendency of opinions can be grasped. If an analyst wants to move increasing opinions in opinion display DB to opinion statistics DB, it can be done by registering new words in opinion display DB to the typical words pattern DB.

In the next section, we discuss a problem about this analysis support system of open-ended questionnaires.

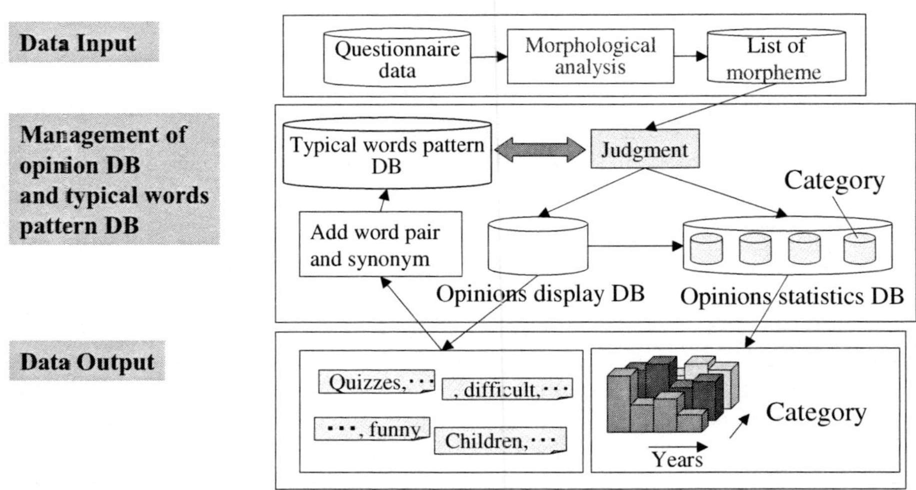

Fig. 1. Outine of the System

2.4 Problem about Analysis Support System of Open-Ended Questionnaires

On the proposed system, an opinion is classified by exactly matching of the opinion's words list with words pair registered in typical words pattern DB. If a known opinion has no words pair in the words list, it can not be classified as the known opinion. Moreover, some known opinions are extracted to opinion display DB due to parameters threshold, in spite of having words pair.

We propose how to reduce the known opinions miss-extracted to opinion display DB, and to classify them to categories in opinions statistics DB.

3 Opinions Classification Method Using Category Based Dictionary

3.1 Approach

On the proposed system, known opinions in opinions statistics DB are classified to some categories by the meaning. In case that an opinion in opinion display DB is known one, there is at least one similar category from the meaning points of views. On the other hand, in case of an unknown opinion, it has no similar category. Then, on the basis of similarity with the categories opinions, we try to remove the known opinions in opinion display DB.

If the meaning of an opinion is considered to be same to a category, the opinion tends to include the category typical words that strongly represent for the content of the category. According to the feature, we judge similarity between an opinion data and categorized opinion data as follows:

- Category typical words involvement
- Words coincidence with category opinions

We propose the method to remove the known opinions from opinion display DB and classify them to opinion categories using typical words involvement degree and words coincidence index between a target opinion and categorized opinions shown in Fig. 2.

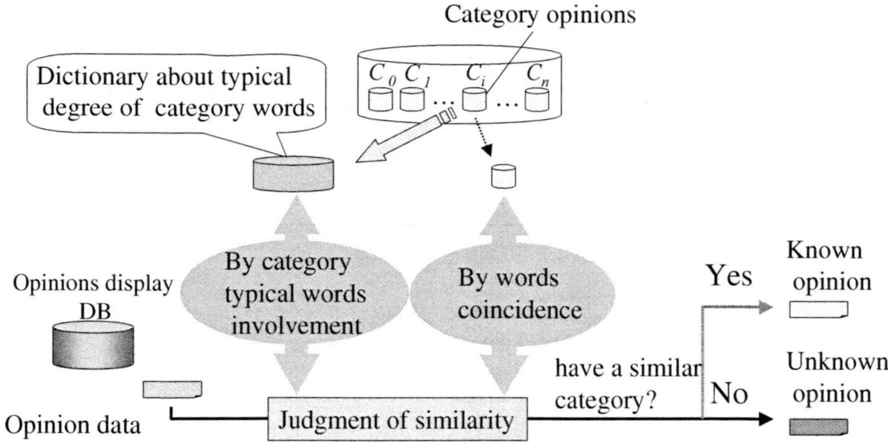

Fig. 2. Overview of the proposed method

3.2 Judgment of Categry Words Involvement

At first, "typical words degree dictionary" is generated, a dictionary which provides typical degrees of all category words. The typical degree of a word in a category is calculated by $tf\text{-}idf$ (term frequency inverted document frequency) which is generally used as an index to decide words typical degrees in documents as follws.

$$tf\text{-}idf = tf \times \log \frac{N}{df} \tag{1}$$

In this paper, tf is defined as the number of occurrence about the target word in the category, N as the number of all categories, df as the number of categories which include the target word. Referring the "typical words degree dictionary", an opinion's typical words involvement degree about a category is decided as average of $tf\text{-}idf$ concerning the opinion's all words.

Typical words involvement degree R_C of the target opinion I is decided as follows.

1. The words list of the target opinion I are $I_w = [w_1, w_2, ...w_n]$.
2. The words used in category C are $C_W = [W_1, W_2, ...W_m]$, and their values of $tf\text{-}idf$ are $C_T = [V_1, V_2, ...V_m]$.
3. In the range $1{\leq}i{\leq}n$, $1{\leq}j{\leq}m$, in case that $w_i = W_j$, the word w_i is given weight $x_i = V_j$. If there is no value $w_i = W_j$, the word w_i is given weight $x_i = 0$
4. By the list of weights $I_x = [x_1, x_2, ...x_n]$ given to the words list I_w of an opinion I, the typical words involvement degree is decided as the follwing expression.

$$R_C = \frac{\sum_{k=1}^{n} x_k}{n} \tag{2}$$

The threshold set for each category is defined using the average of typical words involvement degrees about the category opinions. However, the average values about all combinations of the category opinions are generally much larger than that value about the target opinions. Therefore, we define the threshold as the value that the coefficient $k(0{<}k{<}1)$ is multiplied by the average value of typical words involvement degrees about the category opinions. If the typical words involvement degree is larger than the threshold, typical words are considered to be involved. A judgment example of typical words involvement is shown in Fig. 3.

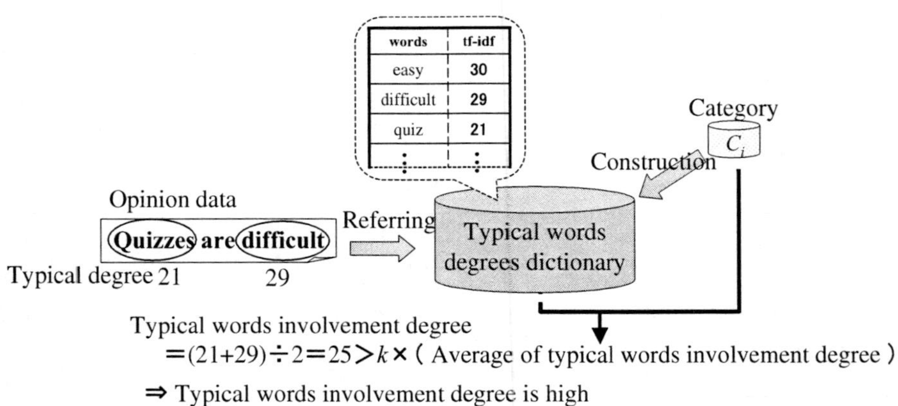

Fig. 3. Juggement of typical words involvement

3.3   Judgment of Words Matching

We introduce the words coincidence index how the target opinion's words concide to the words of category opinions. The words coincidence index is decided as an average of *Jaccard* coefficients with all opinions in a category.

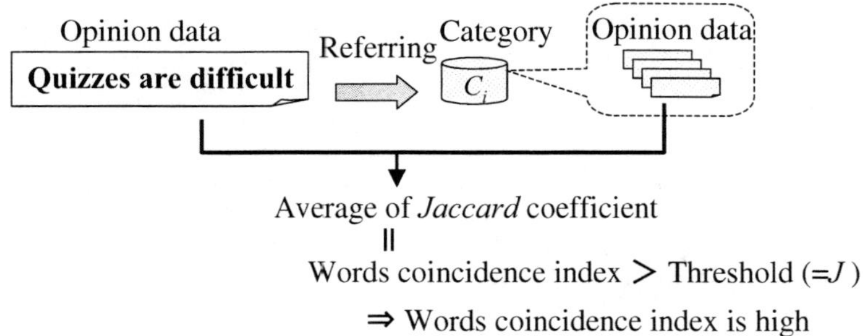

Fig. 4. Judgement of words concidence

Jaccard coefficient is generally used as words matching rate defined as follows.

$$Jaccard \text{ coefficient} = \frac{\text{The number of common words in two opinions}}{\text{The number of total words in two opinions}} \quad (3)$$

If words coincidence index is larger than a threshold which is inputted as a parameter,the opinion is included in the category. An example of judgment of words coincidence is shown in Fig. 4.

4   Experiment

We conducted an experiment of applying the classification method in the following conditions. The overview of the experiment is shown in Fig. 5.

- Targets are 1202 opinions data which were extracted to opinion display DB by the system[3].
- In opinion statistics DB, 2044 opinions data in 53 categories were used.
- Coefficient k was set in the range $0.03 \leq k \leq 0.2$ and threshold J was set at the value of 0.1 or 0.2 as parameters.

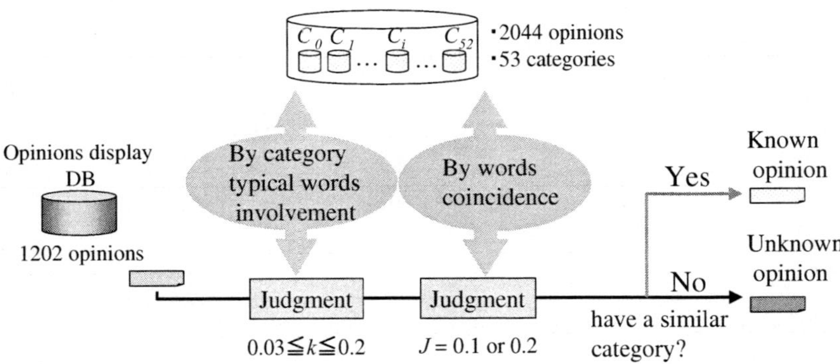

Fig. 5. Overview of the experiment

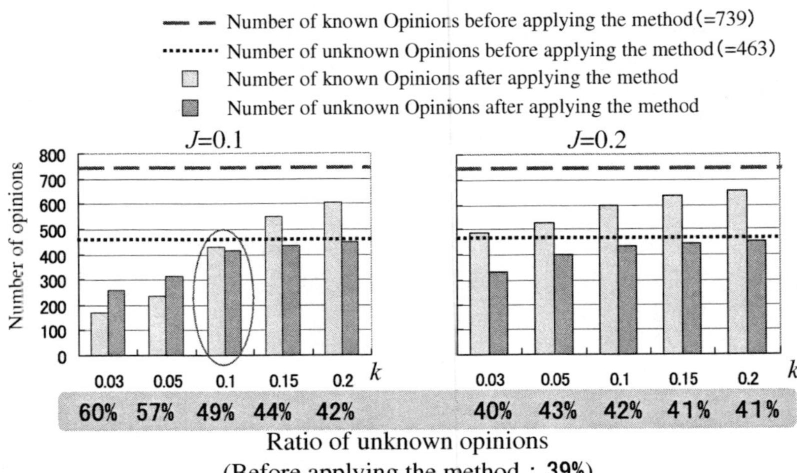

Fig. 6. Changes of the number of opinions by applying the method

In Fig. 6, it is shown the number of opinions in opinion display DB before and after applying the classification method.

The results of the experiment show that reduction of known opinions in opinion display DB by applying the method increased the ratio of unknown opinions, and it could improve efficiency of reading opinions by analyst. Specifically when the both values of k and J are 0.1, as shown in Fig. 6 shows, known opinions can be effectively removed without removing not so many unknown opinions. Through this experiment, we confirmed our method is effective.

5   Conclusion

In this paper, by defining the similarity between opinions data and categories, it is proposed the method of removing the known opinions from opinion display DB using category typical words involvement and words coincidence with category opinions.

Applying the method to the opinions data in opinion display DB by using category opinions data already classified, known opinions can be efficiently removed without miss-removing so many unknown opinions, which concludes the effectiveness of the proposed method. Our future work is about tuning of the parameters for the judgment as opitimum points of views.

References

1. Shono, Y., Takada, Y., Komoda, N., Oiso, H., Hiramatsu, A., Fukaya, K.: Customer Analysis of Monthly-Charged Mobile Content Aiming at Prolonging Subscription Period. In: Proc. of IEEE International Conference on Computational Cybernetics, pp. 279–284 (2004) (in CD-ROM)

2. Montananont, S., Hiramatsu, A., Oiso, H., Komoda, N.: Customer Unsubscription Intention Analysis in Mobile Game by using Structural Equation Modeling. In: Int. Management and Technology Conf. (2004) (in CD-ROM)
3. Hiramatsu, A., Oiso, H., Tamura, S., Komoda, N.: Support System for Analyzing Open-ended Questionnaires Data by Culling Typical Opinions. In: Proc. of 2004 IEEE Int. Conf. on Systems, Man and Cybernetics, pp. 1377–1382 (2004)
4. Lee, C.-S., Kao, Y.-F., Kuo, Y.-H., Wang, M.-H.: Automated Ontology Construction for Unstructured Text Documents. Data & Knowledge Engineering 60(3), 547–566 (2007)
5. Matsumoto, Y., Kitauchi, A., Yamashita, T., Hirano, Y., Matsuda, H., Takaoka, K., Asahara, M.: Japanese Morphological Analysis System ChaSen version 2.2.1 (2000), `http://chasen.aist-nara.ac.jp/chasen/doc/chasen-2.2.1.pdf`

Training of Classifiers for the Recognition of Musical Instrument Dominating in the Same-Pitch Mix

Alicja Wieczorkowska[1], Elżbieta Kolczyńska[2], and Zbigniew W. Raś[1,3]

[1] Polish-Japanese Institute of Information Technology,
Koszykowa 86, 02-008 Warsaw, Poland
alicja@pjwstk.edu.pl
[2] Agricultural University in Lublin,
Akademicka 13, 20-950 Lublin, Poland
elzbieta.kolczynska@ar.lublin.pl
[3] University of North Carolina,
Department of Computer Science, Charlotte, NC 28223, USA
ras@uncc.edu

Abstract. Preparing a database to train classifiers for identification of musical instruments in audio files is very important, especially in a case of sounds of the same pitch, when a dominating instrument is most difficult to identify. Since it is infeasible to prepare a data set representing all possible ever recorded mixes, we had to reduce the number of sounds in our research to a reasonable size. In this paper, our data set represents sounds of selected instruments of the same octave, with additions of artificial sounds of broadband spectra for training, and additions of sounds of other instruments for testing purposes. We tested various levels of added sounds taking into consideration only equal steps in logarithmic scale which are more suitable for amplitude comparison than linear one. Additionally, since musical instruments can be classified hierarchically, experiments for groups of instruments representing particular nodes of such hierarchy have been also performed. The set-up of training and testing sets, as well as experiments on classification of the instrument dominating in the sound file, are presented and discussed in this paper.

1 Introduction

One of the main goals in Music Information Retrieval (MIR) area is a creation of a storage and retrieval system that can automatically index musical input into a database and answer queries requesting specific musical pieces. The system may search for a specified melody, but also it should be able to automatically identify the most dominating musical instruments associated with the musical segment. In knowledge discovery based approach, the system usually divides a musical waveform into segments of equal size right after receiving it as the input data. These segments have to be somehow compared with a very large number of sound objects in a large musical database and checked for their similarity. The problem is greatly simplified by treating a database of singular monophonic sound objects

N.T. Nguyen, R. Katarzyniak (Eds.): New Chall. in Appl. Intel. Tech., SCI 134, pp. 213–222, 2008.
springerlink.com

as a training database for constructing classifiers for music automatic indexing by instruments. The precision and recall of the classifiers, trained on isolated sounds of singular instruments and then tested on polyphonic music, is very low. One way to handle the problem successfully, is trying to improve sound separation algorithm. Another approach is to enlarge the training database by a new set of tuples representing sounds coming from certain pairs (or groups) of instruments with a goal to train new classifiers for recognizing instruments in polyphonic music. The testing we have done on a small group of instruments supports the second approach.

In this paper, the experiments will focus on observing if (and how) mixing the clean musical instrument sound data with other sounds (i.e. adding accompanying sound) may influence the correctness of a classification of instruments, dominating in a polyphonic recording. The clean data represent singular musical instrument sounds of definite pitch and harmonic spectrum. The additions used for testing represent mixes of musical instrument sounds of various levels added to singular monophonic musical instrument sounds; the training additions represent artificial harmonic and noise type sound waves of broadband spectra. Our plan was to establish thresholds (if such thresholds exist) for a level of sounds added to singular instrument sounds which guarantee the highest confidence of classifiers for polyphonic recordings.

The categorization of musical instrument sounds can be used to assist extraction of information requested by users, who are browsing the database of music pieces and looking for the excerpts played by a desired instrument. However, the excerpt played by this instrument may not be available, so in such cases, a piece representing a similar category should be returned by the system.

There are many ways to categorize music instruments, such as by playing methods, by instrument type, or by other generalization concepts. Any categorization process can be represented as a hierarchical schema which is used by a cooperative query answering system to handle failing queries. By definition, a cooperative system is relaxing a failing query with a goal to find its smallest generalization which does not fail. Two different hierarchical schemas, used as models of a decision attribute have been already investigated by authors of this paper (see [10]): Hornbostel-Sachs classification of musical instruments [5], and classification of musical instruments by articulation, i.e. the way the sound is started, played and ended, with the following articulation methods (seen as attribute values): blown, bowed, bowed vibrato, concussive, hammered, lip-vibrated, martele, muted, muted vibrato, picked, pizzicato, rubbed, scraped and shaken. Each hierarchical classification represents a unique hierarchical decision attribute in MIRAI database (http://www.mir.uncc.edu), leading to a construction of new classifiers and the same to a different system for automatic indexing of music by instruments and their types.

In our previous research, we already tried to search for thresholds (describing the loudness level of added sounds) which are most effective for recognition of dominating instrument in sound mix, using 5 linear or 3 logarithmic levels [13]. No such thresholds have been found. This time, we decided to use denser steps for the loudness level, considering also experiments when the accuracy improves or worsens around maximum, to establish the threshold more precisely. Also, since the thresholds may differ for particular instruments or instrument groups, we decided to investigate hierarchical classifications as well. If threshold level can be found (at least for some nodes), the classifiers for each node of a hierarchy can be built basing on the sounds associated with this node, with added sounds from outside the identified class and with loudness level determined by this threshold. We also believe that hierarchical classifiers may outperform the general classifiers at each level of classification, comparing the performance for the groups of classes corresponding to a given level of a hierarchical classifier.

The paper is organized as follows: in Section 2, sound parameterization for instrument identification is briefly described. Section 3 presents sound data used for training and testing in our experiments. Classification results are shown in Section 4. Section 5 summarizes and concludes the paper.

2 Data Parameterization

Digital music (audio) data represent series of samples, where each sample represents instantaneous amplitude value of the recorded sound wave. Therefore, parameterization is needed to perform classification experiments on such sounds, since even the slightest change in the sound wave causes considerable changes in the recorded amplitude samples. Parameterization may describe temporal, spectral, and spectral-temporal properties of the sound. Numerous parameters have been used so far in research on musical instrument sound recognition, including features describing properties of DFT spectrum, wavelet analysis coefficients, MFCC (Mel-Frequency Cepstral Coefficients), MSA (Multidimensional Analysis Scaling) trajectories, and so on [2], [3], [7], [8], [12]. General overview of parameterization and classification of musical instrument sounds is given in [4]. Also, MPEG-7 sound descriptors can be applied for sound parameterization purposes [6]. However, these parameters are not dedicated to recognition of particular instruments in recordings. The choice of parameters for feature vector is an important part of obtaining the data for classification purposes, and the results may vary depending on the parameters and classifiers chosen.

In this research, the feature vector consists of 219 parameters, based on MPEG-7 and other parameters already applied for the recognition of musical instruments from singular sounds, and also in polyphonic (polytimbral) environment [14]. The parameterization was performed for the investigated sounds using 120 ms analyzing frame, sliding along the entire sound, with Hamming window and hop size 40 ms. Long frame was used in order to parameterize low sounds if needed. Most of the calculated parameters represent average value of parameters calculated for consecutive frames of a sound; some of the descriptors are multi-dimensional. This set of features has been already used in previous research [13], [14]; it consists of the following parameters:

- MPEG-7 audio descriptors: *AudioSpectrumSpread, AudioSpectrum Flatness, AudioSpectrumCentroid, AudioSpectrumBasis, LogAttack*

$Time$, $TemporalCentroid$, $HarmonicSpectralCentroid$, $HarmonicSpectral$
$Spread$, $HarmonicSpectralVariation$, $HarmonicSpectralDeviation$;
- other descriptors: $Energy$, $MFCC$, $ZeroCrossingDensity$, $RollOff$, $Flux$,
$AverageFundamentalFrequency$, $Ratio$.

3   Training and Testing Data

In our experiments, we decided to choose 8 instruments, representing aerophones
and chordophones in Hornbostel-Sachs classification:

- B-flat clarinet (aerophone),
- cello - bowed, played vibrato (chordophone),
- trumpet (aerophone),
- flute played vibrato (aerophone),
- oboe (aerophone),
- tenor trombone (aerophone),
- viola - bowed, played vibrato (chordophone),
- violin - bowed, played vibrato (chordophone).

These instruments produce sounds of definite pitch, with spectra of harmonic
type. We have chosen sustained sounds from the octave no. 4 (in MIDI notation) of these instruments, i.e. 12 sounds for each instrument. These sounds
represent instruments to be used for training classifiers. The sounds were taken
from McGill University Master Samples CDs [9]. Sampling rate 44.1 kHz and 16-
bit resolution was chosen to prepare digital audio samples. Sounds were recorded
in .snd format, and left channel of stereo recordings was used.

Apart from singular sounds, mixes with other sounds were used, both for
training and testing purposes. The added sounds were diminished in level. After
rescaling the amplitude of the added sounds, to match the RMS of the main
sound, the level of added sounds was diminished with respect to the main sound
with scaling factor $\sqrt{2}$ (double precision was used for real-valued calculations).
The following levels were used:

- 50%,
- $50\%/\sqrt{2} \approx 35.3553\%$,
- 25%,
- $25\%/\sqrt{2} \approx 17.6777\%$
- 12.5%,
- $12.5\%/\sqrt{2} \approx 8.8388\%$,
- 6.25%.

These levels represent logarithmic diminishing of amplitude, and are perceived
by human hearing system as uniform, since human perception is logarithmic with
respect to changes of stimulus of any type.

Since the amplitude of any musical instrument sound changes in time, the
added sounds were truncated (if needed) to the length of the main sound. Also,
we replaced 0.1 s of the beginning and ending of added sound with silence. Next,

fade-in effect was applied from the end of the silence at the beginning till 1/3 of the sound length, and similarly fade-out from 2/3 of the sound. Thus we ensure that even during transients the main sound is still dominating in the mix.

The training was performed on singular sounds of musical instruments, as mentioned above, and also on the same sounds with added artificial harmonic sound waves of the same pitch and noises, generated using Adobe Audition [1]:

- white noise,
- pink noise,
- triangular wave,
- saw-tooth wave.

All these sounds have broadband spectra, continuous (noises) or harmonic (triangular and saw-tooth wave), strongly overlapping with the main sounds. The frequency values of the generated harmonic waves were rounded to the nearest integers, as below:

- C4 - 262 Hz,
- C#4 - 277 Hz,
- D4 - 294 Hz,
- D#4 - 311 Hz,
- E4 - 330 Hz,
- F4 - 349 Hz,
- F#4 - 370 Hz,
- G4 - 392 Hz,
- G#4 - 415 Hz,
- A4 - 440 Hz,
- A#4 - 466 Hz,
- B4 - 494 Hz.

Eight-second long sounds were generated, since the longest instrumental sound was less than 8 s long.

The testing was performed on the musical instrument sounds mixed with other instruments, of level diminished as described above. Singular sounds were not used for tests, since we know from other experiments that the results in this case are very high, and this was not our subject of experiments.

Additionally, hierarchical classification was performed. In this case, the binary classifier to distinguish between aerophones and chordophones was trained first. Aerophones can be further divided into subclasses: single reed (clarinet), double reed (oboe), lip vibrated (trumpet, trombone), and side blown flute (flute). Chordophones used in our experiments represent one subclass in Hornbostel-Sachs classification. Therefore, we decided to investigate chordophones from our data set, to see if the classification improves, and if thresholds for levels of added sounds can be found. The classifier for chordophones in our research was trained to identify violin, viola, or cello, when one of these instruments dominates in the recording. Obviously, similar classification can be performed for aerophones, but then more instruments should rather be used. The problem which needs to taken into account is that in some cases, if the binary classifier distinguishing aerophones and chordophones yields erroneous results, the next classifier (for the subclass - chordophones in our research) may only yield random results. Therefore, the general classifier (for all classes) can perform better in such cases.

3.1   Training Data

The main training data set consists of 96 singular sounds, representing 4th octave of 8 instruments, as mentioned before. Also, another version of the training set was used, containing both singular sounds, and the same sounds with added noises and artificial harmonic sounds, as described above. These data were used to train classifiers for the recognition of one of these 8 instruments.

Training data for a binary classifier (aerophones and chordopnones) represented the following instruments: clarinet, oboe, trumpet, trombone, and flute sounds (singular and in mixes) represented aerophones, whereas viola, violin and cello sounds (singular and in mixes) represented chordophones. The training data for classifiers dedicated for chordophones contained violin, viola, and cello sounds, singular and in mixes.

3.2   Testing Data

The testing data for the general classifier identifying one of our 8 instruments, as described above, consisted of the sounds of these instruments with added sounds of other 7 instruments. More precisely, for each sound of each instrument, the mix with sounds of the same pitch representing the remaining 7 instruments was prepared. The level of the added sound was calculated as average of this sum, and then modified as in the case of a training set. The same level was used for training and testing (if mixes were used in training set).

In a case of binary classifier distinguishing between aerophones and chordophones, the sounds added to any aerophone sound represented the sum of chordophones for the same pitch (averaged and modified in level, as before). Similarly, the sounds added to any chordophone represented analogous sum of aerophones.

For testing a classifier dedicated to recognize particular chordophones in our experiments, the test data for violin represented mixes with viola and cello, the test data for viola represented mixes with violin and cello, and the test data for cello represented mixes with viola and violin. The sounds added in mixes were modified in loudness level, as before.

4   Experiments and Results

Classification experiments were performed using WEKA software [11]. Support Vector Machine (SMO) classifier was chosen, as appropriate for multi-dimensional

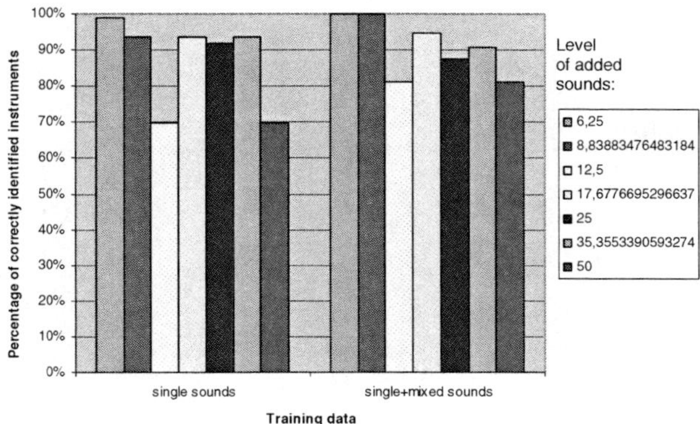

Fig. 1. Correctness of identification of dominating instruments for various types of training data (singular sounds only, or singular and mixed sounds), with testing on sound mixes with various levels of added sounds

data, and already used in similar research. Standard settings of this classifier were used.

General results of correctness of instrument identification for various levels of sounds added to the main instrument are shown in Fig. 1; as we can see, classifier yields higher results if trained on both singular and mixed sounds. Confusion matrices for the training on singular sounds only, and for the training on both singular and mixed sounds (for the same level of added sounds both in the training and test set) is shown in Fig. 2. As we can see, viola was often classified as cello, both in case of training on singular sound only and in training on both singular sounds and mixes. However, the addition of mixed sounds to the training set improves identification of oboe and trombone to 100% for all levels of added sounds, and also improves recognition of flute. However, violin and viola are always difficult to discern for this classifier. Therefore, we expect that the dedicated classifier for chordophones may help to improve this.

The hierarchical classification required the training of a binary classifier first; this classifier should discern between aerophones and chordophones. The contingency tables for this classifier are shown in Fig. 3. As we mentioned before, the next level of classification was performed for chordophones, with 3 instruments to identify: violin, viola, and cello. Contingency tables for this classifier are shown in Fig. 4. Percentage of correctness for both classifiers is shown in Fig. 5. Final correctness of classification can be calculated by multiplying the appropriate correctness values of both classifiers.

Unfortunately, no thresholds for the best performance of classifiers were observed, but generally the lowest levels of added sounds yield best results.

Classified as ->	clarinet		cello		trumpet		flute		oboe		trombone		viola		violin	
clarinet+6.25%	12	12														
clarinet+8.49%	12	12														
clarinet+12.5%	11	10		1						1			1			
clarinet+17.68%	12	12														
clarinet+25%	12	12														
clarinet+35.36%	12	12														
clarinet+50%	11	10		1						1			1			
cello+6.25%			12	12												
cello+8.49%			12	12												
cello+12.5%			12	12												
cello+17.68%			12	12												
cello+25%			12	12												
cello+35.36%			12	12												
cello+50%			12	12												
trumpet+6.25%					12	12										
trumpet+8.49%					12	12										
trumpet+12.5%					12	12										
trumpet+17.68%					12	12										
trumpet+25%					12	12										
trumpet+35.36%					12	12										
trumpet+50%					12	12										
flute+6.25%							12	12								
flute+8.49%							12	12								
flute+12.5%			1	4	1		8	8	2							
flute+17.68%							12	12								
flute+25%							11	12					1			
flute+35.36%							12	12								
flute+50%			1	4	1		8	8	2							
oboe+6.25%									11	12			1			
oboe+8.49%			1						11	12						
oboe+12.5%			3						4	12			1		4	
oboe+17.68%			1						11	12						
oboe+25%									11	12			1			
oboe+35.36%			1						11	12						
oboe+50%			3						4	12			1		4	
trombone+6.25%											12	12				
trombone+8.49%											12	12				
trombone+12.5%			1		2						8	12	1			
trombone+17.68%											12	12				
trombone+25%											12	12				
trombone+35.36%											12	12				
trombone+50%			1		2						8	12	1			
viola+6.25%													12	12		
viola+8.49%			2										10	12		
viola+12.5%			5	3									5	9	2	
viola+17.68%			2	2									10	10		
viola+25%			3	3									9	9		
viola+35.36%			2	2									10	10		
viola+50%			5	3									5	9	2	
violin+6.25%															12	12
violin+8.49%											1		2		9	12
violin+12.5%	1	1								1			4	7	7	3
violin+17.68%											1		2	3	9	9
violin+25%										1			3	8	9	3
violin+35.36%										2	1		2	5	9	5
violin+50%	1	1								1			4	7	7	3

Fig. 2. Contingency table for classifiers trained on singular sounds - left columns, and on both singular and mixed sounds - right columns

A. Classified as ->	aerophone	chordophone
aerophone+6.25%	60	
aerophone+8.49%	59	1
aerophone+12.5%	60	
aerophone+17.68%	60	
aerophone+25%	60	
aerophone+35.36%	56	4
aerophone+50%	54	6
chordophone+6.25%	1	35
chordophone+8.49%		36
chordophone+12.5%		36
chordophone+17.68%	1	35
chordophone+25%	1	35
chordophone+35.36%	3	33
chordophone+50%	2	34

B. Classified as ->	aerophone	chordophone
aerophone+6.25%	60	
aerophone+8.49%	60	
aerophone+12.5%	60	
aerophone+17.68%	60	
aerophone+25%	60	
aerophone+35.36%	59	1
aerophone+50%	59	1
chordophone+6.25%		36
chordophone+8.49%		36
chordophone+12.5%	5	31
chordophone+17.68%	2	34
chordophone+25%		36
chordophone+35.36%	3	33
chordophone+50%	9	27

Fig. 3. Contingency tables for the aerophones/chordophones classifier for the training on singular sounds only (table A) and on both singular and mixed sounds (B)

A. Classified as ->	cello	viola	violin
cello+6.25%	12		
cello+8.49%	12		
cello+12.5%	12		
cello+17.68%	12		
cello+25%	12		
cello+35.36%	12		
cello+50%	12		
viola+6.25%		12	
viola+8.49%		12	
viola+12.5%		12	
viola+17.68%		12	
viola+25%	1	11	
viola+35.36%		11	1
viola+50%	1	10	1
violin+6.25%			12
violin+8.49%			12
violin+12.5%			12
violin+17.68%			12
violin+25%			12
violin+35.36%		2	10
violin+50%		3	9

B. Classified as ->	cello	viola	violin
cello+6.25%	12		
cello+8.49%	12		
cello+12.5%	11	1	
cello+17.68%	12		
cello+25%	11	1	
cello+35.36%	11	1	
cello+50%	11	1	
viola+6.25%		12	
viola+8.49%		12	
viola+12.5%		11	1
viola+17.68%		12	
viola+25%		12	
viola+35.36%		12	
viola+50%	1	11	
violin+6.25%			12
violin+8.49%			12
violin+12.5%		2	10
violin+17.68%		1	11
violin+25%		1	11
violin+35.36%		1	11
violin+50%		3	9

Fig. 4. Contingency tables for the chordophone classifier, for the training on singular sounds only (table A) and on both singular and mixed sounds (B)

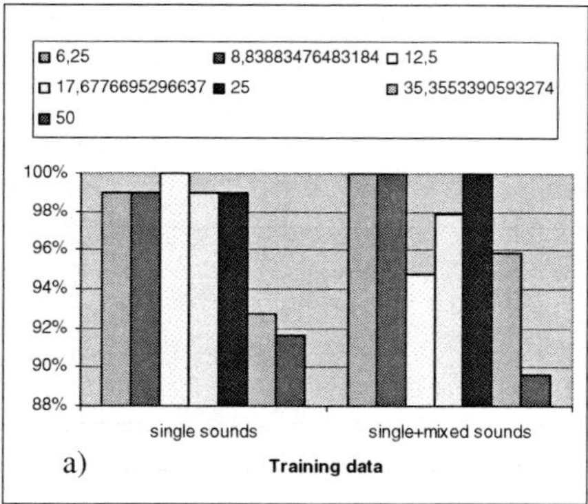

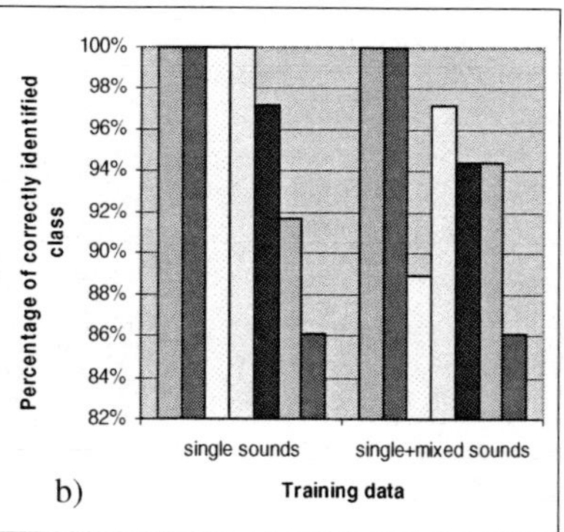

Fig. 5. Correctness of identification of dominating instruments for the aerophones/chordophones classifier (a), and for the chordophone classifier (b)

5 Conclusions

These experiments were performed on selected instruments, representing aerophones and chordophones. We consider continuation of our experiments using all instruments from these classes, since the selected instruments do not show the whole picture of the problem. In a case of a few instruments, the testing data can easily represent sounds of all other instruments, covered by a given classifier. In a case of numerous instruments, we should rather use artificial sounds, or other mixes, because it is rather unrealistic to prepare all possible mixes with other instruments. When collecting data, we may also consider sounds played with various articulation. They may represent one class (i.e. instrument), or separate classes for each articulation method. Also, we can perform similar experiments to identify articulation, since, for example, sounds of viola and violin played pizzicato can be considered more similar that sound of violin played vibrato and pizzicato. Additionally, more attributes describing sound samples may be needed, especially for lower levels of hierarchical classification. It is possible that the most efficient levels of added sounds can be found for some nodes of hierarchical classifier, or maybe we can find new classes of instruments, for which we are able to find such thresholds, and thus introduce a new decision attribute to our database, and new classification.

Acknowledgments. This work was supported by the National Science Foundation under grant IIS-0414815, and also by the Research Center of PJIIT, supported by the Polish National Committee for Scientific Research (KBN).

References

1. Adobe Systems Incorporated: Adobe Audition 1.0 (2003)
2. Aniola, P., Lukasik, E.: JAVA Library for Automatic Musical Instruments Recognition. AES 122 Convention, Vienna, Austria (May 2007)
3. Brown, J.C.: Computer identification of musical instruments using pattern recognition with cepstral coefficients as features. Journal of the Acoustical Society of America 105, 1933–1941 (1999)
4. Herrera, P., Amatriain, X., Batlle, E., Serra, X.: Towards instrument segmentation for music content description: a critical review of instrument classification techniques. In: Int. Symp. on Music Information Retrieval ISMIR 2000, Plymouth, MA (2000)
5. Hornbostel, E.M.V., Sachs, C.: Systematik der Musikinstrumente. Ein Versuch. Zeitschrift fur Ethnologie 46(4-5), 553–590 (1914)
6. ISO/IEC JTC1/SC29/WG11: MPEG-7 Overview (2004),
   `http://www.chiariglione.org/mpeg/standards/mpeg-7/mpeg-7.htm`
7. Kaminskyj, I.: Multi-feature Musical Instrument Sound Classifier w/user determined generalisation performance. In: Proceedings of the Australasian Computer Music Association Conference ACMC, pp. 53–62 (2002)
8. Martin, K.D., Kim, Y.E.: Musical instrument identification: A pattern-recognition approach. 136-th meeting of the Acoustical Society of America, Norfolk, VA (1998)
9. Opolko, F., Wapnick, J.: MUMS - McGill University Master Samples. CD's (1987)
10. Ras, Z.W., Zhang, X., Lewis, R.: MIRAI: Multi-hierarchical, FS-Tree Based Music Information Retrieval System. In: Kryszkiewicz, M., Peters, J.F., Rybinski, H., Skowron, A. (eds.) RSEISP 2007. LNCS (LNAI), vol. 4585, pp. 80–89. Springer, Heidelberg (2007)
11. The University of Waikato: Weka Machine Learning Project. Internet (2007),
    `http://www.cs.waikato.ac.nz/~ml/`
12. Wieczorkowska, A.: Towards Musical Data Classification via Wavelet Analysis. In: Ohsuga, S., Raś, Z.W. (eds.) ISMIS 2000. LNCS (LNAI), vol. 1932, pp. 292–300. Springer, Heidelberg (2000)
13. Wieczorkowska, A., Kolczyńska, E.: Quality of Musical Instrument Sound Identification for Various Levels of Accompanying Sounds. In: Ras, Z.W., Tsumoto, S., Zighead, D. (eds.) Mining Complex Data, Post-proceedings. LNCS/LNAI (2007)
14. Zhang, X.: Cooperative Music Retrieval Based on Automatic Indexing of Music by Instruments and Their Types. Ph.D thesis, Univ. North Carolina, Charlotte (2007)

Part 4
Intelliengence Technologies in Optimization and Combinatorial Problems

Impact of Fuzzy Logic in the Cooperation of Metaheuristics

J.M. Cadenas, M.C. Garrido, and E. Muñoz

Dept. Ingeniería de la Información y las Comunicaciones, Facultad de Informática, Universidad de Murcia, 30100-Espinardo, Murcia, Spain
jcadenas@um.es, carmengarrido@um.es, enriquemuba@dif.um.es

Summary. Algorithm selection problem is a common problem when we solve optimization problems. To cope with it we have proposed a hybrid system of metaheuristics that intelligently combines different strategies using a coordinator based on Fuzzy Logic. In this paper we study the impact of Fuzzy Logic in the behaviour of this hybrid system. In order to do that we perform some test to study the impact of an important parameter, the $\alpha - cut$ used in the fuzzy engine of the system, demonstrating how the variations on this parameter may change the performance of the system with different kind of instances.

Keywords: Meta-heuristic, Cooperative System, Fuzzy Rules, Data Mining.

1 Introduction

Optimization problems have focused the interest of the research community for a long time. For that reason a large amount of strategies have been developed in order to solve them in a reasonable period of time finding solutions with a near optimum quality. However, when we try to solve different instances of an optimization problem we can find algorithm selection problem [14], which tries to decide which algorithm has to be used to solve an instance of a problem, trying to maximize a measure of performance. This problem is undecidable [7], and the most common approach to solve it is to measure the performance of a set of algorithms over a set of instances and use the one with the best performance. Nevertheless, this approach seems to be too rigid and it would be more interesting to search for more tolerant strategies that adapt better to the changing conditions of the problems.

An interesting way of obtaining flexible mechanisms is using hybrid systems, which allow us to solve complex problems, very hard to solve using less tolerant approaches. Consequently, combining intelligently different strategies using a hybrid system we can tackle the algorithm selection problem. But to obtain an "intelligent" combination of strategies that achieves good results for all type of instances and problems we need a tolerant approach, as the one provided by "Soft Computing", specially by Fuzzy Logic. On the other hand this increase in tolerance may produce some precision loss, but we can sacrifice it in order

N.T. Nguyen, R. Katarzyniak (Eds.): New Chall. in Appl. Intel. Tech., SCI 134, pp. 225–234, 2008.
springerlink.com

to obtain a more robust system which could face the changes of the problems and instances. The use of different strategies together with the methodologies provided by Fuzzy Logic for building hybrid systems, will give us reasoning mechanisms and search methods which will allow us to combine domain knowledge with experimental data in order to obtain new computation tools to solve complex problems that are very difficult to solve with less tolerant approaches. In [9], Kuncheva shows a study and comparative of combination methods of fuzzy and odd nonfuzzy classifiers. The work demonstrates that better results are obtained with fuzzy combination methods and, though Kuncheva does not want to establish that the fuzzy methods are better in general, she wants to clarify that the fuzzy alternatives must not be forgotten.

In this paper, we study the impact of Fuzzy Logic in the behaviour of a hybrid system developed with the aim of solving optimization problems. In section 2, we present the design of this hybrid system, which is based on a Data Mining and Knowledge Discovery. As we use Fuzzy Logic to model some components of the system, in section 3 we show the impact of these components in the improvement of the found solutions. To finish, in section 4 we present the conclusions and work to develop.

2   A Cooperative Meta-heuristic System

2.1   Related Works

Several studies have shown that heuristics and meta-heuristics are successful tools for providing reasonably good solutions (excellent in some cases) using a moderate number of resources. An interesting trend in this area is to obtain hybrid strategies which cooperate in a parallel way in order to solve a problem, and two fields that follow this approach are *parallel meta-heuristics* and *hybrid meta-heuristics*, where the same or different metaheuristics are parallelized in order to reduce its execution time or even improve its results.

Many efforts have been focused on these fields, and we can find different implementations. First appeared synchronous implementations, where information is shared in regular intervals, such as [11]. More recently asynchronous implementations showed up, such as [6], and, according to the reports provided in [5], they obtain better results than synchronous. It has been pointed out that these approaches obtain better results than independent methods, but previous studies, [6], show that if the access to shared information is not restricted they can experiment premature convergence problems. This seems to be owe to the stabilization of the shared information produced as a result of the intense exchange of the best solutions. Trying to cope with this problem in [13] is proposed a cooperative strategy that uses memory to control this effect. Here a coordinating agent, modeled by a set of fuzzy rules defined by the user, monitors a set of solver agents and sends orders to them about how they have to continue, implementing each agent the same meta-heuristic.

It has also been noticed that those strategies based on an unique meta-heuristic does not cover all the possibilities, thus we find two interesting challenges:

- To find ways of controlling the information exchange.
- To combine different meta-heuristics.

In this paper we face both of them and propose a similar structure to the one in [13] where a coordinator modeled by a set of fuzzy rules gets information about the performance of the different meta-heuristics and sends orders to them. The main differences are that it combines a set of different meta-heuristics and that the rules are obtained as the result of a knowledge extraction process.

2.2   Design of the System

There exist many ways to carry out the cooperative execution of different meta-heuristics. The one that seems to be more appropriate to us is a multi-agent system, in which each meta-heuristic is an agent that has to solve the problem while coordinates itself with the rest of meta-heuristics. In this schema, an approach to perform the cooperation is the use of a coordinating agent which will control and modify the behaviour of the agents, [2]. The coordinator will have two fundamental tasks: to gather information on the performance of each of the meta-heuristics and to send orders to modify their search behaviour, [2].

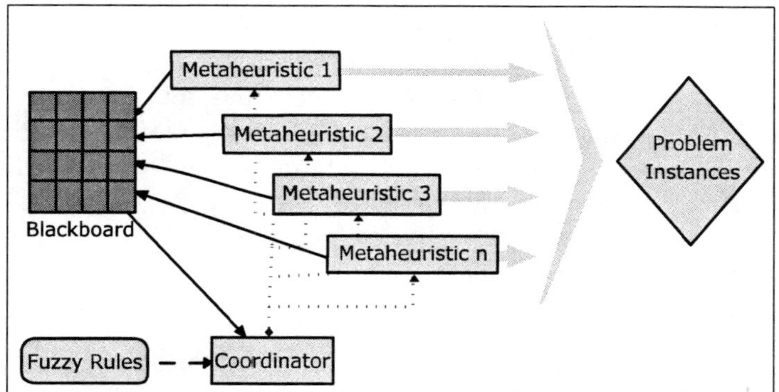

Fig. 1. Multi-agent Meta-heuristic System

To perform the communication among the different meta-heuristics we will use an adapted blackboard model. In this model each agent controls a part of the blackboard where periodically writes the best solution it has found. The coordinator then, consults the blackboard in order to monitor the behaviour of each meta-heuristic, and decides which actions have to be taken to improve the performance.

The problem that arises immediately is how to describe the coordinator, in such a way that it can modify the behaviour of the meta-heuristics efficiently and soften the problems showed before, related with the behaviour of the meta-heuristics when taken individually.

We propose to give intelligence to the coordinator using a set of fuzzy rules, since they allow to represent data in a way very similar to human reasoning, and that will do more comprehensible the system and besides will allow to easily incorporate expert knowledge into the system as ad hoc rules. These rules will give intelligence to the coordinator and this will arise as a result of the methodology proposed on [1]. In this methodology, first, we apply a knowledge extraction process from which we will obtain the set of fuzzy rules. After that, the rules are implemented and constitute the coordinator of a completely operative system.

The knowledge extraction process is supervised and divided in three phases. It starts with the Data Preparation phase, where a database which contains useful information for data mining is obtained. Next, Data Mining phase is applied, in which the model of the coordinator of the system is obtained using data mining techniques. And the process ends with the Model Evaluation phase, where the efficiency of the set of fuzzy rules that model the coordinator is tested.

2.3   Obtaining a Set of Fuzzy Rules for the Coordinator

In this section we briefly describe the knowledge extraction process. Its details are shown in [1].

In the first phase, Data Preparation, the first thing that has to be done is select the meta-heuristics that are going to be used. With them, a broad set of training instances has to be solved, using different configurations of parameters and gathering any interesting information, including data about the instance being solved, the parameters of each meta-heuristic, the final solution and the intermediate ones. With this information a database is created and is advisable the use of a preprocess before starting with the next phase, Data Mining.

In Data Mining phase, a data mining technique has to be chosen and applied to different subsets of the database in order to fill a template of fuzzy rules, which is shown later.

Model of the coordinator

The fuzzy rule template is composed of two high level rules:

- IF [($weight_1 * d_1$ OR ... OR $weight_n * d_n$) IS *enough*] THEN change the current solution of the worst meta-heuristic.
- IF [($weight_1 * d_1$ OR ... OR $weight_n * d_n$) IS *high* AND (time IS *THigh* OR *TVeryHigh*)] THEN *changeParameters* of Meta-heuristic.

where:

- d_n is the difference between the benefit obtained by the meta-heuristic n and the one that is being studied divided by the best.

- the weights were obtained during the knowledge extraction process. The weights for each metaheuristic are a function of the error obtained by each one of them. This error is classified with the fuzzy sets: *zero, verysmall, acceptable* and *bad*. After inferring the error obtained by every meta-heuristic, each weight is calculated as the frequency of instances solved by each metaheuristic with an error classified as acceptable or lesser with respect to the instances solved by all the metaheuristics with this error. This frequency is biased according to the error, being *zero* the best and *acceptable* the worst.
- *Enough* is a fuzzy set with trapezoidal membership function defined as follows:

$$\mu(x, a, b, c, d) = \begin{cases} 0 & x \leq a \ or \ x \geq d \\ (x-a)/(b-a) & x \in (a, b] \\ (d-x)/(d-c) & x \in [c, d) \\ 1 & x \in [b, c] \end{cases}$$

where a, b, c, d are 0.005, 0.01, 1, 1 respectively. This membership function, as the rest of membership functions described, was determined using trial and error.
- *High* is a fuzzy set with trapezoidal membership function where a, b, c, d are 0.05, 0.1, 1, 1, respectively.
- *THigh* is a fuzzy set with trapezoidal membership function where a, b, c, d are 0.4, 0.5, 0.7, 0.8, respectively.
- *TVeryHigh* is a fuzzy set with trapezoidal membership function where a, b, c, d are 0.7, 0.8, 1, 1, respectively.
- *ChangeParameters* is a function that change the parameters of a meta-heuristic for a new set of parameters which showed good performance during the process of knowledge extraction. In order to do that, for each kind of instance, we obtain an order over the different parameters based on the error that obtain and assign parameters according to this order.
- A rule is considered to be fired if its activation is bigger than an $\alpha - cut$, that can be configured.

The first rule tries to indicate how the position in the search space of the meta-heuristic with the worst behaviour can be changed for a position near to another meta-heuristic with a better behaviour. In order to change the solution of the worst meta-heuristic, we can take into account the following situations:

- The meta-heuristic that receives the solution is based on trajectories. The best solution obtained among the meta-heuristics that have fired the rule is sent to it.
- The meta-heuristic that receives the solution is based on populations. There are different options:
  ◇ The meta-heuristic that sends the solution is based on trajectories. It has to send a set of solutions consisting of the replication of its best solution.

 ⋄ The meta-heuristic that sends the solution is based on populations. It has
   to send a set of solutions consisting of the best members of its population.
 ⋄ Different meta-heuristics has to send solutions. They have to send a set
   of solutions where each meta-heuristic choose its solutions as said before
   and they are combined paying attention to their weights.

On the other hand, the second rule shows how the parameters of the different
meta-heuristics have to be modified. That way if a meta-heuristic is obtaining
solutions with a benefit smaller than the rest, and it has been a long time since
its parameters have been changed, then we can change its parameters for a new
set using new values obtained during Data Mining phase.

System Prototype for Knapsack Problem

In order to test this idea we applied the knowledge extraction process to obtain
a prototype of the system for solving knapsack problem, [3]. The prototype is
composed of three meta-heuristics, a genetic algorithm, a tabu search and a
simulated annealing. In Data Preparation phase we solved 2000 instances using
these algorithms with different parameter configurations. In Data Mining phase
we used a fuzzy decision tree, FID 3.4 [8] to fill the template, obtaining different
sets of parameters for each meta-heuristic and their weights, 0.45 for the Genetic
Algorithm, 0.3 for the Simulated Annealing and 0.25 for the Tabu Search.

The prototype of the system was implemented synchronously, that is, every communication is carried out at a given moment, previously specified. At
this moment each meta-heuristic writes its current solution and the coordinator
checks which actions have to be performed. It is important to highlight the difference between the learning phase and the execution phase. In the learning phase
we obtain the model of the system coordinator, being performed only once. On
the other hand, in the execution phase we apply the system, previously modeled, to solve instances of the problem, without the need of applying knowledge
extraction once again.

In order to obtain the fuzzy engine used to model the coordinator we used the
tool XFuzzy 3.0 [12]. With this tool we modeled the different rules and obtained
a fuzzy engine that was finally slightly modified to obtain an appropriate engine
to our purpose.

3 Impact of Fuzzy Sets in the System

In this section we want to test the benefits of using fuzzy rules in our system.
That way, we will test that as the system is partially controlled by fuzzy sets
we can obtain, for each $\alpha - cut$, a different set of rules, and thus, better results
depending on instance type. To test that we executed different experiments using
a system with crisp rules and systems using different values for the $\alpha - cut$ that
controls the firing of the rules.

3.1 Test Database

To carry out the tests we solved a database of instances composed of 20 instances where changed both size (number of objects) and the way they were generated. Regarding size we can find four different sizes: 500, 1000, 1500 and 2000 objects. As for the way the instances are generated there exist five types:

- Spanner: These instances are constructed in such a way that all their items are multiple of a small set of items called key. That key was generated using three distributions:
    - Uncorrelated,
    - Weakly correlated,
    - Strongly correlated.
- Profit ceiling: In these instances all the benefits are multiple of a given parameter d.
- Circle: These instances are generated in such a way that the benefits are a function of the weights, having its graph an elliptic representation.

These instances were different from the instances used in the process of knowledge extraction. That way we can validate the model of the coordinator using a test database different from the training database.

3.2 Methodology

In order to compare the systems we decided that each one had to be executed during 120 seconds stopping each 250 milliseconds to execute the coordination. Each instance was solved 10 times, and the results show the averages. All tests were executed on an Intel core2 Quad 1.66Ghz with 2GB of Memory.

3.3 Results

The results of the tests can be seen on table 1, that shows the average error ratio, after 10 executions, obtained in the resolution of the different types of instances by each system, being highlighted the system which obtained the best performance. In the first part of the table the average error ratio is shown for each type of instance, and in the second for each instance size.

We can check the better performance of the fuzzy systems, as only in one case the crisp system outperforms them. That way, for every type of instance the fuzzy systems obtained better results than the crisp one, and only for instances of size 1500 the crisp system obtained the best performance. With regard to the different $\alpha - cut$ values none can be said to be the best. However, by changing it we can obtain different behaviors and thus, the system becomes more tolerant and flexible when we try to solve different instances. Thus, we have found an interesting parameter for the system configuration. Because if we change $\alpha - cut$ we can obtain a better behaviour depending on the type of the instance being solved.

Table 1. Tests results

	Crisp	$\alpha = 0.5$	$\alpha = 0.6$	$\alpha = 0.7$	$\alpha = 0.75$	$\alpha = 0.8$	$\alpha = 0.9$
unc span	0,463	0,404	**0,33**	0,362	0,431	0,361	0,514
wea span	2,647	2,611	2,595	**2,556**	2,606	2,582	2,669
str span	0,913	0,917	0,913	0,920	**0,899**	0,923	0,919
pceil	0,114	**0,11**	0,115	0,116	0,116	0,116	0,114
circle	7,821	7,738	7,966	7,834	7,893	**7,724**	7,803
500	2,238	**2,102**	2,286	2,164	2,144	2,192	2,14
1000	2,745	2,748	2,816	2,766	2,833	**2,711**	2,850
1500	**3,205**	3,259	3,244	3,294	3,25	3,262	3,218
2000	3,771	3,677	3,572	3,564	3,719	**3,541**	3,811

4   Conclusions and Work to Develop

During the development of this paper we have shown the construction of a hybrid, centralized, cooperative system and how the use of fuzzy technologies can improve its performance due to the elasticity that provide us.

The system is based on a multi-agent system, where each meta-heuristic is an agent that has to solve the problem while cooperates with the rest. In order to accomplish this coordination we have used a coordinating agent which controls and modifies the behaviour of the agents during their execution. To add intelligence to the coordinator we have applied a knowledge extraction process obtaining a set of fuzzy rules. With these rules the coordinator can:

- Change the position in the search space of a meta-heuristic which is obtaining poor results for another position near to the position of a meta-heuristic with a better performance.
- Change the parameters of a meta-heuristic intelligently if it persists in having bad results.

With the model of the coordinator we built a synchronous prototype of the system, that was used to solve a set of 20 instances, using crisp rules and different values for the $\alpha - cut$ that controls the firing of the fuzzy rules. After this comparison we could observe how the fuzzy engine improved the performance of the system and became an important parameter. This parameter should be studied and configured for each type of instance.

It is important to outline that if we increment the number of meta-heuristics cooperating the knowledge extraction process will take more time, but the performance of the system will not decrease, because each meta-heuristic is independent from the others. Only the coordinator will be a bit slower, however, its execution time compared with the execution time of the meta-heuristics is negligible.

Regarding future work we have outlined that the $\alpha - cut$ is an important parameter for the design of the coordinator of our system because it can change its behaviour. For that reason we propose to obtain a decision tree that, added to the coordinator, will decide which $\alpha - cut$ has to be used with each instance. But the $\alpha - cut$ is not the only parameter from the template that needs tuning, for that reason we want to obtain models (based on decision trees) that will choose the parameter values that have to be used for each instance.

On the other hand, we can outline that the system has been applied to a very simple problem, knapsack problem, which is considered one of the "easiest" NP problems. For that reason we expect to apply this process to more complex problem, as the p-median, the p-hub median or the protein structure comparison.

It has also been noticed that the cost of the knowledge extraction process can be too large. To cope with this problem we propose two approaches:

- To use Active Learning to reduce the time expended in Data Preparation phase, as with this technique we can reduce the instances that need to be solved to generate the database.
- To use a system based on Online Learning. That way we will have an initial system with a "bad" performance that will be improving as it solves more instances.

Acknowledgements

The authors thank the "Ministerio de Educación y Ciencia" of Spain and the "Fondo Europeo de Desarrollo Regional" (FEDER) for the support given to this work under the project TIN2005-08404-C04-02. And the "Fundación Séneca, Agencia de Ciencia y Tecnología de la Región de Murcia", which give support under the "Programa Séneca".

References

1. Cadenas, J.M., Garrido, M.C., Hernández, L.D., Muñoz, E.: Towards a definition of a Data Mining process based on Fuzzy Sets for Cooperative Metaheuristic systems. In: International Conference on Information Processing and Management of Uncertainty in Knowledge-Based Systems, IPMU 2006, Paris, pp. 2828–2835 (2006)
2. Cadenas, J.M., Garrido, M.C., Liern, V., Muñoz, E., Serrano, E.: Un prototipo del coordinador de un Sistema Metaheurístico Cooperativo para el Problema de la Mochila. In: V congreso español sobre Metaheurísticas, Algoritmos Evolutivos y Bioinspirados, MAEB 2007, Tenerife, Spain, pp. 811–818 (2007)
3. Cadenas, J.M., Garrido, M.C., Muñoz, E.: A Cooperative System of Metaheuristics. In: 7th International Conference on Hybrid Intelligent Systems, HIS 2007, Kaiserslautern, Germany (2007)
4. Cohoon, J., Martin, W., Richards, D.: A multi-population genetic algorithm for solving the k-partition problem on hyper-cubes. In: Richar, K., Velw, L.B. (eds.) Fourth International Conference on Genetic Algorithms, Morgan Kaufmann Publishers, San Mateo. CA (1991)
5. Crainic, T.G., Toulouse, M.: Parallel Strategies for Metaheuristics. Handbook of Metaheuristics. Kluwer Academic Publisher, Dordrecht (2003)

6. Crainic, T.G., Gendreau, M., Hansen, P., Mladenovic, N.: Cooperative parallel variable neighborhood search for the p-median. Journal of Heuristics 10, 293–314 (2004)
7. Guo, H.: A Bayesian Approach for Automatic Algorithm Selection. In: IJCAI 2003 Workshop on AI and Autonomic Computing, Mexico, pp. 1–5 (2003)
8. Janikow, C.Z.: Fuzzy decision trees: Issues and methods. IEEE Transaction System, Man, and Cybernetics, Part B 28(1), 1–14 (1998)
9. Kuncheva, L.I.: 'Fuzzy' vs 'Non-fuzzy' in combining classifiers designed by boosting. IEEE Transactions on Fuzzy Systems 11(6), 729–741 (2003)
10. Le Bouthillier, A., Crainic, T.G.: A cooperative parallel meta-heuristic for the vehicle routing problem with time windows. Computers and Operations Research 32(7), 1685–1708 (2003)
11. Lee, k., Lee, s.: Efficient parallelization of simulated annealing using multiple markov chains: An application to graph partitioning. In: International Conference on Parallel Processing, Michigan, USA, pp. 177–180 (1992)
12. Moreno-Velo, F.J., Baturone, I., Sánchez-Solano, S., Barriga, A.: XFUZZY 3.0: A Development Environment for Fuzzy Systems. In: International Conference in Fuzzy Logic and Technology, Leicester, England, pp. 93–96 (2001)
13. Pelta, D., Cruz, C., Sancho-Royo, A., Verdegay, J.L.: Using memory and fuzzy rules in a cooperative multi-thread strategy for optimization. Information Sciences 176(13), 1849–1868 (2006)
14. Rice, J.R.: The algorithm selection problem. Advances in Computers 15, 65–118 (1976)
15. University of Waikato. Weka, Data Mining with Open Source Machine Learning Software in Java. URL: http://www.cs.waikato.ac.nz/ml/weka/

An OCL-Based CSP Specification and Solving Tool

Samira Sadaoui, Malek Mouhoub, and Xiaofeng Li

University of Regina Regina, SK, Canada S4S 0A2
{sadaouis,mouhoubm,li235}@cs.uregina.ca

Abstract. Due to the importance of solving discrete combinatorial problems in so many fields, constraint programming was developed. One of the most challenging concerns is the representation of these problems through the well known Constraint Satisfaction Problem (CSP) framework. In this paper we present a tool that facilitates the specification of constraint applications via CSP. Based on the Object Constraint Language (OCL), the tool provides a CSP template that can be easily specialized to describe a wide range of constraint problems. More precisely, we enhance OCL with new keywords to be able to express constraints and requirements of CSP. From the CSP specification, our tool automatically generates the constraint and solution graphs. Afterwards, we demonstrate the usability and efficiency of our tool on large size problems.

1 Introduction

The CSP paradigm is very powerful for representing and solving problems under constraints [7]. Many real-world applications, such as scheduling and planning problems, natural language processing and business applications, are considered as instances of CSP. In general a CSP is known to be an NP-hard problem, therefore a backtrack search algorithm of exponential time cost is needed to find a complete solution. In practice in order to overcome this difficulty, constraint propagation techniques have been proposed [7, 10, 14, 15] to reduce the size of the search space before and during the backtrack search. Although several solving tools have been developed to efficiently solve constraint problems [18, 5, 6, 8], using these tools is still challenging because of the following reasons:

- When requirements and constraints of CSP are formalized with specialized formal models or constraint programming languages, users will have difficulty to understanding and learning them.
- Constraint propagation techniques represent a real challenge. Most users lack the knowledge about the complex CSP solving algorithms which require strong Artificial Intelligence foundations. In addition, users cannot decide which algorithms best suit their specific problems.

Consequently, we propose in this work a tool to facilitate the specification and solving of CSPs. More precisely, we enhance the specification language OCL [20] to express constraints and requirements of CSPs. Based on these extensions, our tool provides a CSP template that can be easily instantiated to describe a wide range of constraint applications. Indeed users just need to enter the values of their variables and constraints in the OCL specification. Our tool will then automatically generate the constraint graph

N.T. Nguyen, R. Katarzyniak (Eds.): New Chall. in Appl. Intel. Tech., SCI 134, pp. 235–244, 2008.

to visualize the problem and therefore increase its understandability, and also the corresponding solution(s) displayed in both text and graphic modes. With our tool, users do not need to learn specific constraint programming languages to solve their problems. We perform here several tests on instances of two known problems, the N-Queens and the map coloring. The results are appealing and demonstrate the efficiency of our tool to deal with large size problems. In order to describe CSPs, we extend OCL with several new keywords that we subsequently associate with the existing OCL expressions. Our tool is designed with a generic CSP architecture that specifies any CSP application. Furthermore to find solutions for a given CSP, this architecture is coupled with a generic constraint library representing any CSP solver including ILOG[18], Prolog[8], Java Cream[6] and Choco [5]. In this paper, our tool is experimented with the Choco solver. There are some specialized languages to describe CSP constraints, such as CDL [12] and CCEL [16]. However these formal languages are difficult to learn and use. Moreover, these languages are platform and language dependent. In contrast, as an expression language, OCL is not tied to any programming language and environment. We propose here to use OCL as a well-known language with a simple syntax to make it easier for non-expert users to specify constraints problems through CSP.

The rest of this paper is structured as follows. Section 2 gives an overview of the CSP framework. Section 3 presents the CSP template based on the proposed OCL extensions. It also illustrates the template instantiation through two examples. Section 4 shows the CSP architecture as well as the friendly graphical user interface of the tool. Section 5 demonstrates the usability and performance of our tool. Section 6 concludes with a summary and future work.

2   CSP Backgound

A CSP involves a list of variables defined on finite domains of values and a list of relations or constraints restricting the values that the variables can simultaneously take [10, 13, 14, 15, 7]. A solution to a CSP is a set of assigned values to variables that satisfy all the constraints. Generally, the domain of a variable is a finite set of discrete values. A constraint is an arbitrary relation over a set of variables. It can be represented by mathematical or logical formulas. Let us illustrate the CSP framework with the following example.

Example 2.1 (3-Map Coloring). *"Given a graph with six nodes and eight edges, we want to assign one color from a set of three colors (red, blue and green) to each node such that no two adjacent nodes have the same color".*

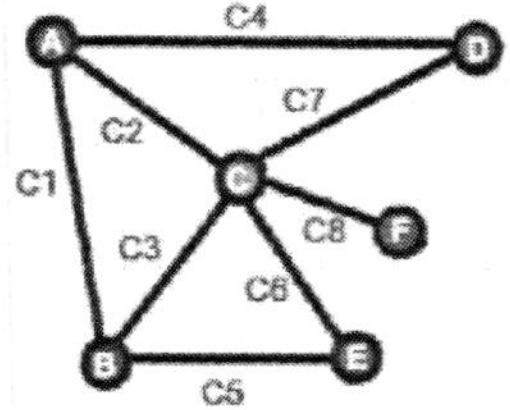

As illustrated above, via the CSP framework, this problem can be described with six variables, A to F, and eight constraints, C_1 to C_8, where C_1 means "A <> B", C_2 means "A <> C", and so on. Here all the variables have the same domain: {red, blue, green}. This problem has twelve solutions. For instance, A, E and F have color red, B and D have color green, and C has color blue.

The basic way to solve a CSP is the backtrack search (BT) which explores the search space in a systematic manner. BT incrementally attempts to extend a partial solution towards a complete one, by repeatedly choosing a value for another variable [10, 7]. The late detection of inconsistency is the disadvantage of BT. Constraint propagation approach uses BT with local consistency algorithms. This allows the early detection of inconsistencies. Local consistency is used before and during BT phase to prune earlier later failure. In binary CSPs (where constraints are unary or binary relations) various local consistency techniques have been proposed [13, 14, 7]: (i) node consistency which checks the consistency according to the unary constraint of each variable, (ii) arc consistency which checks the consistency between any two variables, (iii) path consistency which checks the consistency between any subset of three variables ..., etc. The key AC algorithm was developed by Mackworth[14] called AC-3 over twenty years ago and remains one of the easiest to implement and understand today. AC3-FC (forward-checking), and AC3-LA (look-ahead) are two backtrack search algorithms using constraint propagation via arc consistency [10]. Forward-checking is the easiest way to prevent future conflicts. It performs a restricted form of arc consistency between the current variable (the variable that is being assigned a value) and the future ones (the variables that will be assigned a value). Look-ahead does more than forward-checking by further detecting the conflicts between future variables and therefore allows more branches of the search tree that will lead to failure to be pruned earlier than with forward checking.

3   OCL Extension-Based CSP Template

OCL is used for specifying constraints for the UML diagrams [20]. It provides a textual description of constraints on attributes, operations and classes. A constraint in OCL is a set of rules that govern an object and its evaluation always delivers a value. In this paper to be able to describe a given CSP and its solution rules, we extend OCL with new keywords as shown in figure 1. Basically we have three types of solutions: (i) all complete solutions, (ii) one complete solution, (iii) best solution which satisfies most constraints of a CSP. Best solution is useful when CSPs do not have a complete solution (over constrained CSPs) or when users do not want to wait for a long time to obtain a complete solution. Hence for complex applications, a best solution within a specific period of time might be users' favourite.

OCL uses invariant, pre and postcondition expressions to describe constraints of object-oriented systems. In OCL, these expressions are associated with classes or operations. In this paper, we improve these expressions to be associated with special objects denoted by the new keywords of figure 1. The syntax of these expressions is given below.

- **Invariants.** Here invariants are associated with three objects, a given CSP, a variable or a constraint: **context** (*CSP/Vars[i]/Cons[i]*) **inv**: expressions
- **Pre and Post conditions.** Here these expressions are only associated with a given constraint. Preconditions must be true when the constraint is going to be met, and postconditions are true after the constraint has just ended. The syntax is as follows:

CSP Description

- **CSP**: CSP application.
- **Name**: name for a CSP, a variable or a constraint.
- **VarNum**: variable number in a CSP.
- **ConsNum**: constraint number in a CSP.
- **Vars**: variable collection in a CSP where Vars[1] denotes the first variable, and so on.
- **Value**: initial value of a variable.
- **Domain**: domain of a variable. For instance, Domain = [1, 4] expresses that the domain is an integer set from 1 to 4; Domain = {blue, green, yellow} denotes an enumerated type.
- **Cons**: constraint collection in a CSP where Cons[1] denotes the first constraint, and so on.

Solution Rule Description

- **TimeLimited**: time needed to solve a CSP. The default value 0 means that there is no deadline.
- **SolutionType**: **AllSolutions** means all complete solutions, **OneSolution** means one complete solution, **BestSolution** means the best solution within TimeLimited.

Fig. 1. OCL extensions for CSPs

context (*Cons[i]*) **inv**:
    pre: expressions
    post: expressions

Based on the proposed OCL extensions, we present the generic CSP template, shown in the left part of figure 2, that models constraint applications through the CSP framework. In the right part of figure 2, we instantiate the CSP template to generate the specification of the 3-map coloring problem given in example 2.1. Here we just show three variables and two constraints. In the following we give an example to show the usability of pre and post conditions.

Example 3.1. *"Mike, Tom and Jack have in total ten apples. Mike can eat more than two apples, Tom and Jack more than six. Jack has to eat two apples before Tom eats. After Jack is full, Tom can eat more than three apples".*

This problem is represented with three variables, Mike, Tom and Jack (represented respect. by vars[1], vars[2] and vars[3]), and three constraints given in figure 3. We may notice that there are only two solutions for this problem, for instance, Mike eats 3 apples, Tom 4 and Jack 3.

4   A Tool for CSPs

The main functions of our tool are illustrated in figure 4 including: CSP specification in OCL, constraint graph generation, CSP solver calling, and solution(s) display. As shown in figure 5, we design our tool with a generic architecture (modeled with the UML class diagram) representing any CSP application. The architecture compiles the CSP specification using a given constraint library in order to produce the solutions. We design

--**CSP and Solution Rules**
 context *CSP* **inv**:
  Name =--CSP name
  VarNum =--variable number
  ConsNum =--constraint number
  TimeLimited =--time deadline
  SolutionType =--solution type

--**Variables and Domains**
 context *Vars[1]* **inv**:
  Name =--Vars[1]'s name
  Domain =--Vars[1]'s domain
  Value =--Vars[1]'s initial value
•••
 context *Vars[n]* **inv**: •••

--**Constraints on Variables**
 context *Cons*[1] **inv**:
  Name =--*Cons*[1]'s name
  --*Cons*[1]'s formulas
•••
 context *Cons*[m] **inv**: •••

--**CSP and Solution Rules**
 context *CSP* **inv**:
  Name = Map Coloring Problem
  VarNum = 6
  ConsNum = 8
  TimeLimited = 0
  SolutionType = AllSolutions
--**Variables and Domains**
 context *Vars[1]* **inv**:
  Name = A--vertex A
  Domain = {red, blue, green}
  Value = red--initial coloring of A is red
 context *Vars[2]* **inv**:
  Name = B--vertex B
  Domain = {red, blue, green}
  Value = red-- initial coloring of B is red
 context *Vars[3]* **inv**:
  Name = C--vertex C
  Domain = {red, blue, green}
  Value = red--initial coloring of C is red
•••
--**Constraints on Variables**
 context *Cons*[1] **inv**:
  Name = C_1
  Vars[1] <> *Vars*[2]
 context *Cons*[2] **inv**:
  Name = C_2
  Vars[1] <> *Vars*[3]
•••

Fig. 2. CSP template and its instantiation for the 3-map coloring problem

–Constraints on Variables
context *Cons*[1] **inv**:
  Name = C1
  Vars[1] + *Vars*[2] + *Vars*[3] = 10
context *Cons*[2] **inv**:
  Name = C2
  Vars[1] > 2
context *Cons*[3] **inv**:
  Name = C3
  pre: *Vars*[3] >= 2
  inv: *Vars*[2] + *Vars*[3] > 6
  post: *Vars*[2] > 3

Fig. 3. OCL constraints for example 3.1

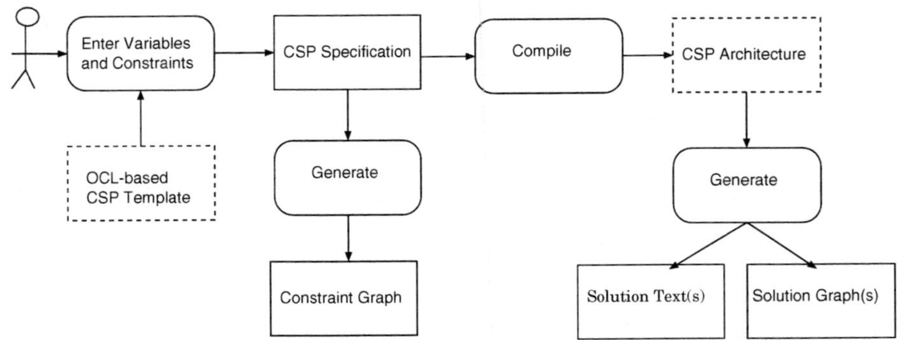

Fig. 4. A Specification and Solving Tool for CSPs

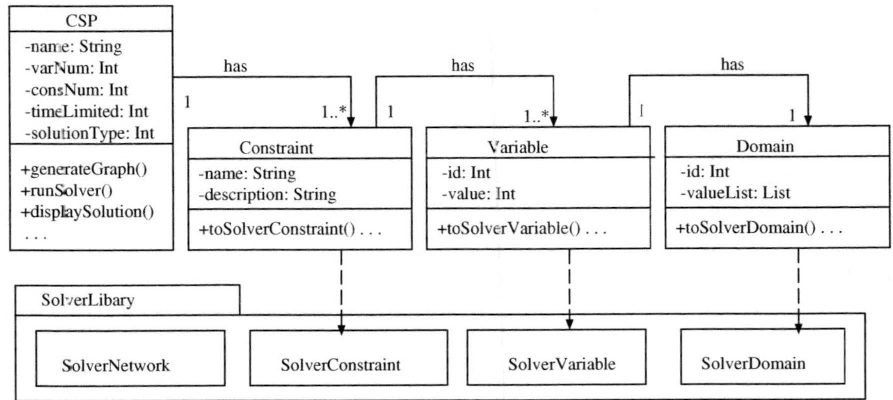

Fig. 5. Generic CSP Architecture

any constraint solver with the generic package called SolverLibrary that contains four template classes: SolverVariable, SolverDomain, SolverConstraint and SolverNetwork. The rest of the CSP architecture consists of the following classes: Variable records all the variable information and it is then converted to SolverLibrary::SolverVariable; Domain records domain description of a variable and it is then converted to Solver-Library::SolverDomain. Because values of CSP variables are discrete numbers, we assume here all the values of variables are natural numbers; Constraint records constraint information of a CSP and it is then converted to SolverLibrary::SolverConstraint; CSP records CSP general information and solution rules. Performing the operation CSP::generateGraph(), a constraint graph will be generated directly from the CSP specification. After calling CSP::runSolver(), a given CSP solver is used to produce solutions according to the rules given by the user. The operation CSP::displaySolution() visualizes solution(s) in both text and graphic modes.

In this paper, we experiment our tool with the Choco solver [5] which is based on the backtrack search strategies described in Section 2. Since Choco is open source, we

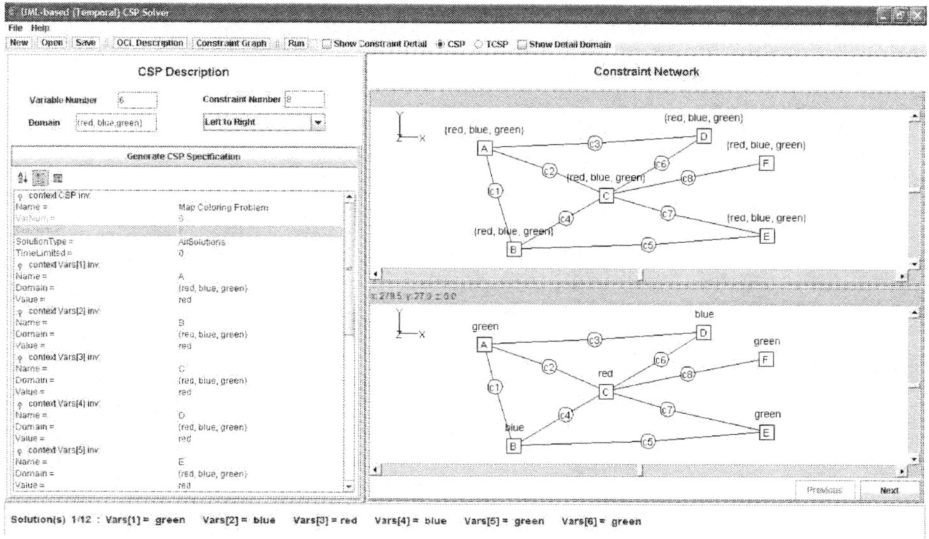

Fig. 6. Specification and constraint network of the 3-map coloring

can therefore extend its class library with the required solving algorithms. As shown in figure 6, our tool comes with a friendly graphical user interface that facilitates the description of constraint applications. Eclipse 3.12 is used as the implementation environment for its support for Java SDK 1.5. Several open source Java libraries have been used, including L2FProd.com to implement the graphical interface, JGraphT (a Java graph-editor) that we extended to draw both the constraint and solution graphs.

We now apply our tool to the 3-map coloring problem as shown in figure 6. After the user chooses the problem type (here CSP), and enters the number of variables (here 6) and constraints (here 8), and the default domain of all variables (here {red, blue, green}), the first version of the CSP specification is generated from the CSP template. The user has then to complete it by providing the solution type (here AllSolution), variable and constraint values. From the CSP description, our tool produces the constraint network composed of the constraint graph and the twelve solutions visualized in text and graphic modes (pressing the button "Next" displays the next solution). We may also notice that the 3-map coloring can be easily changed to, for instance the 5-map coloring, by just adding the new colors to the default domain of the variables. The corresponding constraint network will be automatically re-generated. The Choco code of the 3-map coloring specification is given below.

```
//Choco code of the 3-map coloring
Problem pb = new Problem();
IntDomainVar[] intVar = new IntDomainVar[6];
int[] d = {1,2,3}; // 1 = red, 2 = green, 3 = blue
//Generating variables
intVar[0] = pb.makeEnumIntVar("A", d);
intVar[1] = pb.makeEnumIntVar("B", d);
```

```
intVar[2] = pb.makeEnumIntVar("C", d);
intVar[3] = pb.makeEnumIntVar("D", d);
intVar[4] = pb.makeEnumIntVar("E", d);
intVar[5] = pb.makeEnumIntVar("F", d);
//Generating constraints
pb.post(pb.neq(intVar[0],intVar[1]));
pb.post(pb.neq(intVar[0],intVar[2]));
pb.post(pb.neq(intVar[0],intVar[3]));
pb.post(pb.neq(intVar[1],intVar[2]));
pb.post(pb.neq(intVar[1],intVar[4]));
pb.post(pb.neq(intVar[2],intVar[3]));
pb.post(pb.neq(intVar[2],intVar[4]));
pb.post(pb.neq(intVar[2],intVar[5]));
//Calling the Choco solver
solver = pb.getSolver();
pb.solveAll();
```

5   Evaluation

In order to evaluate the performance in response time of our tool, we conduct several tests on instances of the map coloring and N-Queens problems. The experiments are performed on a PC Pentium 4 with Windows XP. Let us first introduce the N-Queens problem.

Example 5.1 (N-Queens). *"Given any integer N, the problem is to place N queens on N distinct squares in an N×N chess board satisfying the constraints that no two queens should threaten each other, that is, no two queens can be placed on the same row, same column, or same diagonal".*

We show below the generic constraints of the N-Queens where we use the global constraint in C1, C2 and C3 to denote that all the queens are mutually not on the same row nor on the same diagonal.

```
--Constraints on Queens
context Cons[1] inv:
  Name = C1
  Vars→forAll (Vars[i], Vars[j]| i<>j implies Vars[i]<>Vars[j])
context Cons[2] inv:
  Name = C2
  Vars→forAll (Vars[i], Vars[j]| i<>j implies Vars[i]-Vars[j] <> i-j)
context Cons[3] inv:
  Name = C3
  Vars→forAll (Vars[i], Vars[j]| i<> j implies Vars[i]-Vars[j] <> j-i)
```

The specification of the particular case of the 4-Queens is illustrated in figure 7 where we use binary constraints in the constraint network. The 4-Queens has two solutions; for instance the second generated solution is shown in figure 7.

The results of the experiments are shown in table 1. The column Choco Code corresponds to the time in seconds needed to generate the Choco code while the column Solution shows the time in seconds needed to find one complete solution. For small

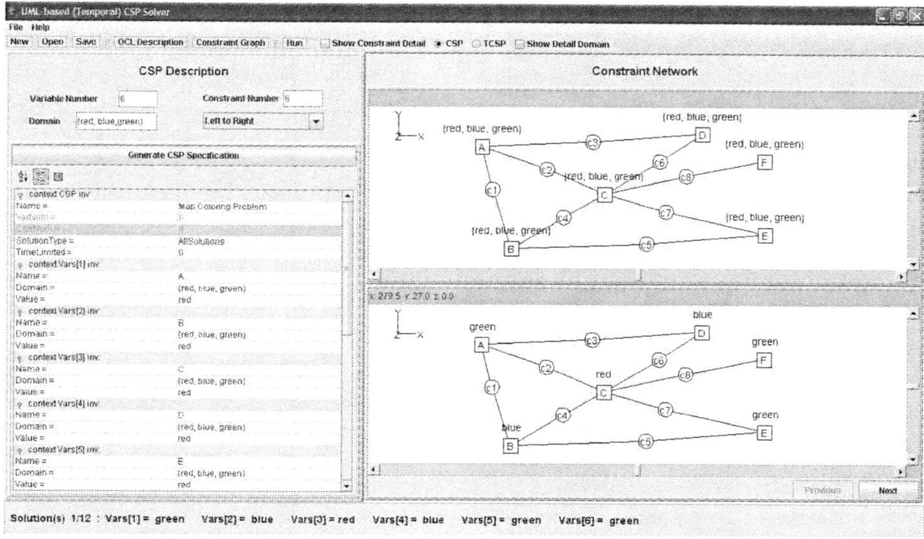

Fig. 7. Specification and constraint network of the 4-Queens

Table 1. Test results for the Map Coloring and N-Queens

Example	# Constraints	# Variables	Choco Code	OneSolution
Map Coloring	6	8	0.312	0.01
Map Coloring	100	400	0.337	16
8-Queens	4	8	0.317	0.032
16-Queens	8	16	0.320	0.062
64-Queens	32	64	0.338	22.15

instances, most of the time is spent on the code generation part. This time is in general independent of the number of variables|constraints and the type of the problem. The actual solving part however increases exponentially when the number of variables grows (for instance the map coloring with 400 variables and 100 constraints). This is justified by the fact that we are dealing with hard problems that require an exponential time cost algorithm to find a complete solution.

6   Conclusion and Future Work

Today, there is a significant need to solve more complex combinatorial problems with more intelligent and time efficient solving techniques. A more challenging task is the description of these problems which involves complex constraints. In this paper, we presented a tool for CSP specification, constraint and solution graphs generation. Distributed constraint programming [21] is an interesting approach that uses agent technology to solve more complex CSPs. In this case, we are interested in first modeling the distributed constraint problems with OCL and agent-UML [1] and then solving them by extending the Choco solver with distributed CSP solving techniques.

References

1. Bauer, B., Muller, J.P., Odell, J.: An extension of UML by protocols for multi-agent interaction. In: The Fourth International Conference on MultiAgent Systems (ICMAS 2000) (2000)
2. Bessière, C.: Arc-consistency and arc-consistency again. Artificial Intelligence 65, 179–190 (1994)
3. Bessière, C., Freuder, E., Regin, J.C.: Using inference to reduce arc consistency computation. In: IJCAI 1995, Montréal, Canada, pp. 592–598 (1995)
4. Bessière, C., Régin, J.C.: Refining the basic constraint propagation algorithm. In: Seventeenth International Joint Conference on Artificial Intelligence (IJCAI 2001), Seattle, WA, pp. 309–315 (2001)
5. Choco (2006), http://choco.sourceforge.net/
6. Cream: Class Library for Constraint Programming in Java, http://bach.istc.kobe-u.ac.jp/cream
7. Dechter, R.: Constraint Processing. Morgan Kaufmann, San Francisco (2003)
8. Diaz, D., Codognet, P.: Design and implementation of the gnu prolog system. Journal of Functional and Logic Programming 6 (2001)
9. Fowler, M.: UML distilled: A brief guide to the standard object modeling language, 3 (2003)
10. Haralick, R.M., Elliott, G.L.: Increasing tree search efficiency for Constraint Satisfaction Problems. Artificial Intelligence 14, 263–313 (1980)
11. Van Hentenryck, P.: Constraint Satisfaction in Logic Programming. The MIT Press, Cambridge (1989)
12. Klarlund, N., Koistinen, J., Schwartzbach, M.I.: Formal design constraints. In: OOPSLA 1996: Proceedings of the 11th ACM SIGPLAN conference on Object-oriented programming, systems, languages, and applications, New York, NY, USA, pp. 370–383. ACM Press, New York (1996)
13. Kumar, V.: Algorithms for Constraint Satisfaction Problems: A survey. AI Magazine (1992)
14. Mackworth, A.K.: Consistency in networks of relations. Artificial Intelligence 8, 99–118 (1977)
15. Mackworth, A.K., Freuder, E.: The complexity of some polynomial network-consistency algorithms for constraint satisfaction problems. Artificial Intelligence 25, 65–74 (1985)
16. Meyers, S., Duby, C.K., Reiss, S.P.: Constraining the structure and style of object-oriented programs. Technical paper, Brown University USA (1993)
17. Mohr, R., Henderson, T.: Arc and path consistency revisited. Artificial Intelligence 28, 225–233 (1986)
18. ILOG Solver (2006), http://www.ilog.com/products/solver/
19. Wallace, R.J.: Why AC-3 is almost always better than AC-4 for establishing arc consistency in CSPs. In: IJCAI 1993, Chambery, France, pp. 239–245 (1993)
20. Warmer, J., Kleppe, A.: The object constraint language, vol. 2. Addison-Wesley, Reading (2003)
21. Yokoo, M., Hirayama, K.: Algorithms for distributed constraint satisfaction: A review. Journal of Autonomous Agents and Multi-agent Systems 3(2) (2000)
22. Zhang, Y., Yap, R.H.C.: Making ac-3 an optimal algorithm. In: Seventeenth International Joint Conference on Artificial Intelligence (IJCAI 2001), Seattle, WA, pp. 316–321 (2001)

Effects of At-Home Nursing Service Scheduling in Multiagent Systems

Hiroshi Date and Tokuro Matsuo

Department of Informatics, Graduate School of Engineering,
Yamagata University, 4-3-16, Jonan, Yonezawa, Yamagata, 992-8510, Japan
matsuo@yz.yamagata-u.ac.jp, date2007@e-activity.org
http://veritas.yz.yamagata-u.ac.jp/

Abstract. In this paper, we propose a method of discovering the best combination between helpers and elders under much restriction based on multi-agent systems. Scheduling problem is defined as task assignment problem. Computational costs increase exponentially according to the number of restrictions. We present a new allocation algorithm which uses agent based experienced technique to detect an efficient allocation. The agent technology is one field of the artificial intelligence. It is promising in the negotiation among users who have various roles because of the problem solving. Our proposing method is based on constraint satisfaction problem. We also derive the optimal solution with reducing the solution space. To use our proposed calculation method, elderly person can serve an effective service allocation since business managers can improve their tasks efficiently. We show an example that we found efficient combination from a situation in which five elders and five helpers are given several parameters. By using our proposed method, a business manager for home-care can deliver appropriate services in both city and rural town.

1 Introduction

Recently, many developed countries are rapidly aging society. Accordingly, in medical care, reemployment is catching the society's attention. Japan is also rapidly aging and has the problem about aging society. In addition, Japan is currently experiencing a declining birthrate. After 1971 - 1974's demographic changing rate of year shows cleary that Japan's birthrate decreasing year by year. That shows proportion of elder population to total population growing rapidly. Accelerating this trend and Japanese society facing the serious concerns of rapid aging and very low birthrate.

At once, acceleration of demographic aging and increasing number of households that only elderly person. In addition, cohabitation ratio of living with an aged person has been falling. National Livelihood Survey undertaken by Ministry of Health, Labor and Welfare in 2005 shows that the rate in which an aged person take care of other elder person is increasing. A lot of caretakers are required by many people who cannot take care of themselves. In the future, half-care becomes more important in Japan. Half-care becomes popular in medical field since elderly person serves half-care services without changing his/her environment. Efficiently preparing adequate human resources becomes important problem. Caretaker delivering business is rapidly increasing and some companies try starting it.

N.T. Nguyen, R. Katarzyniak (Eds.): New Chall. in Appl. Intel. Tech., SCI 134, pp. 245–254, 2008.
springerlink.com

In this paper, this scheduling problem defines as view of task assignment problem. Computational costs increase exponentially according to the number of restriction. To detect an efficient allocation, we propose new allocation algorithm that uses experienced technique based on agents. The agent technology is one field of the artificial intelligence. It is promising in the negotiation among users who have various roles because of the problem solving. The algorithm proposes with this thesis is based on constraint satisfaction problem. We also derive the optimal solution by reducing the solution space. We show an example that we found efficient combination from a situation in which five elders and five helpers are given several parameters. By using our proposed method, a business manager for home-care can deliver appropriate services in both city and rural town.

The rest of this paper consists of the following six parts. In section 2, we show related work. In Section 3, we show preliminaries on several terms. In Section 4, we consider an example based on the Constraint Satisfaction Problems. In Section 5, we discuss about our method's adequacy and availability. Finally, we present our concluding remarks and future work.

2   Related Work

2.1   Nurse Scheduling Problem

Nurse Scheduling Problem is enumerated as an example of the multi-attribute effect problem. The nurse scheduling is a problem of allocating working shift in many nurse to fill many constraints. But it is difficult to meet all constraints. Yamaguchi's research group evaluated the value with the highest importance degree as much as possible [1] [2].

We propose a mechanism of at-home care. At-home care is a form in which caretaker visits the elder's house. Therefore, commute time should be taken the scheduling into consideration. The elder selects an appropriate helper based on his/her preferences. In existing research, such condotion is not employed since the situation becomes more complex.

2.2   Constraint Satisfaction Problem

Constraint Satisfaction Problem (CSP) is a method which finds some answer. The answers fill many constraints. CSP gives value to each parameter which fill all constraints. The value is selected from domain which consists of finite set of discrete-value [3] [4].

The constraint is defined relation which is decided between components of objects and attributes.

CSP is expressed by graph which consists of node that shows variable and arc that is stickled label of restriction adhere. Such a graph is considered to be a knowledge representation. That is suitable for the description of a static problem. The advantage of the approach that uses the network is easy to structuralize knowledge, and is a thing can be efficiently managing knowledge.

In general, Constraint Satisfaction Problems is to solve the restriction network. That is equal to thing to give value to all variables in the network and shows that shows obtain a single solution or all solutions.

3 Preliminaries

3.1 Model

In this section, we give some definition of model used in mechanism we proposed. Then, we define a person who needs care is as elder, and define caretaker is as helper.

Elders offer helper to nursing care through an intermediary. Elder tells his/her budget and helper's quality level for each care content to intermediary when negotiate. After, intermediary acquaint helper who a deal acceptable with elder.

- $W = \{w_1, w_2, ..., w_i, ..., w_l\}$ is set of care contents.
- $H = \{h_1, h_2, ..., h_j, ..., h_m\}$ is set of helpers.
- $E = \{e_1, e_2, ..., e_k, ..., e_n\}$ is set of elders. Each elder has budget and helper's skill quality level for each care contents.
- We describe set of elder's budgets as B.
  $B = \{b_{1,1}, b_{1,2}, \cdots, b_{1,l}, b_{2,1}, \cdots, b_{k,1}, \cdots, b_{k,i}, \cdots, b_{n,1}, \cdots, b_{n,l}\}$
- We describe set of quality that elder expects of helper as Q.This quality is helper's skill.
  $Q = \{q_{1,1}, q_{1,2}, \cdots, q_{1,l}, q_{2,1}, \cdots, q_{k,1}, \cdots, q_{k,i}, \cdots, q_{n,1}, \cdots, q_{n,l}\}$
- We describe the set for employed helper's income as P.
  $P = \{p_{1,1}, p_{1,2}, \cdots, p_{1,l}, p_{2,1}, \cdots, p_{j,1}, \cdots, p_{j,i}, \cdots, p_{m,1}, \cdots, p_{m,l}\}$
- We describe set of helper's skills for each care content as S.
  $S = \{s_{1,1}, s_{1,2}, \cdots, s_{1,l}, s_{2,1}, \cdots, s_{j,1}, \cdots, s_{j,i}, \cdots, s_{m,1}, \cdots, s_{m,l}\}$
- $C_{k,i}$ is set for income paid by the employing elder for count i, and $CT(k)$ is the total amount paid by that elder for each content.

$$CT(k) = \sum_{i=1}^{l} C_{k,i}$$

- $D_{a,b} = D_{b,a}(a, b \in E)$ is the distance between elder a's home and elder b's home.
- $D_{k,0} = D_{0,k}(k \in E)$ is the distance between elder k's home and intermediary's office.
- Intermediary bears each helper's traveling expenses. It is1\$/hour. $HC(j)$ is each helper's traveling expenses and TC is total traveling expenses.

$$TC = \sum_{j=1}^{m} HC(j)$$

- $T_j \leq 8$ is helper j's working time which also includes commuting time.
- $UB_{k,i}$ is monetary utility for elder k receiving nursing care i.
  $UB_{k,i} = $ (Total price requested by elder) $-$ (Total price paid by elder)
- $UQ_{k,i}$ is qualitative utility that elder k receiving nursing care i.
  $UQ_{k,i} = $ (Sum of helper's skill that elder was given nursing each care) $-$ (Sum of helper's skill that elder requested that each care)
- $x\%$ of all of the elder's payment is intermediary's profit. Intermediary pays helpers using the rest of the money as a salary. We describe intermediate's total profit as A.

$$A = \sum_{k=1}^{n} CT(k) \cdot \frac{x}{100} - TC$$

- Helpers complete each task within one hour. After that the helper heads for the following elder's home.
- Time-span is one day.
- All elders live within the range in which the helper is able to commute to and from.
- Helpers will depart for the first house from the office and will return to the office after all work is done.

3.2   Exact Algorithm

Exact search Algorithm searches all combinations. Because of this, it ensures the result's perfection. But there is a possibility the calculation may not be completed in the allocated time. It takes amount of time if we use this algorithm in our research. Therefore, it is unsuitable for complex calculations. In order to make working schedule for helpers, we need a method to reduce computation time.

Our proposing method use process that compare restriction to another restriction and remove unnecessary parameter to solve matter of time.

3.3   Relaxation Method

CSP is NP complete problem. We do not have efficient algorithm to solve CSP because it is very difficult to do so in the time allocated [3] [4]. Relaxation method is defined as a process that removes unnecessary elements. Using this method, we can refine our research space and this can greatly reduce computing time because there is no need to search for all possibilities.

4   Combinatorial Algorithm

4.1   Combinatorial Assumption

We show assumption about our proposing method. Suboptimal solutions meet the following assumptions.

Assumption1: Solutions maximize the benefits of intermediary.

Assumption2: If it have some combinations which equall of the benefits of intermediary, we choose the combination that become maximize the elder's utility.

Each elder's Utilities are financially $\succ$ quality. If the answer was not decided in Financially, the answer decides based on quality.

Intermediary and helpers must be able to satisfy elders because they are client. It means utility of elders don't become minus.

Assumption3: Helper works continuously up to 8 hours, which is decided by the law.

4.2   Combinatorial Steps

Step 1: In this step, we make tables of elders of whom each helper can takes charge. We are comparing B_{ki} with P_{ji} and select combinations of $B_{ki} \geq P_{ji}$. Similarly, we are comparing Q_{ki} with S_{ij} and select combinations of $Q_{ki} \leq S_{ij}$.

Step 2: We make the table of the set of $B_{ki} \geq P_{ji}$ and $Q_{ki} \leq S_{ij}$. If the set is the empty set, we turn up the set of $B_{ki} \geq P_{ji}$.

Step 3: We select the combination. That is all helper's working time within eight hours from table 3.

If the set of $B_{ki} \geq P_{ji}$ and $Q_{ki} \leq S_{ij}$ is empty set, we use the set of $B_{ki} \geq P_{ji}$. If, the set of $B_{ki} \geq P_{ji}$ is empty set, we use set of $Q_{ki} \leq S_{ij}$.

This process is choosing combinations that can take charge of all elders are satisfied. If there are some combinations, we choose the total of helper's working time is the fewest one.

4.3   Example

We think above tables pattern. We give parameters $0 \leq B_{ki}, P_{ij} \leq 10, 1 \leq Q_{ki}, S_{ij} \leq 3$. In Step 1, we comparing B_{ki} with P_{ji} and choose sets of $B_{ki} \geq P_{ji}$. Similarly, we comparing Q_{ki} with S_{ij} and choose sets of $Q_{ki} \leq S_{ij}$.

We can choose Set of helpers who are each elders can employ for financially and Set of helpers who are each elders can employ for quality in this process(table 6, 7).

Table 1. B_{ki}

B_{ki}	e_1	e_2	e_3	e_4	e_5
w_1	2	6	3	4	5
w_2	4	5	2	10	3
w_3	9	2	7	6	5

Table 2. P_{ij}

P_{ij}	h_1	h_2	h_3	h_4	h_5
w_1	1	2	3	4	5
w_2	4	5	6	2	3
w_3	2	6	5	6	8

Table 3. Q_{ij}

Q_{ki}	e_1	e_2	e_3	e_4	e_5
w_1	1	2	1	2	2
w_2	2	2	1	3	1
w_3	3	1	3	3	2

Table 4. S_{ij}

S_{ij}	h_1	h_2	h_3	h_4	h_5
w_1	1	1	1	2	2
w_2	2	2	3	1	1
w_3	1	2	2	3	3

Table 5. D_{ab}

$D_{10} = D_{01} = 1$	$D_{50} = D_{05} = 1$	$D_{15} = D_{51} = 1$	$D_{34} = D_{43} = 1$
$D_{20} = D_{02} = 2$	$D_{12} = D_{21} = 1$	$D_{23} = D_{32} = 2$	$D_{35} = D_{53} = 1$
$D_{30} = D_{03} = 1$	$D_{13} = D_{31} = 1$	$D_{24} = D_{42} = 2$	$D_{45} = D_{54} = 1$
$D_{40} = D_{04} = 1$	$D_{14} = D_{41} = 1$	$D_{25} = D_{52} = 2$	

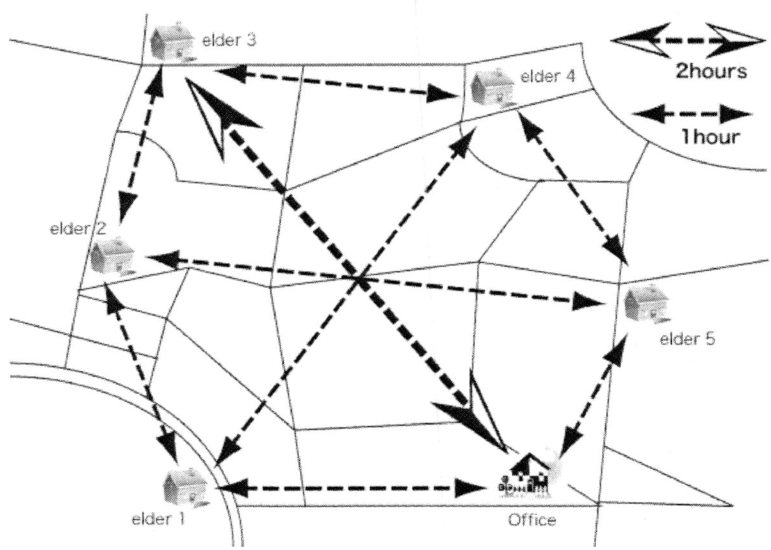

Fig. 1. Address map

Table 6. Set of helpers each elders can employ for financially

	e_1	e_2	e_3	e_4	e_5
w_1	$\{h_1, h_2\}$	$\{h_1, h_2, h_3, h_4, h_5\}$	$\{h_1, h_2, h_3\}$	$\{h_1, h_2, h_3, h_4\}$	$\{h_1, h_2, h_3, h_4, h_5\}$
w_2	$\{h_1, h_4, h_5\}$	$\{h_1, h_2, h_4, h_5\}$	$\{h_4\}$	$\{h_1, h_2, h_3, h_4, h_5\}$	$\{h_4, h_5\}$
w_3	$\{h_1, h_2, h_3, h_4, h_5\}$	$\{h_1\}$	$\{h_1, h_2, h_3, h_4\}$	$\{h_1, h_2, h_3, h_4\}$	$\{h_1, h_3\}$

Table 7. Set of helpers each elders can employ for quality

	e_1	e_2	e_3	e_4	e_5
w_1	$\{h_1, h_2\}$	$\{h_1, h_2, h_3, h_4, h_5\}$	$\{h_1, h_2, h_3\}$	$\{h_1, h_2, h_3, h_4\}$	$\{h_1, h_2, h_3, h_4, h_5\}$
w_2	$\{h_1, h_4, h_5\}$	$\{h_1, h_2, h_4, h_5\}$	$\{h_4\}$	$\{h_1, h_2, h_3, h_4, h_5\}$	$\{h_4, h_5\}$
w_3	$\{h_1, h_2, h_3, h_4, h_5\}$	$\{h_1\}$	$\{h_1, h_2, h_3, h_4\}$	$\{h_1, h_2, h_3, h_4\}$	$\{h_1, h_3\}$

In Step 2,we make table 8 which set of helpers each elders can employ for totally based on table 6, 7.

Afterwards, we choose combinations to which elders can receive requested nursing as many as possible.

In Step 3, we decide optimal answers from helper's working time.

h_4's working time is over eight hours only by caring elders who can be nursed only to h_4. Therefore, his/her working combination is decided. Reflect the result of this, h_5's working time is over eight hours too only by caring elders who can be nursed only to h_5 and we repeat this process.

Table 8. Set of helpers each elders can employ for totally

	e_1	e_2	e_3	e_4	e_5
w_1	$\{h_1, h_2\}$	$\{h_4, h_5\}$	$\{h_1, h_2, h_3\}$	$\{h_4\}$	$\{h_4, h_5\}$
w_2	$\{h_1\}$	$\{h_1, h_2\}$	$\{h_4\}$	$\{h_3\}$	$\{h_4, h_5\}$
w_3	$\{h_4, h_5\}$	$\{h_1\}$	$\{h_4\}$	$\{h_4\}$	$\{h_3\}$

All elders can take care of requested caring except e_1's w_1, e_2's w_2, e_3's w_1 after above selection process(Table 9).

Table 9. Set of helpers each elders can employ

	e_1	e_2	e_3	e_4	e_5
w_1	$\{h_1, h_2\}$	$\{h_5\}$	$\{h_1, h_2, h_3\}$	$\{h_4\}$	$\{h_5\}$
w_2	$\{h_1\}$	$\{h_1, h_2\}$	$\{h_4\}$	$\{h_3\}$	$\{h_5\}$
w_3	$\{h_5\}$	$\{h_1\}$	$\{h_4\}$	$\{h_4\}$	$\{h_3\}$

Table 9 shows us solution has $2 \cdot 2 \cdot 3 = 12$ patterns. We can find optimal solution in searching this patterns.

We calculate the benefits of intermediary. $x\%$ of all of the elder's payment is intermediary's profit. Intermediary pays helpers using the rest of the money as a salary. Therefore, optimal solution maximizes income of Intermediary. Income of Intermediary is determined by below expression.

$$\textbf{Intermediary}'\textbf{sIncome} = \sum_{k=1}^{n} CT(k) - TC \qquad (1)$$

e_1's w_1 takes charge of e_1, e_2's w_2 takes charge of e_1 and e_3's w_1 takes charge of e_2.

In this case, the benefits of intermediary become maximize in 40\$. Therefore,this combination is optimal answer.

In this case, total number of combination case are $5! \cdot 5! \cdot 5! = 1,728,000$ cases. It has possibilities that it cannot find optimal solutions in allowed time because it takes much computation time. When it has large calculation amount in like this case, the full search method takes so much time. Our example uses few elders, helpers and care contents. But it has so many combinations in reality. Therefore, full search algorithm is impractical method in these combinatorial-problem.

However, our proposed method is possible to reduce the computation time compared with full search method. It says optimal solution can be found in allowed time and we can seek the solution faster than full search algorithm. This protocol suggests that it is more useful algorithm than full search algorithm when we choose optimal solution of the problem in Constraint Satisfaction Problem.

Table 10. Optimal combination

	e_1	e_2	e_3	e_4	e_5
w_1	$\{h_1\}$	$\{h_5\}$	$\{h_3\}$	$\{h_4\}$	$\{h_5\}$
w_2	$\{h_1\}$	$\{h_1\}$	$\{h_4\}$	$\{h_3\}$	$\{h_5\}$
w_3	$\{h_5\}$	$\{h_1\}$	$\{h_4\}$	$\{h_4\}$	$\{h_3\}$

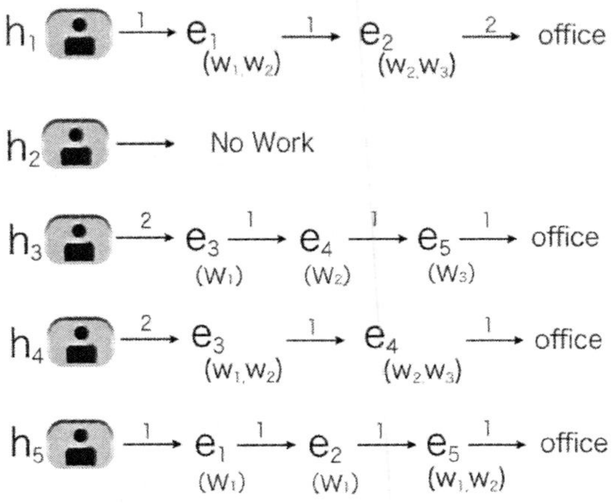

Fig. 2. Each helper's route

5 Discussion

We simulate showing the high speed of our method against full search method.

We generate Q_{ki}, B_{ki}, S_{ji} and P_{ij} with the actual experiment at random. Afterwards, we measure CPU-Time when we pair helper and elder. We run this process 1000 times and take the average. The parameter used for the actual experiment is as follows.

- Number of elders : 10, 20, 30, 40, 50
- Number of helpers : 5, 10
- Our code is implement in C and run on a 1.83 GHz processor core Mac mini with 1GB memory under Mac OS X 10.4.11.

The experiment result is as follows.

Computation time increase very much when number of helper increase in both method. These tables show our method is faster than Full Search method. Because the computing time of our technique is less, our method is effective to solve optimal answer.

If the number of helpers increases any further, it is thought that the optimal solution is not requested in Full search method. Because, it is thought that the number of helpers is more in the reality.

We remove combinations which cannot become optimal answer. After this proceeding, we search all candidates of combination. In the above-mentioned example, we use

Table 11. 5 helpers

elder	10	20	30	40	50
Time of Full Search(second)	0.002305575	0.003001249	0.004195484	0.00624500	0.007173199
Time of Our method(second)	0.000009850	0.000018285	0.00027946	0.00038785	0.00050778

Table 12. 10 helpers

elder	10	20	30	40	50
Time of Full Search(second)	64.814293000	65.999648000	84.438852000	129.235965000	163.957217000
Time of Our method(second)	0.000014801	0.00027276	0.000.40930	0.000055564	0.000071575

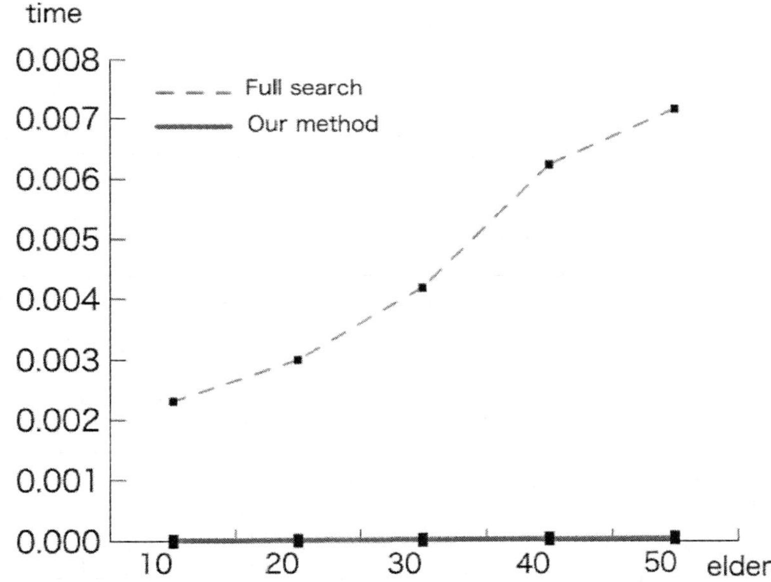

Fig. 3. Result of the experiment

three restraints that monetary, qualitative factors and time. In reality, we have more restraint for combinatorial problem. Even if the restriction increases, our method is useful because the computing time can be shortened by using the method which refines the solution space before searching answer.

In addition, if the set of $B_{ki} \geq P_{ji}$ and the set of $Q_{ki} \leq S_{ji}$ are empty set, the elders have no helper who can employ. The elders compromise his/her request of care in this case. They are three alternative to this problem which are raising his/her budget of care, lower his /her request of helper's skill and he/she given up to receive nursing. In this paper, we solve this problem by using the method that the elder compromises technically. In other words, elders decide the helper who employ it within the range of his/her budget. This thing is similar to the thing that people can shop only within their budget and this is correct in economics.

6 Conclusion

In this paper, we proposed technique for discovering the best combination under many restriction conditions. In the above-mentioned example, we simplified restraint to illustrate. To use our proposed calculation method, elderly person can serve an effective service since business managers can improve their task efficiently. Additionally, the benefits of intermediary become maximum in our method. Time to make it persist in the calculation and the distribution by using our technique can be shortened, and that time in can be spared to other work. Therefore, our method is useful to solve Constraint Satisfaction Problem like at-home nursing service.

Our future work include that extend time interval and add the restriction. As for helpers, the time that hangs at each work is different. Additionally, the time takes to each elderly person's house to go is variously too. We think method of solving this question and develop our method.

Multi-issue Negotiation protocols represent a promising field since most negotiation problems in the real world involve interdependent multiple issues. This protocol is very useful for doing the negotiation with a lot of points under discussion. [5] Constraint Satisfaction Problem like at-home nursing service has various points under discussion. Therefore, it is thought that the technique uses Multi-issue Negotiation is very effective to solve the problem. We innovate this protocol to our method, and develop the technique.

References

1. Yamaguchi, T., Ito, T., Shintani, T.: Implementing a Nurse Scheduling System by Search Method Using Local Optima on VCSP. In: The 22nd Conference Proceedings of Japan Society for Software Science and Technology (2005)
2. Yamaguchi, T., Ozono, T., Ito, T., Shintani, T.: An Implementation of a Nurse Scheduling System with a Team Structure Using CSP. In: The 19th Annual Conference of Japanese Society for Artificial Intelligence (2005)
3. Kanoh, H.: Approximate Alrorithm for Constraint Satisfaction Problems. Journal of JSAI, 359–365 (1997)
4. Nishihara, S.: Fundamentals and Perspectives od Constraint Satisfaction Problems. Journal of JSAI, 351–358 (1997)
5. Fujita, K., Ito, T., Hattori, H., Klein, M.: An Approach to Implementing A Threshold Adjusting Mechanism in Very Complex Negotiations: A Preliminary Result . The Proceedings of The 2nd International Conference on Knowledge, Information and Creativity Support Systems (2007)

Greedy and Exact Algorithms for Invitation Planning in Cancer Screening

Marco Gavanelli[1], Michela Milano[2], Sergio Storari[1], Luca Tagliavini[1],
Paola Baldazzi[3], Marilena Manfredi[3], and Gianfranco Valastro[4]

[1] Department of Engineering - University of Ferrara Via Saragat, 1 – 44100 – Ferrara, Italy
    {marco.gavanelli,sergio.storari,luca.tagliavini}@unife.it
[2] DEIS - University of Bologna Viale Risorgimento, 2 – 40136 – Bologna, Italy
    mmilano@deis.unibo.it
[3] Department of Public Health - Sanitary Agency of Bologna Via Montebello,
    6 – 40136 – Bologna, Italy
    {paola.baldazzi,marilena.manfredi}@ausl.bo.it
[4] NOEMALIFE SpA Via Gobetti, 52 – 40129 – Bologna, Italy
    gvalastro@noemalife.com

Abstract. Cancer screening is a method of preventing cancer by early detecting and treating abnormalities. One of the most critical screening phase is invitation planning since screening resources are limited and there are many people to invite. For this reason, smart resource allocation approaches are needed.

In the paper, we propose and compare two solutions for smart invitation plan definition, one based on greedy approaches and one based on Constraint Programming techniques that enable the definition of the optimal invitation plan.

1 Introduction

Cancer screening is a process finalized to the prevention of the illness from its starting phases. Early diagnosis of tumors is fundamental, because a timely intervention makes the healing easier and reduces the risk of death. In fact, uterus cancer, breast cancer and many other tumors are preventable and fully curable if they are early diagnosed.

We focused on cervical cancer screening, that enables the identification of tumors in the cervix. To reduce the mortality related to this kind of tumor it is necessary to ensure periodically pap-test screening [2] for the entire female population with age between 25 and 64 years.

The screening process is managed by the screening center manager and consists of several phases. First of all, the involved patients (composing the so-called *target population*) are identified, by excluding, e.g., people that have already a cancer and residents in other areas. Given the target population, the next step is to create an invitation plan for the screening examination. This plan should be coherent with the time availability of the centers in which the screening examinations are performed. Once this planning phase is finished, invitations are sent by mail to the target population, and usually patients are visited at the screening center in the scheduled time slots. The screening process then proceeds in different ways, depending on the result of the screening examination.

N.T. Nguyen, R. Katarzyniak (Eds.): New Chall. in Appl. Intel. Tech., SCI 134, pp. 255–264, 2008.
springerlink.com

Among all these phases, the most complex one, from an organizational point of view, is the invitation planning since the number of patients is typically high and the pap-test center resources (time and personnel) are limited. For this reason, we need smart resource allocation approaches, exploiting optimization technology.

Baker and Atherill [1] studied, by means of simulations of queue theory, the order of patients to be invited. The order is then optimized by means of a sort of hill climbing algorithm; the objective is to minimize the dissatisfaction of the patients (modelled as a function of the waiting time), and the server idle time. Other authors [3, 5] used a weighted sum of the patients average waiting time and server idle time. In [1], authors analyzed datasets of pap-test invitations in order to identify probabilistic models of patient attendance and appointment rebooking.

In this paper, we describe the research activity carried on within an industrial project of the Emilia Romagna region of Italy for handling the invitations of cervix cancer screening in the Bologna district. In particular, we explain how invitation plans are currently generated and we propose two solutions to improve the efficiency of the process. The first is based on greedy algorithms: we show two algorithms and the corresponding results. The second is based on Constraint Programming techniques that provide the optimal invitation plan.

Performance evaluations have been conducted on exact and heuristic solutions by means of simulations on different scenarios involving different groups of women and different pap-test center resources.

2   Pap-Test Invitation Management

The definition of the pap-test invitation plan is a very complex task since it involves many women and consequently requires a lot of resources. Involved information include: pap-test center resources, last pap-test examination date, screening history of each woman in the target population, women addresses.

Pap-test center resources are represented by time periods offered each day for the execution of the examinations. These time periods can change each month so each center regularly communicates its monthly agenda to the screening center manager. Usually the time assigned for each pap-test execution is 10 minutes. For this reason, given a pap-test invitation at time T, the next one can be scheduled at time T plus 10 minutes. Moreover, given the pap-test duration, the number of patients that can be invited a day D in a pap-test center is the number of 10 minute slots contained in its available time period.

The last pap-test examination date is important because the next expected pap-test should be performed three years after the last one.

The screening history of a woman is the collection of all the events happened during her screening process (e.g., received/refused invitations, pap-test results). Depending on these events, the woman is classified into three main priority levels: High Priority (HP), Normal Priority (NP) and Low Priority (LP). A woman is classified as High Priority when during her screening history a high risk event has occurred (e.g., if a tumor was found and treated in the last two years). Normal Priority is associated to women that have accepted the last pap-test invitation and results were normal. Low priority is assigned to women who have not accepted the last pap-test invitation. The screening protocol prescribes to track such women and retry the pap-test invitation several times. Statistics show that LP women have very low probability of accepting an invitation: typically less than 30% of the invited women show up. For this reason, overbooking is a common practice: in our instance, the examination duration is reduced down to 3 minutes for LP patients. The assigned priority is one of the most important parameters for the definition of the pap-test invitation plan because usually a fixed percentage of the time-periods available in the pap-test centers is allocated for each priority level.

During the round, centers might be early or late on calls. In the first case, the center can be excluded from the invitation plan. In the second case, overbooking is performed.

The address of a woman is important because she should be invited in the nearest pap-test center in order to increase the probability of showing up.

Given the information described above, the definition of the monthly invitation plan is made in several steps. The screening center manager receives from the pap-test center the availability agenda for the next month expressed in minutes. A list of women to invite is identified by filtering the target population by choosing among the target population only those women whose invitation expires before a certain deadline. The overall time availability is subdivided in slots of 10 minutes each. A percentage of slots is then assigned to each priority level (default percentages are: 50% for High priority, 30 for Normal priority and 20 for Low priority). The manager tries heuristically to match the availability of the resources and the number of patients:

- If the number of slots is much higher than the number of patients to invite, the manager moves the invitation expiration deadline to include as many patients as possible without anticipating too much their invitations.
- If the number of slots is not enough, the manager decides if it is necessary to perform overbooking on some priority classes or postpone some invitations to the next month with a time tolerance.

If a reasonable solution could not be obtained despite the heuristic fixes, the manager contacts the pap-test centers asking for additional time availability.

3   Greedy Approach

The invitation planning activity, shown in the previous section, relies heavily on trial-and-error, is very error prone, and does not guarantee optimality (or even near-to optimality). Its only chances of success stand in the manager's experience.

We developed two greedy algorithms to support the screening center manager in the definition of the pap-test invitation: Priority-Date and Weighted.

The Priority-Date greedy algorithm schedules the women considering two aspects: the expected invitation date and the priority.

Women in the target population are divided into three different lists depending on their priority. Women in the same priority list are then ordered w.r.t. their expected date.

In each day, available slots are subdivided in three groups according to a percentage associated to each priority level (as described in Section 2). Each group represents the maximum number of slots that can be used for each priority level.

For each priority list, women are extracted from the top of the list and assigned to slots reserved for the corresponding priority. If for some priority the allotted time slots in a month exceed the number of patients of the same priority, the remaining slots are assigned to women of lower priorities.

The Weighted greedy algorithm tries to balance the two aforementioned criteria in order to limit the introduced delays and to give importance to high priority classes. In fact, Priority-Date tends to provide extreme solutions, in which high priority classes are scheduled too eagerly, and low priority patients can be given significant delay.

We give to each patient a weight that depends on her associated delay and priority:

$$W = delay(Patient) \cdot p(Patient)$$

where $delay(Patient)$ is a function that returns the delay of the $Patient$ invitation with respect to the expected examination date and $p(Patient)$ is a coefficient associated to the priority level of $Patient$ (the highest the coefficient, the highest the importance given to the delay). Moreover, as in the Priority-Date algorithm, the user can state that in each day some slots are reserved for patients of a specific priority.

The patients are then ordered according to their weights. Given the ordered list, the algorithm starts the assignment from the first day of the month and associates to a slot reserved for a particular priority level the patient of the corresponding priority with the highest weight. The slots non assigned for this priority level are associated to the women with the highest objective function values independently from their priorities.

3.1   Experiments on Greedy Algorithms

In order to test the proposed algorithms and highlight their pros and cons, we set up a simulation with very difficult conditions (more women to invite than the available time). The instance spans over 5 months, and involves 2400 women with expected invitation dates randomly generated with uniform distribution. Out of the 2400 women, 1150 were given low priority, 950 normal and 300 high. The pap-test center has a daily time availability of 50 minutes (5 pap-test examinations of 10 minutes or 16 if we consider overbooking with 3 minutes for each examination), 7 days a week.

As shown in Table 1, the Priority-Date algorithm, configured with default parameters (50% of time for high priority, 30% for normal priority and 20% for low priority), gives too much importance to the high priority women introducing significant delays

Table 1. Max number of delay/anticipation days

Algorithm	Priority	Max Anticipation	Max Delay
Priority-Date	HP	22	1
Weighted	HP	0	16
Priority-Date	NP	5	9
Weighted	NP	0	16
Priority-Date	LP	0	75
Weighted	LP	2	52

for the low priority women (up to 75 days of delay). The introduction of an objective function in the weighted greedy algorithm represents an evolution of the Priority-Date one, capable of reducing the delays for low priority women (up to 52) as shown in Table 1. It also introduces, for each day, a better allocation of the available slots by balancing priorities and delays in the objective function.

The problem of this greedy algorithm is that it cannot identify an optimal invitation plan as it only discovers local optima. Consider for instance a day in which low priority patients are subject to overbooking, and a free slot of 10 minutes. We can accommodate either 1 high priority woman with a time delay of one day or 3 low priority women with a time delay of one day each. Consider to assign to the high, normal and low priority levels respectively a weight of 10, 7 and 4 in the objective function. The algorithm orders patients according to their weight: first the high priority woman whose weight is $10 \cdot 1 = 10$, then the three low priority patients whose weight is $4 \cdot 1 = 4$ each. Indeed, even if the delay of low priority women rises up to two days they are still ordered after the high priority woman. The weighted algorithm then selects the first patient in the list, assigns the slot of 10 minutes to the high priority woman, thus delaying the three low priority women of one day. This solution costs $10 + 4 \cdot 2 + 4 \cdot 2 + 4 \cdot 2 = 34$.

Looking globally to our list, we observe that the one generated is not the optimal solution as reserving the 10 minutes slot for inviting the 3 low priority women and delaying the high priority invitation of one day has a lower cost $10 \cdot 2 + 4 + 4 + 4 = 32$.

For this reason we used artificial intelligence techniques and Constraint Programming for identifying the optimal invitation plan (the plan that has the lowest sum of all the woman objective function values). This approach is described in details in Section 4.

4   Constraint Programming

The greedy algorithms presented in Section 3 provide reasonable solutions in a very short time. The generated appointment schedules were submitted to the final users, that deemed them acceptable. However, due to the combinatorial nature of the problem, a greedy algorithm in general does not provide the optimal solution, and it never proves optimality.

We decided to experiment with optimization algorithms, in order to find the optimal solution, and to compare the quality of the solutions given by optimal and greedy algorithms. The aim was to evaluate the viability of an Artificial Intelligence module, exploiting a complete algorithm, in the appointment scheduling application.

Constraint Programming (CP) languages are devoted explicitly to the solution of hard combinatorial problems. Initially born as a rib of Logic Programming, CP was then extended also to the object-oriented paradigm. Modern CP languages contain libraries and solvers for different domains. Popular instances are CP(FD), in which the unknowns range on Finite Domains, and CP($\mathcal{R}$), in which variables range on the set of real numbers. The corresponding solvers are based on tree search enriched with propagation algorithms reaching Arc-Consistency (and its generalizations) for the FD domain, and on (Integer) Linear Programming for the domain of the reals.

We first experimented the viability of a CP(FD) model, but it did not provide optimal solutions in reasonable time. We then applied a CP($\mathcal{R}$) solution, exploiting an integer linear programming model, that opens the way to efficient solvers based on linear programming enriched with a branch and bound strategy.

4.1   CP($\mathcal{R}$) Model

At a first sight, one could think to associate a decision variable AD_i, representing the appointment date, to each patient. Unluckily, the number of patients could be large, and many of them share same category and expected date, so the search space can contain an exponential number of symmetric solutions obtained by permuting patients with same features, that gives a well known combinatorial explosion of the search space [4]. Symmetric solutions can be pruned by adding the constraints $AD_i \leq AD_j$ whenever $i \leq j$. However, the number of variables is still very large. Therefore, we decided to classify the patients into groups, each group being identified by a expected date and a category, and associate a variable to each group.

Suppose we have ng groups and nd days. For each group of patients g, and for each possible invitation day id, we define a positive decision variable $I_{g,id} \geq 0$, representing the number of patients from group g invited in day id.

For each decision variable there is a *cost* associated to such assignment. For the group of patients g the cost depends on the category and on the introduced delay with respect to the expected day $ed(g)$. Categories with higher priorities will contribute with a higher cost than low-priority categories. The cost depends on the delay through a nonlinear function. If the invitation date coincides with the expected date, the cost is zero; the same holds if the invitation date is before the expected date, provided that the anticipation is limited: there exists a parameter α defining the maximal number of days a patient can be called in advance. The protocol required delays not to be higher than 40 days; we defined a parameter δ (that defaults to 40). Delays superior to δ or patients called more than α days before their expected date contribute to the total with a very high cost M. A delay between 0 and δ contributes with a cost proportional to the number of days of delay, multiplied to the priority coefficient $p(g)$ of the group g. The objective function is then:

$$min \sum_{g=1}^{ng} \sum_{id=1}^{nd} I_{g,id} cost(g, id - ed(g))$$

were the cost is defined as

$$cost(g, d) = \begin{cases} 0 & if \ \alpha \leq d \leq 0 \\ p(g) \cdot d & if \ 0 < d \leq \delta \\ M & if \ d < \alpha \vee d > \delta \end{cases}$$

The constraints (1) impose that the total capacity of the day is not exceeded. *capacity* is the total number of minutes available for visits in a given day; *duration* is the duration of a visit, and it depends on the category and on the day (which enables the user to define detailed policies for overbooking, varying the visit durations).

$$\sum_{g=1}^{ng} I_{g,id} \, duration(g, id) \leq capacity(id) \quad \forall id \in 1..nd \tag{1}$$

Note that an instance could be infeasible if the number of days is not enough to accommodate all patients; in such a case a constraint solver does not provide a solution, but simply returns failure. To provide the manager a reasonable answer also in this case, we avoid infeasibility by introducing an additional day with unlimited capacity and with a high cost M to accommodate all patients.

Each patient should be invited exactly once, stated as constraint (2), where $|g|$ is the number of patients belonging to the group g.

$$\sum_{id=1}^{nd} I_{g,id} = |g| \quad \forall g \in 1..ng \tag{2}$$

Finally, the percentage of time devoted to visiting patients of each category should be respected. Actually, in order to fully exploit the power of the optimizer, the problem should not be too constrained (otherwise, if there are no freedom degrees, the optimal solution boils down to the same solution given by a greedy algorithm). We decided to guide the optimization process toward the specifications of the final user as follows. We ask the user to impose a capacity per day per group of patients, $CapPerc(c, id)$. We check whether the total allotted time for each category is enough for visiting all the patients in that category. If the allotted time is enough, we impose that in each day the number of patients of category c is at most the one specified by the user (3). Otherwise (if the total time is not enough for that category), for each day we impose that the number of patients of category c is at least the one prescribed by the user (4).

$$\forall id \in 1..nd, \forall c \in 1..nc \ s.t.$$
$$\sum_{id'=1}^{nd} CapPerc(c, id') < \sum_{id''=1}^{nd} duration(c, id'') \implies \tag{3}$$
$$\sum_{g=1}^{ngc(c)} I_{g,id} duration(g, id) \geq CapPerc(c, id)$$

$$\forall id \in 1..nd, \forall c \in 1..nc \ s.t.$$
$$\sum_{id'=1}^{nd} CapPerc(c, id') > \sum_{id''=1}^{nd} duration(c, id'') \implies \tag{4}$$
$$\sum_{g=1}^{ngc(c)} I_{g,id} duration(g, id) \leq CapPerc(c, id)$$

The model consisting of the objective function, constraints (1), (2), (3), (4) and the integrality constraint for each variable $I_{g,id}$ is solved through branch and bound exploiting a linear relaxation for bound computation. The branch and bound algorithm solves the problem to optimality and proves the solution is optimal.

5   Experiments

We selected a series of experiments to compare the quality and the runtime of the greedy algorithms with respect to the use of the CP($\mathcal{R}$) solver. In the experiments, we used an instance with 204 patients to be scheduled in a period of one month, with random expected day. The patients are divided into three categories: 28 patients HP, 74 NP and 104 LP. The visiting time is 10 minutes without overbooking, while it is reduced to 3 minutes in case of overbooking (only for LP patients). The availability of the screening centre is 50 minutes per day, which is not enough to visit all the patients without overbooking, thus some of the patients have to be be moved to the following month.

In Figure 1 we show the distribution of the difference between expected day and invitation day for each of the categories for the weighted greedy algorithm detailed in

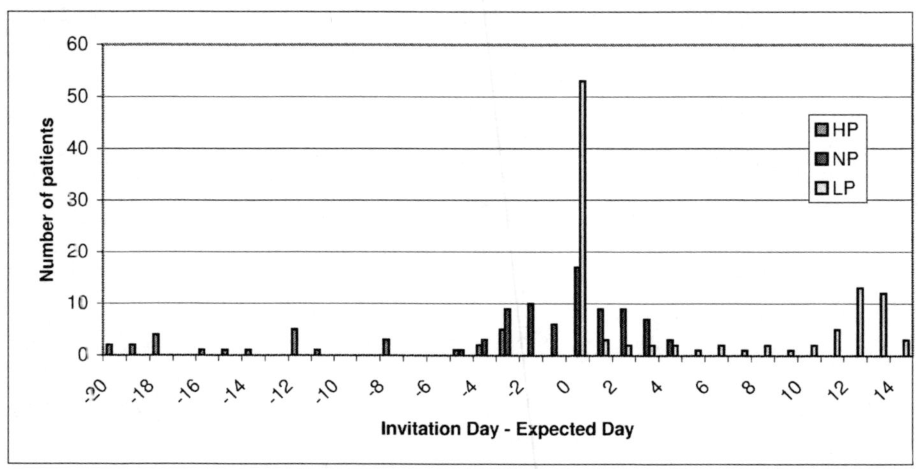

Fig. 1. Distribution of the patients: weighted greedy algorithm

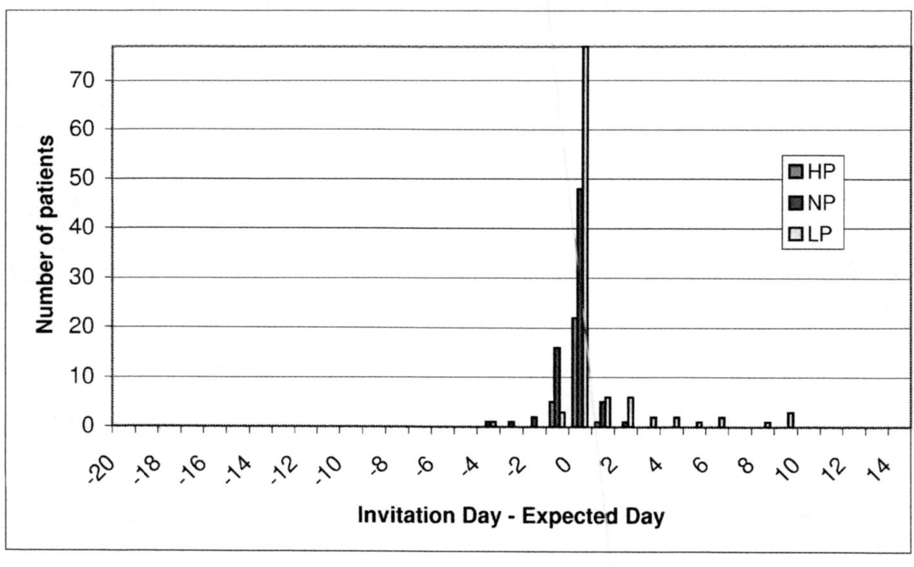

Fig. 2. Distribution of the patients: optimal solution

Section 3. In abscissa we represent the difference expected day - invitation day, i.e., negative numbers represents anticipation with respect to the optimal invitation date, while positive numbers represent delay. In ordinate, we have the number of patients (for each category) that has such an anticipation/delay. The algorithm gives high priority to high risk patients, which are anticipated, with respect to their ideal date, up to 20 days. Correspondingly, delays are introduced for lower priority patients. This shows that there

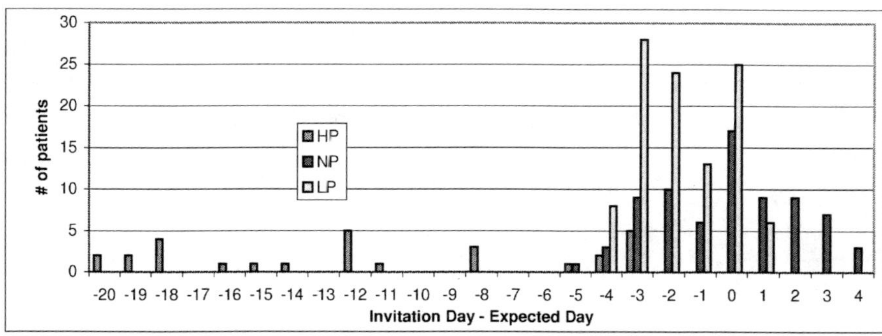

Fig. 3. Distribution of the patients: weighted greedy algorithm with overbooking

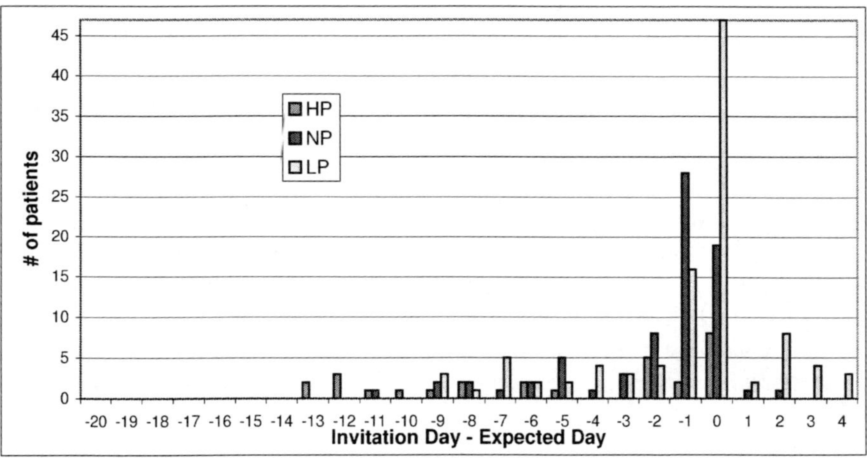

Fig. 4. Distribution of the patients: optimal solution with overbooking

is room for improvement: intuitively, some of the early patients could be swapped with patients that are delayed.

Figure 2 shows the distribution in the optimal solution. Both anticipations and delays are drastically reduced: no patient is anticipated more than 4 days or delayed more than 9 days. The values of objective function in the two situations synthesize the same information visually presented in the graph: the greedy solution has cost 2037, while the optimal cost is almost an order of magnitude better: 325.

The same can be said in the case with overbooking, as shown in Figures 3 and 4. The corresponding costs are 558 for the greedy solution and 153 for the optimal one.

We used ILOG CPLEX 9.0 as solver; it was able to find the optimal solution in a very small time on an Intel Celeron CPU 2.4 GHz, 512MB RAM computer. In order to test the scalability of the algorithm, we experimented with a higher number of patients, up to 20,000. The algorithm scales very well: all the instances were solvable within one minute, which is by far acceptable for an algorithm that is run once every month. The scalability can be easily explained: the unknowns in our models do not depend directly on the number of patients, which can be large, but the number of groups, that cannot grow beyond the number of possible days multiplied by the number of categories.

6   Conclusions

In the paper we have proposed greedy and exact algorithms for the invitation plan generation for cancer screening.

Invitation plans, generated during experiments performed with different patient and resource configurations, were submitted to the final users, that deemed them acceptable, in any case better than the current hand-generated plans.

Clearly, the choice between a greedy and an optimal algorithm should take into account issues related to scalability, efficiency and solution quality. Small instances (up to hundreds of patients to be scheduled in a month time horizon) can be effectively solved via the exact approach proposed in this paper. When the time horizon raises up to several months we can either face the overall instance with a greedy approach or we can decompose it by dividing the time horizon in monthly slices and solve each sub-instance with the exact algorithm.

Acknowledgments. This work has been partially funded by the SPRING (Screening PRotocol INtelligent Government) project, partially financed by Emilia-Romagna Region (Italy) under (PRRIITT 3.1.A), and by the MIUR PRIN projects n. 2005-015491 and n. 2005-011293. We would like to thank Dr. Natalina Collina of the Sanitary Agency of Bologna for her contribution on the description of the screening process. We also thank Evelina Lamma and Paola Mello for their useful suggestions.

References

1. Baker, R., Atherill, P.: Improving appointment scheduling for medical screening. IMA Journal of Management Mathematics 13, 225–243 (2002)
2. Cervical cancer screening in the Emilia Romagna region of Italy,
   `http://www.regione.emilia-romagna.it/screening/`
3. Ho, C.J., Lau, H.S.: Minimising total cost in scheduling outpatient appointments. Management Sci. 38, 1750–1764 (1992)
4. Puget, J.F.: On the satisfiability of symmetrical constraint satisfaction problems. In: Komorowski, J., Raś, Z.W. (eds.) ISMIS 1993. LNCS, vol. 689, pp. 350–361. Springer, Heidelberg (1993)
5. Rohleder, T.R., Klassen, K.J.: Using client-variance information to improve dynamic appointment scheduling performance. Omega 28(10), 293–302 (2000)

A Comparison of Three Meta-heuristics for a Closed-Loop Layout Problem with Unequal-Sized Facilities

Hadi Panahi, Masoud Rabbani, and Reza Tavakkoli-Moghaddam

Department of Industrial Engineering, Faculty of Engineering,
University of Tehran, P.O. Box: 11155/4563, Tehran, Iran
{hadipanahi,mrabani,tavakoli}@ut.ac.ir

Abstract. This paper presents a novel mathematical model of a closed-loop layout problem with unequal-sized facilities. This problem belongs to a class of combinatorial optimization and NP-hard problems. Obtaining an optimal solution for this complex, large-sized problem in reasonable computational time by using traditional approaches and is extremely difficult. Therefore, we propose three well-known meta-heuristics, namely genetic algorithm (GA), ant colony optimization (ACO), and simulated annealing (SA), to solve the closed-loop layout problem. These algorithms report near-optimal and promising solutions in a short period of time because of their efficiency. The computational results obtained by these algorithms are compared with the results reported by the Lingo 8.0 software package. Finally among our three proposed meta-heuristics, the output of SA is better than other two algorithms and the Lingo.

Keywords: Closed-loop layout problem, unequal-sized facilities, Genetic algorithm, Ant colony optimization, Simulated annealing.

1 Introduction

Pattern of the flow of materials may have different shapes. But it must have conformity with one of the general flow patterns. Some of general flow patterns are: straight line, U shape, closed loop, zigzag and undefined. Straight line pattern is used when production is simple or short or number of the products components is few or number of the machines is few. When general handling systems are in one side of the factory or it is necessary to use common facilities at the first and the last step of production, so that the product should be returned to the starting point, U shaped pattern is chosen. Closed loop pattern is usually applied when the product is needed to be returned exactly to the starting point or send and receive units must be at same place or it is needed to use a special machine twice. Zigzag pattern is used when production line is long due to available space. Undefined pattern is used when the goal is to establish shortest path between related units and maximum use of the plant area [1].

Existence of an efficient material flow results in an increase in production efficiency, better use of plant area, simplification of a material handling system,

N.T. Nguyen, R. Katarzyniak (Eds.): New Chall. in Appl. Intel. Tech., SCI 134, pp. 265–278, 2008.
springerlink.com   © Springer-Verlag Berlin Heidelberg 2008

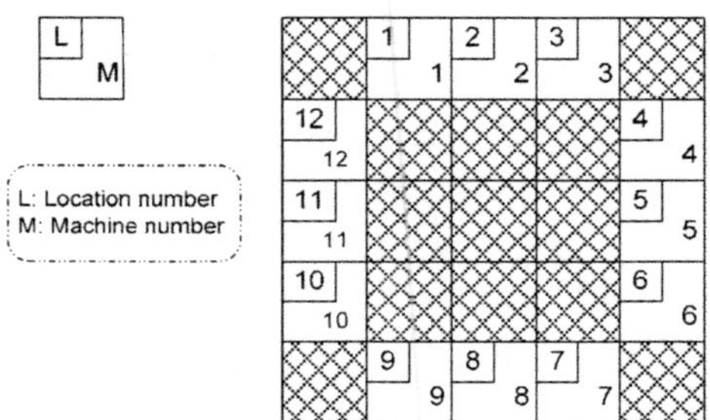

Fig. 1. Typical closed loop layout

decrease of the production time, decrease of work-in-process inventories, and the like [1]. In this paper, we focus on the closed-loop flow pattern. Fig. 1 shows an illustration of a layout based on closed loop flow pattern.

This figure depicts a layout based on closed-loop flow pattern. Facilities (e.g., machines, departments, machining centers, etc.) cannot be assigned to the hatched locations. In each square, the number in the right-down corner represents the number of the facility assigned to the related location that is surely not fixed and the number on the left-up corner represents the location number that is always fixed. The problem considered in this paper is to find the optimal assignment of facilities to the locations by minimizing the material handling function between facilities. This belongs to a class of non-deterministic polynomial hard (NP-hard) problems. It must be emphasized that sizes of facilities are unequal. In other words, it is assumed that facilities are rectangular with unequal sizes.

Exact methods, such as a branch-and-bound procedure, have been applied to solve a closed-loop unidirectional cyclic and bidirectional circular layout problems [2,3]. A number of meta-heuristic algorithms were proposed to solve loop layout design problem [4,5]. Bennell1, Potts, Whitehead [6] proposed a local search for a loop layout problem.

It must be emphasized that the assumptions of proposed mathematical model are different from other previous works, so for evaluation, we can not use the computational results of previous works. In this paper, we propose meta-heuristic methods based on genetic algorithms (GA) combined by a local search technique, namely 2-Opt, and ant colony optimization (ACO) and simulated Annealing to solve the given problem. To show the efficiency of the proposed methods, the experimental results obtained by these algorithms are compared with the results reported by the Lingo 8.0 software package in terms of the objective function value (OFV) and computational time. We conclude that the one of the algorithms (SA) reaches to the optimum solution between other algorithms and even Lingo and the processing time is reduced saliently.

2   Mathematical Model

Fig. 2 illustrates the alteration facilities centers during each given layout.

As shown in Fig. 2, centers of the facilities are changing in every given layout. So the distance between facilities is changing due to 2 factors: a) location of the facility

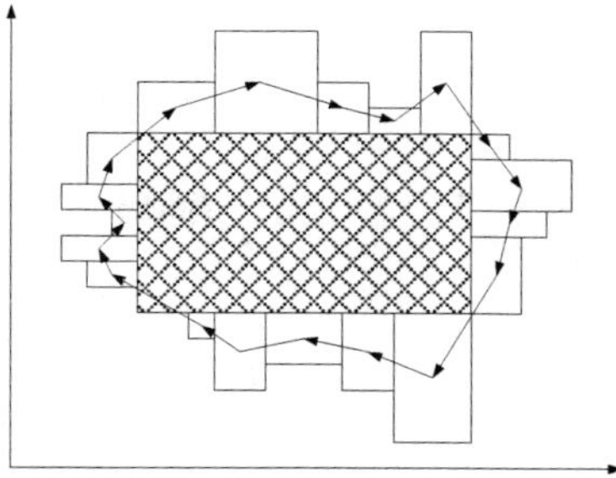

Fig. 2. Alteration of facilities center's

b) size of the facility. It is necessary to notice that as it can be seen from Fig. 1, we have assumed that the shape of the total layout is square like shape and facilities gather around it. It may be asked that is square shape is appropriate for this problem or not because for solving this problem it is assumed that the number of facilities on parallel sides is equal. In order to answer this question we answer in this way:

Every natural number which is representative of the number of facilities it can be encountered as below:

$$n = \begin{cases} (I): 4k \\ (II): 4k+1 \\ (III): 4k+2 \\ (IV): 4k+3 \end{cases}$$

In case (I), there is no problem because there is the same number of the facilities on each side.

In case (III), we will have 2 sides with k facilities and 2 sides with k+1 facility, so a rectangular shape is appropriate for this case.

In cases (II) and (IV), we apply a dummy facility in order to transform them to case (II) or (IV). Therefore in the frequency matrix, there will be a column and a row with zero values because of the dummy facility.

The parameters used in this paper are as below:

$f(i,j)$: Frequency between facilities i and j.

$C(i,j)$: Cost of transportation between two facilities which in this paper is assumed to be $[1]_{n \times n}$.

The variable notations used in this paper are as below:

$l(i,1)$: Linear position of center of facility i.
$l(i,2)$: Phenomenal position of center of facility i.
$d(i,1)$: Length of facility i.
$d(i,2)$: Width of facility i.

$$x(i,j) = \begin{cases} 1 & \text{If facility } i \text{ is assigned to location } j; \\ 0 & \text{Otherwise.} \end{cases}$$

$l(i,j)$: Free variable. $i=1,2,\dots,n$ $j=1,2$

$$\text{Min } Z = \sum_{i=1}^{n}\sum_{j=1}^{n}\left(c\,(i,j)\times f\,(i,j)\right)\times\left(\left|l\,(i,1)-l\,(j,1)\right|+\left|l\,(i,2)-l\,(j,2)\right|\right) \tag{1}$$

$$\sum_{i=1}^{n} x(i,j)=1 \qquad j=1,2,\dots,n \tag{2}$$

$$\sum_{i=1}^{n} x(i,j)=1 \qquad i=1,2,\dots,n \tag{3}$$

$$l(1,1)=\sum_{i=1}^{n}\left(x(i,1)\times d(i,1)\right)/2\,. \tag{4}$$

$$l(1,2)=\sum_{i=1}^{n}\left(x(i,2)\times d(i,2)\right)/2\,. \tag{5}$$

$$l(i,1)=\sum_{j=1}^{n}\left(x(j,i)\times d(j,1)\right)/2+\sum_{j=1}^{n}x(j,(i-1))\times d(j,1)/2+l(i-1,1) \quad 1\prec i\prec\frac{n}{4}+1. \tag{6}$$

$$l(i,2)=\sum_{j=1}^{n}x(j,i)\times d(j,2) \qquad 1\prec i\prec\frac{n}{4}+1. \tag{7}$$

$$l(n/4+1,1)=\sum_{i=1}^{n}(x(i,n/4+1)\times d(i,2))/2+\sum_{i=1}^{n}x(i,n/4)\times d(i,1)+l(n/4,1)+l(n/4,1) \tag{8}$$

$$l(n/4+1,2)=-\sum_{i=1}^{n}(x(i,n/4+1)\times d(i,1))/2\,. \tag{9}$$

$$l(i,1)=\sum_{j=1}^{n}(x(j,i)\times d(j,2))/2+\sum_{j=1}^{n}(x(j,n/4)\times x(j,1))/2+l(n/4,1) \qquad \frac{n}{4}+1\prec i\prec\frac{n}{2}+1\,. \tag{10}$$

$$l(i,2)=-\sum_{j=1}^{n}(x(j,i)\times d(j,2))/2-\sum_{j=1}^{n}(x(j,i-1)\times d(j,1))/2+l(i-1,2) \qquad \frac{n}{4}+1\prec i\prec\frac{n}{2}+1. \tag{11}$$

$$l(n/2+1,1)=\sum_{j=1}^{n}(x(j,n/4)\times d(j,1))/2-\sum_{j=1}^{n}(x(j,n/2+1)\times d(j,1)) \tag{12}$$

$$l(n/2+1,2)=-\sum_{j=1}^{n}(x(j,n/2)\times d(j,2))/2-\sum_{j=1}^{n}(x(j,n/2+1)\times d(j,n/2))/2+l(n/2,2) \tag{13}$$

$$l(i,1) = -\sum_{j=1}^{n}(x(j,i)\times d(j,1))/2 - \sum_{j=1}^{n}(x(j,i-1)\times d(j,1))/2 + l(i-1,1) \qquad \frac{n}{2}+1 \prec i \prec 3\times\frac{n}{4}+1 \quad . \tag{14}$$

$$l(i,2) = -\sum_{j=1}^{n}(x(j,i)\times d(j,2))/2 - \sum_{j=1}^{n}(x(j,n/2)\times d(j,2))/2 + l(n/2,2) \qquad \frac{n}{2}+1 \prec i \prec 3\times\frac{n}{4}+1. \tag{15}$$

$$l(3\times n/4+1,1) = -\sum_{j=1}^{n}(x(j,3\times n/4+1)\times d(j,2)). \tag{16}$$

$$l(3\times n/4+1,2) = -\sum_{j=1}^{n}(x(j,n/2)\times d(j,2))/2 + l(n/2,2) + l(3\times n/4+1,1) \quad . \tag{17}$$

$$l(i,1) = -\sum_{j=1}^{n}(x(j,i)\times d(j,2))/2 \qquad 3\times\frac{n}{4}+1 \prec i \prec n+1 \quad . \tag{18}$$

$$l(i,2) = \sum_{j=1}^{n}(x(j,i-1)\times d(j,1))/2 + \sum_{j=1}^{n}(x(j,i-1)\times d(j,1))/2 + l(i-1,2) \qquad 3\times\frac{n}{4}+1 \prec i \prec n+1. \tag{19}$$

Eq. (1) is the objective function of the closed-loop layout problem. For simplification, we assume that C matrix is a constant matrix, in which all arrays are 1. Eq. (2) guarantees that each facility can be assigned to just one position. Eq. (3) ensures that only one facility can be assigned to each location. Eqs. (4) and (5) determine the linear and phenomenal position of the center of a facility assigned to the first position of the first side. Eqs. (6) and (7) determine the linear and phenomenal position of the center of facilities assigned to the other positions of the first side.

Eqs. (8) and (9) determine the linear and phenomenal position of the center of a facility assigned to the first position of the second side. Eqs. (10) and (11) compute the linear and phenomenal position of the center of facilities assigned to the other positions of the second side. Eqs. (12) and (13) compute the linear and phenomenal position of the center of a facility assigned to the first position of the third side. Eqs. (14) and (15) determine the linear and phenomenal position of the center of facilities assigned to the other positions of the third side. Eqs. (16) and (17) determine the linear and phenomenal position of the center of a facility assigned to the first position of the forth side. Eq. (18) and (19) compute the linear and phenomenal position of the center of facilities assigned to the other positions of the forth side.

3   Genetic Algorithm

Genetic algorithms are known as evolutionary algorithms and stochastic search techniques [7]. It explores the solution space by the aid of concepts which are derived from natural genetics and evolution theory [8-10]. In recent years GA is applied for solving layout problems as a meta-heuristic algorithm [11-15]. The steps of genetic algorithm which is used in this article are as follows:

- Generating first population randomly.
- Evaluation of the population and calculation of objective function for every solution.
- Local search with 2-opt.
- Selection with Roulette wheel.
- Crossover.
- Inserting elitist parents in the new population.
- Mutation.
- Check for a stop Criterion or go to step 2.

3.1 String Representation

There are many representations for formulating this problem, two major representations are described in this paper and compared with each other.

- Due to Fig. 1, if we want to show blank spaces in the chromosome, therefore a sample chromosome will be the same as Fig. 3.

b	5	b	6	b	9	b	b	10	b	b	11	8
b	b	7	2	b	b	1	b	3	4	12	b	

Fig. 3. A possible representation of the chromosome

"b" represents blank spaces. Due to this representation we would get that by running GA many infeasible solutions would be created and just a few percent of each generation would be feasible. But a good point of this representation is diversified search on the solution space. So we if we choose this method we may have to put a correction step (which would very complicated for this special problem) or not to use correction step and just give more time to GA in order to search solution space.

- If we do not show blank spaces so a sample chromosome for mentioned representation will be the same as Fig. 4.

7	4	9	10	3	6	2	1	8	11	12	5

Fig. 4. Representation of the selected chromosome

The benefit of this representation is that just feasible solutions are generated. In this paper, second representation is applied.

3.2   Selection Method

In this paper, a Roulette wheel method is applied for the selection phase.

3.3   Crossover Operator

Due to the type of the chromosome, Uniform order-based crossover is applied. Two parents are selected at random and then a pattern is created at random too. At first offspring 1 is filled in the way that in every place where there is 1 in the pattern, the content of the parent 1 is taken, so up to now first offspring is not filled completely, for rest of the alleles wherever there was zero in the pattern, they are filled in the in the order which they have appeared in the second allele. Crossover operator rate is set between 0.9-0.95.

3.4   Mutation Operator

The mutation operator which is used in this paper is swapping which selects a parent at random and selects 2 alleles of it at random and swaps the content of them. Mutation operator rate is set to 0.01.

3.5   Stop Criterion

Stop criterion is running specific numbers of generation.

3.6   Local Search

In this paper, a local search, based on a 2-Opt algorithm, is applied to improve the chromosomes created in the first generation (a pool of chromosomes, namely, population). Using this algorithm for large-sized problems is very time consuming, so an upper limit is a number of replacements to reduce the computational time.

4   Ant Colony Optimization

The ACO algorithm was first developed by Dorigo et.al [16]. This algorithm is a multi-agent meta-heuristic algorithm, which is broadly applied in combinatorial optimization and other problems [16]. The ACO simulates the way real ants find the shortest way between a food source and their nest. Ants communicate with each other by pheromone and exchange information about selecting the best path. The path which is chosen by more ants gains more chance to be chosen by other ants and if they choose that path, they increase the probability of that path to be chosen by next ants.

String solutions are empty at the first iteration and a pheromone matrix is initialized to the equal values. This matrix is updated considering the objective function value for each agent. Description of the progress in the current iteration is as follows:

Solutions are made with consideration to the information produced by the pheromone matrix updated at the end of latter iteration. An agent has two choices to allocate each facility to each place in order to build solutions of each iteration:

(i) Place with the maximum pheromone concentration is chosen with the probability of q_0, a predefined probability number, this process is known as exploitation and/or,

(ii) One of the places using a stochastic distribution with a probability $(1-q_0)$, denoted as p_{ij}. This process is known as biased exploration [17].

The above processes work together simultaneously. For each facility random numbers using uniform distribution is generated, if the random number was less than q_0 then first procedure is taken, else second procedure is taken. If the first procedure is taken, then the facility or better say the appropriate element of the solution string is assigned to the place with the highest pheromone concentration due to the pheromone matrix. But if the second procedure is taken, any one of the places is chosen with a normalized pheromone probability (pheromone probability normalized to 1) calculated as below:

$$p_{ij} = \frac{\tau_{ij}}{\sum_{k=1}^{k} \tau_{ik}}, / \quad j = 1,...,k \cdot \tag{20}$$

where, p_{ij} is the normalized pheromone probability for facility i belongs to location (place) j. The assessment of each solution is measured in terms of the value of objective function. The objective function is the same as GA. Like other meta-heuristics, ACO employs local search procedures with a hope to achieve better solutions [17,18].

Usually local search is applied on a few percent of the total population; applying local search on all of the total population in each iteration may consume a lot of time. In this paper, local search is applied on a predefined percent of best solutions. There are various algorithms for local search. In this paper, a swapping procedure is applied for local search. After applying local search it must be checked if it has improved the current solution or not. If it had improved the solution then local search is accepted else local search is rejected. Then it is time to update the pheromone matrix. This updating helps the next generation of ants to choose best way. We employ the best L agents from R agents; they simulate the pheromone trail splashing by assigning a number τ_{ij} considering each solution's specifications.

Pheromone trails are updated by using Eq. (21).

$$\tau_{ij}(t+1) = (1-\rho)\tau_{ij}(t) + \sum_{l=1}^{L} \Delta\tau_{ij}^{l} \quad i = 1,...,N \quad j = 1,...,K \cdot \tag{21}$$

where, ρ is the steadfastness of trail and therefore $(1-\rho)$ is the evaporation rate. Higher value of ρ means that the information achieved in the past iterations is forgotten faster and $\Delta\tau_{ij}^{l}$ is calculated below:

$$\Delta\tau_{ij}^{l} = \begin{cases} 1/F_l & \text{If facility } j \text{ is assigned to place } i. \\ 0 & \text{Otherwise.} \end{cases} \tag{22}$$

4.1 String Representation

The string representation of the ACO algorithm is as the same as the GA that is used in this paper. There are three main steps at any iteration:

- Generation of new R solutions via the last updated pheromone trail matrix.
- Executing local search procedure.
- Updating pheromone trail matrix.

The algorithm will perform above steps till a predefined number of iterations is met. The flowchart of the proposed ACO algorithm is shown in Fig. 5.

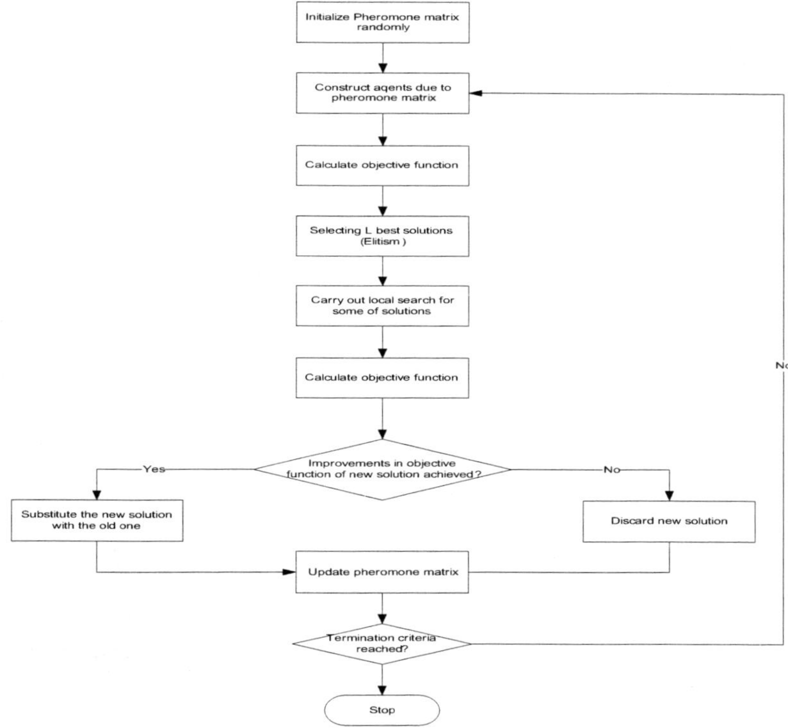

Fig. 5. Flowchart of ACO algorithm

5 Simulated Annealing

Simulated annealing algorithm was first introduced by Kirkpatrick et al. [19], in which it is used to solve large, combinatorial optimization problems [20]. It has the capability of jumping out of the local optima for global optimization. This capability is achieved by accepting with probability neighboring solutions worse than the current solution. The acceptance probability is determined by a control parameter (*Temperature*), which decreases during the SA procedure [21].

The main steps of the proposed SA algorithm used in this paper are as follows:

Initializing T
Initializing p
Initializing ΔT
While (not(end))
$$q = Neighbor(p)$$
$$\Delta C = C(q) - c(p)$$
$$if \ \Delta C \prec 0 \ then \ p = q$$
$$else \ if \ rand(0,1) \prec e^{-\frac{\Delta C}{T}} \ then \ p = q$$
$$T = \alpha(T)$$
$$if \ T \prec \Delta T \ then \ end(while)$$
$$repeat \ (while)$$

String representation
The string representation of SA is as the same as GA used in this paper.

Neighborhood generation
The strategy, which is used in this paper to generate a neighbor point, is swapping.

Cooling Temperature
The coefficient α is constant so temperature is cooled by a linear equation.

6   Computational Results

Three proposed meta-heurisitcs are coded and programmed in MATLAB 7 language and all experiments are performed on a Pentium IV 1.66MHZ. The test problems generated randomly are classified to small, medium, and large sizes. These test problems are illustrated in Table 1. The computational results obtained by the above-mentioned methods are shown in Tables 2, 3, and 4.

Table 1. Test problems

Size	n	Frequency	Dimension
Small	20	U[1,10]	U[1,10]
Small	28	U[1,10]	U[1,10]
Medium	48	U[1,10]	U[1,20]
Medium	60	U[1,10]	U[1,20]
Large	80	U[1,20]	U[1,10]
Large	100	U[1,20]	U[1,10]

Table 2. Results of GA and Lingo for a closed-loop layout problem

Facility number	Population	Generation	Objective Function Value		Error (%)	Time (s,)	
			Proposed GA	Lingo		Proposed GA	Lingo
20	20	20	46089	44324	3.98	0.219	3000
20	20	40	45653	44324	2.99	0.515	3000
20	20	100	44722	44324	0.89	1.125	3000
20	20	5000	43947	44324	-0.85	45	3000
20	40	20	44846	44324	1.17	0.469	3000
20	40	40	44525	44324	0.51	0.984	3000
20	40	100	44553	44324	0.45	2.187	3000
28	40	40	117620	110442	6.49	12.45	3600
28	40	1000	1.16E+05	110442	5.15	50.57	3600
28	60	1000	117960	110442	6.80	76.50	3600
28	60	2000	117140	110442	6.06	136.1	3600
28	60	4000	116012	110442	5.04	180	3600
28	80	500	117908	110442	6.76	64.07	3600
28	90	500	116490	110442	5.47	180	3600
40	100	500	693645	654179	6.03	117.1	3600
40	100	1000	684690	654179	4.66	237.4	3600
40	200	1000	688790	654179	5.29	604.8	3600
60	60	100	3013400	2812540	7.14	60.92	3600
60	60	200	3064800	2812540	8.96	122.4	3600
60	60	500	3043735	2812540	8.22	312.5	3600
60	80	100	2984034	2812540	6.09	74.32	3600
60	100	100	3013136	2812540	7.13	99.383	3600
60	100	200	3067768	2812540	9.07	190.5	3600
80	80	50	7.53E+06	6926210	8.66	40.25	3600
80	240	30	7.46E+06	6926210	7.72	121.3	3600
80	240	200	7.45E+06	6926210	7.58	602.3	3600
100	100	100	14248540	12965200	9.89	308.8	3600
100	120	100	13995000	12965200	7.94	387.8	3600

Table 3. Results of ACO & Lingo for a closed-loop layout problem

n	r	q	Elitist agents	ρ	Objective Function Value		Error (%)	Time (s)		Runs	pls
					ACO	Lingo		ACO	Lingo		
20	20	0.98	5	0.01	46342	44324	4.552838	18.4530	3000	700	0.2
20	20	0.97	5	0.01	45896	44324	3.546611	18.3280	3000	700	0.2
20	20	0.96	5	0.01	46309	44324	4.478386	18.0780	3000	700	0.2
20	40	0.99	10	0.01	45369	44324	2.357639	36.8280	3000	700	0.2
20	40	0.98	10	0.01	45955	44324	3.679722	36.5930	3000	700	0.2
20	80	0.98	20	0.01	44929	44324	1.364949	72.6880	3000	700	0.2
28	60	0.97	12	0.01	117715	110442	6.585357	18.6870	3600	200	0.01
28	60	0.98	12	0.01	117100	110442	6.028504	18.7970	3600	200	0.01
28	60	0.98	15	0.01	117030	110442	5.965122	18.8430	3600	200	0.01
28	60	0.98	20	0.01	117520	110442	6.408794	18.5310	3600	200	0.01
40	80	0.98	20	0.01	690300	654179	5.521577	88.8750	3600	300	0.2
40	80	0.98	20	0.01	686833	654179	4.9916	87.2660	3600	300	0.1
60	60	0.98	15	0.01	3010600	2812540	7.042033	208.0150	3600	500	0.1
60	80	0.98	20	0.01	3025065	2812540	7.556337	277.1720	3600	500	0.1
60	80	0.98	25	0.01	3019493	2812540	7.358224	277.1292	3600	500	0.1
80	80	0.98	20	0.01	7553100	6926210	9.050982	277.2500	3600	300	0.2
80	160	0.98	40	0.01	7460048	6926210	7.707505	549.8120	3600	300	0.1
100	80	0.98	20	0.01	14099131	12965200	8.745958	206.3440	3600	200	0.01
100	120	0.98	24	0.01	14195000	12965200	9.485392	268.7970	3600	200	0.01

Table 4. Results of SA and Lingo for a closed-loop layout problem

n	T_0	α	ΔT	Objective Function Value		Error (%)	Time (s)	
				SA	Lingo		SA	Lingo
20	1000	0.95	0.001	44110	44324	-0.48281	0.156	3000
20	1000	0.9	0.001	45559	44324	2.786301	0.11	3000
20	1000	0.9	0.001	44110	44324	-0.48281	0.11	3000
20	1000	0.9	0.0001	44163	44324	-0.36323	0.109	3000
20	1000	0.95	0.0000001	44110	44324	-0.48281	0.219	3000
20	2000	0.9	1E-07	44149	44324	-0.39482	0.141	3000
28	2000	0.95	0.00000001	109620	110442	-0.74428	0.375	3600
28	2000	0.94	0.00000001	109690	110442	-0.6809	0.328	3600
28	2000	0.93	0.00000001	109739	110442	-0.63653	0.266	3600
28	3000	0.93	0.00000001	109730	110442	-0.64468	0.265	3600
28	2000	0.95	0.001	109700	110442	-0.67185	0.234	3600
28	2000	0.95	0.0001	109724	110442	-0.65011	0.25	3600
28	3000	0.95	0.001	109842	110442	-0.54327	0.234	3600
28	2000	0.96	0.001	109780	110442	-0.59941	0.266	3600
40	4000	0.9	0.00000001	650590	654179	-0.54863	0.734	3600
40	4000	0.9	0.0000001	650146	654179	-0.6165	0.672	3600
40	4000	0.9	0.000001	652407	654179	-0.27087	0.609	3600
40	4000	0.95	0.0000001	647124	654179	-1.07845	1.328	3600
40	5000	0.95	0.0000001	647460	654179	-1.02709	1.313	3600
50	5000	0.95	0.00000001	2818275	2812540	0.203908	1.422	3600
50	4000	0.95	0.00000001	2824856	2812540	0.437896	1.406	3600
50	4000	0.96	0.00000001	2816300	2812540	0.133687	1.735	3600
60	4000	0.97	0.00000001	2812800	2812540	0.009244	2.328	3600
60	4000	0.98	0.00000001	2806414	2812540	-0.21781	3.578	3600
60	5000	0.98	0.00000001	2806628	2812540	-0.2102	3.609	3600
60	5000	0.97	0.00000001	2810500	2812540	-0.07253	2.328	3600
80	5000	0.95	0.00000001	6953000	6926210	0.386792	2.453	3600
80	5000	0.96	0.00000001	6936000	6926210	0.141347	3.141	3600
80	7000	0.96	0.00000001	6934600	6926210	0.121134	3.094	3600
80	7000	0.98	0.00000001	6914251	6926210	-0.17266	6.5	3600
80	7000	0.97	0.00000001	6921344	6926210	-0.07025	4.14	3600
100	9000	0.98	0.0000000001	12963000	12965200	-0.01697	11.313	3600
100	9000	0.98	0.00000000001	12960866	12965200	-0.03343	12.157	3600
100	9000	0.98	0.000000000001	12956000	12965200	-0.07096	13.078	3600

7 Conclusions

We presented a novel, new mathematical model of a closed-loop layout problem. This model was solved by three meta-heuristic algorithms, namely genetic algorithm (GA) combined with a local search method (e.g., namely 2-Opt), ACO, and SA. The related results were compared with the Lingo 8.0 software package in order to show the efficiency of our three proposed algorithms. We concluded that the proposed SA works more efficiently than the Lingo, GA, and ACO. It approaches to the optimal solution in a short time compared with the Lingo, and in most of the cases the result of the SA was better than the Lingo. Note that during solving the Lingo solves medium and large-sized problems; we set a one-hour time limit. Indeed, the Lingo reports just a lower bound and it cannot reach to a global optima. Further, the 2-Opt algorithm was applied for only medium and large-sized problems in order to improve the quality of solutions.

References

1. El-Baz, M.A.: A genetic algorithm for facility layout problems of different manufacturing environments. Computers & Industrial Engineering 47, 233–246 (2004)
2. Bozer, Y.A., Rim, S.C.: A branch and bound method for solving the bidirectional circular layout problem. Appl. Math. Modeling 20(5), 342–351 (1996)
3. Öncan, T., Altınel, İ.K.L.: Exact solution procedures for the balanced unidirectional cyclic layout problem. Eur. J. of Operational Research (to appear, 2007)
4. Nearchou, A.C.: Meta-heuristics from nature for the loop layout design problem. Int. J. Production Economics 101, 312–328 (2006)
5. Junjae, C., Peters, B.A.: A simulated annealing algorithm based on a closed loop layout for facility layout design in flexible manufacturing systems. International J. of Prod. Research 44(13), 2561–2572 (2006)
6. Bennell1, J.A., Potts, C.N., Whitehead, J.D.: Local search algorithms for the min-max loop layout Problem. J. of the Operational Research Society 53, 1109–1117 (2002)
7. Goldberg, D.: Genetic algorithms in search, optimization, and machine learning. Addison-Wesley, New York (1989)
8. Kazerooni, M., Luonge, L., Abhary, K.: Cell formation using genetic algorithms. Int. J. of Flexible Automation and Integrated Manufacturing 3(3-4), 283–299 (1995)
9. Michalewicz, Z.: Genetic algorithms + data structures = evolution programs. Springer, Berlin (1992)
10. Tavakkoli-Moghaddam, R., Shayan, E.: Facilities layout design by genetic algorithms. Computers and Industrial Engineering 45(3-4), 527–530 (1998)
11. Al-Hakim, L.: On solving facility layout problems using genetic algorithms. Int. J. of Production Research 38(11), 2573–2582 (2000)
12. Chan, K.C., Tansri, H.: A study of genetic crossover operations on the facility layout problem. Computers and Industrial Engineering 26(3), 537–550 (1994)
13. Gau, K.Y., Meller, R.D.: An iterative facility layout algorithm. Int. J. of Production Research 37(16), 3739–3758 (1999)
14. Hamamoto, S.: Development and validation of genetic algorithm-based facility layout a case study in the pharmaceutical industry. Int. J. of Production Research 37(4), 749–768 (1999)

15. Islier, A.A.: A genetic algorithm approach for multiple criteria facility layout design. International Journal of Production Research 36(6), 1549–1569 (1998)
16. Levine, J., Ducatelle, F.: Ant colony optimization and local search for bin packing and cutting stock problems. J. of the Operational Research Society 55, 705–716 (2004)
17. Gambardella, L.M., Dorigo, M.: An ant colony system hybridized with a new local search for the sequential ordering problem. INFORMS Journal on Computing 12(3), 237–255 (2000)
18. Gambardella, L.M., Taillard, E., Agazzi, G.: MACS-VRPTW: A multiple ant colony system for vehicle routing problems with time windows. In: Corne, D., Dorigo, M., Glover, F. (eds.) New Ideas in Optimization, pp. 63–76. McGraw-Hill, London (1999)
19. Kirkpatrick, S., Gelatt, C.D., Vecchi, M.P.: Optimization by simulated annealing. Science 220, 671–680 (1983)
20. Wang, T.Y., Wu, K.B., Liu, Y.W.: A simulated annealing algorithm for facility layout problems under variable demand in cellular manufacturing systems. Computers in Industry 46, 181–188 (2001)
21. Baykasoglu, A., Gindy, N.N.Z.: A simulated annealing algorithm for dynamic layout problem. Computers & Operations Research 28, 1403–1426 (2001)

A Genetic Algorithm with Multiple Operators for Solving the Terminal Assignment Problem

Eugénia Moreira Bernardino[1], Anabela Moreira Bernardino[1],
Juan Manuel Sánchez-Pérez[2], Juan Antonio Gómez-Pulido[2],
and Miguel Angel Vega-Rodríguez[2]

[1] Department of Computer Science, School of Technology and Management,
Polytechnic Institute of Leiria, 2400 Leiria, Portugal
{eugenia,anabelab}@estg.ipleiria.pt
[2] Department of Technologies of Computers and Communications, Polytechnic School,
University of Extremadura, 10071 Cáceres, Spain
{sanperez,jangomez,maveca}@unex.es

Abstract. In recent years we have witnessed a tremendous growth of communication networks resulted in a large variety of combinatorial optimization problems. One of these problems is the terminal assignment problem. In this paper, we propose a genetic algorithm employing multiple crossover and mutation operators for solving the well-known terminal assignment problem. Two sets of available crossover and mutation operators are established initially. In each generation a crossover method is selected for recombination and a mutation method is selected for mutation based on the amount fitness improvements achieved over a number of previous operations (recombinations/mutations). We use tournament selection for this purpose. Simulation results with the different methods implemented are compared.

Keywords: Terminal Assignment Problem, Evolutionary Algorithms, Genetic Algorithm.

1 Introduction

The literature on telecommunication network problems has quickly grown. This is mainly due to the dramatic growth in the use of the Internet [1], [2]. Terminal assignment (TA) is an important issue in telecommunication networks optimization to increase their capacity and reducing the cost of them. The task here is to assign a given collection of terminals to a given collection of concentrators.

The TA problem is a NP-complete combinatorial optimization problem. This means that we cannot guarantee to find the best solution in a reasonable amount of time. The intractability of this problem is a motivation for the pursuits of Genetic Algorithms (GAs) that produce approximate, rather than exact, solutions. GAs are evolutionary algorithms (EAs) that use the principle of natural selection to evolve a set of solutions toward an optimum solution. An EA is a subset of evolutionary computation, a generic population-based metaheuristic optimization algorithm [3]. The EAs have been used successfully for the solution of hard numerical and combinatorial optimization problems. An EA uses some mechanisms inspired by biological evolution: mutation, recombination, natural selection and survival of the fittest. Candidate solutions to the optimization problem play the role of individuals in

N.T. Nguyen, R. Katarzyniak (Eds.): New Chall. in Appl. Intel. Tech., SCI 134, pp. 279–288, 2008.
springerlink.com

a population, and the cost function determines the environment within which the solutions survive [3].

In this article we present a GA with multiple operators to solve the TA problem. We use a method similar to tournament selection to select the mutation and crossover operators based on the fitness improvements achieved over a number of previous operations.

The paper is structured as follows. In Section 2 we present the TA problem; in Section 3 we present the previous work; in Section 4 we describe the proposed GA; while in Section 5 the studied examples; in Section 6 we discuss the computational results obtained by the GA, finally, in Section 7 we report about the conclusions.

2   The Terminal Assignment Problem

The objective of the TA problem is to minimize the link cost to form a network by connecting a given set of terminals to a given set of concentrators [4], [5]. The TA problem involves determine what terminals will be serviced by each concentrator. The terminals and concentrators have fixed and known locations. The capacity requirement of each terminal is known and may vary from one terminal to another. Each concentrator is limited in the amount of traffic that it can accommodate. The capacities of all concentrators and the cost of linking each terminal to a concentrator are also known. The problem is to identify for each terminal the concentrator to which it should be assigned, under three constraints, in order to minimize the total cost. The three constraints imposed in this article for solving the TA problem are [5]: (1) each terminal must be connected to one and only one concentrator; (2) the aggregate capacity requirement of the terminals connected to any one concentrator cannot exceed the capacity of that concentrator; (3) guarantee the balanced distribution of terminals among concentrators.

To represent the TA problem we use the following components: (1) a set N of n distinct terminals; (2) a set M of m distinct concentrators; (3) a vector C with the capacity required for each concentrator; (4) a vector T with the capacity required for each terminal; (5) a vector CP with the location (x,y) of each concentrator; (6) a vector CT with the location (x,y) of each terminal.

3   Previous Work

Some interesting approaches for the TA can be found in the literature. Atiqullah and Rao [6] proposed Simulate Annealing (SA) to find the optimal design of small-scale networks. Pierre et al. [7] proposed SA to find solutions for packet switched networks. Glover and Ryan [8] and Koh and Lee [9] adopted Tabu Search (TS) to find an appropriate design of communications networks. Abuali et al. [4] proposed a Greedy Algorithm and a Hybrid Greedy-GA for solving the TA problem. Khuri and Chui [5] proposed a GA with a penalty function as an alternative method for solving the TA problem and compare the results with the Greedy Algorithm. Salcedo-Sanz and Yao [1] proposed two different GAs using Hopfield Neural Network and compare the results with the GA. Xu et al. [10] proposed TS and compare the results with the

GA and the Greedy Algorithm. Salcedo-Sanz et al. [11] proposed to solve TA problem with Groups Encoding: the Wedding Banquet Problem. Yao et al. [2] proposed Hybrid GAs and compare the concentrator-based and terminal-based representations.

4   The Proposed GA

GAs are inspired by the natural process of reproduction [12]. Metaphors as chromosomes and population stand for solutions and solution set, respectively. Analogously, a single variable is often indicated as a gene. Mechanisms as recombination and mutation give rise to new offspring by manipulating the current population of solutions. In particular, mutation applies to a single solution (chromosome) while crossover creates new solutions from a pair of solutions selected in the current population. Following a standard Darwinistic approach, selection extracts the most promising individuals in the current population. The main features of a GA are the following [13]: (1) chromosomal representation - the correspondence between chromosomes and solutions; (2) initial population - an initial set of solutions or chromosomes; (3) fitness function - the function used to evaluate the quality of candidate solutions; (4) selection - a mechanism to select promising chromosomes in conjunction with fitness function; (5) crossover and mutation - mechanisms to generate new solutions from the currently selected chromosomes.

The first step for the GA implementation involves choosing a representation for the problem. In this work, the solutions are represented using integer vectors. We use the terminal-based representation (Fig. 1). Each position in the vector corresponds to a terminal. The value carried by position i of the chromosome specifies the concentrator that terminal i is to be assigned to.

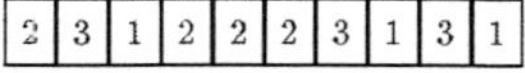

Fig. 1. Terminal Based Representation

The initial population can be created randomly or in a deterministic form. The deterministic form is based in the Greedy Algorithm proposed by Abuali et al. [4]. This algorithm assigns terminals to the closest feasible concentrator. This kind of assignment can lead to infeasible solutions even if a feasible solution exists. The algorithm can fail to produce a feasible solution when: (1) the total terminal capacity requirement is higher than the total concentrator capacity; (2) there is not a feasible solution to the problem instance; (3) the algorithm can't reach a feasible solution [5].

It's necessary to evaluate how good a potential solution is relative to other potential solutions. The fitness function is responsible for performing this evaluation and returning a positive number (fitness value) that reflects how optimal the solution is. The fitness function is based on the fitness function used in [1]. The objective is to minimize the fitness function. The fitness function is based on: (1) the total number of terminals connected to each concentrator - the objective is to guarantee the balanced distribution of terminals among concentrators; (2) the distance between the concentrators and the terminals assigned to them - the objective is to minimize the distances between concentrators and terminals assigned to them; (3) the penalization if a solution is not feasible - the objective is to penalize the solutions when the total capacity of one or more concentrators is overloaded.

$$fitness = 0,9 * \sum_{c=0}^{M-1} bal_c + 0,1 * \sum_{t=0}^{N-1} dist_{t,c(t)} + Penaliza$$

$$bal_c = \begin{cases} 10 & if\left(total_c = round\left(\frac{N}{M}\right)+1\right) \\ 20*abs\left(round\left(\frac{N}{M}\right)+1-total_c\right) \end{cases}$$

$$c(t) = concentrator\ of\ terminal\ t$$
$$t = terminal$$
$$c = concentrator$$

$$Penaliza = \begin{cases} 0 \\ 500 \end{cases} \quad if\ (Feasible)$$

$$total_c = \sum_{t=0}^{N-1} \begin{cases} 1 \\ 0 \end{cases} \quad if\ (c(t)=c)$$

$$dist_{t,c(t)} = \sqrt{(CP[c(t)]x - CT[t]x)^2 + (CP[c(t)]y - CT[t]y)^2}$$

For selection, we implement three selection methods: Roulette, Tournament y Tournament with Elitism (the elite size is 20% of the population size).

For recombination, we implement eight crossover operators. One point, 2-point, 4-points and uniform are very well-known and widely used in practice. In "reciprocal translocation", randomly located and arbitrary-length chromosomal segments are exchanged between parent chromosomes, as illustrated in Fig. 2.

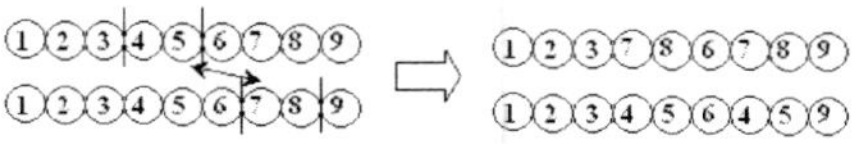

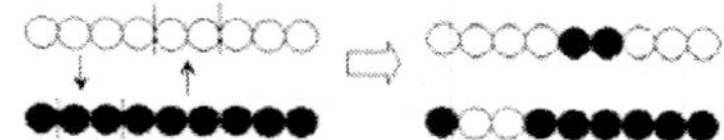

Fig. 2. Reciprocal translocation

In "exchange positions", one chromosomal segment is randomly selected for each parent and exchanged between parent chromosomes, as illustrated in Fig. 3.

Fig. 3. Exchange positions

In "exchange terminals of two concentrators", two concentrators are randomly selected and the terminals of that concentrators are exchanged. In "multiple", a set of available crossover operators is established initially. The crossover set contains the seven crossover operators applicable to the problem. In the initialization phase, each crossover operator has the same probability of being selected. From thereon and after every recombination is assigned a fitness value to the respective crossover operator based on its contribution to individuals fitness. This fitness value is used for recombination selection. The recombination operators are selected using a tournament selection. The program chooses three random operators. The operator with the higher fitness will win. After a predefined number of recombinations (NUM_PREV_OPS_RECOMB) the probabilities of each crossover operator are updated based on their contribution in the last recombinations.

Multiple Recombination:

```
FitIni1 = fitness(individual1)
FitIni2 = fitness(individual2)
if (first time)
  initialize numOperationsR
  initialize fitPreviousOperationsR
  initialize fitnessActualR
  num_operations_totalR = 0
else
  If (num_operations_totalR = NUM_PREV_OPS_RECOMB)
     num_operations_totalR = 0
     For i=1 to NUM_OPERATORS do
        fitnessActualR[i] = fitPreviousOperationsR[i]/
                      numOperationsR[i]
     initialize numOperationsR
     initialize fitPreviousOperationsR
num_operations_totalR = num_operations_totalR + 1
operator = random(NUM_RECOMB_OPERATORS)
for op = 1 to 3 do
    op = chooseRandomOperator()
    if (fitnessActualR[op] > fitnessActualR[operator])
         operator = op
switch(operator)
    case 1: recomb = Recombination1Point()
    case 2: recomb = Recombination2Points()
    case 3: recomb = Recombination4Points()
    case 4: recomb = RecombinationUniform()
    case 5: recomb = RecombinationTranslocation()
    case 6: recomb = RecombinationExchangePositions()
    case 7: recomb = RecombinationConcentrators()
run(recomb, individual1, individual2)
fitPreviousOperationsR[operator]=
   fitPreviousOperationsR[operator]+max(fitIni1,fitIni2)
   - max(fitness(individual1),fitness (individual2))
numOperationsR[operator] = numOperationsR[operator] + 1
```

For mutation, we implement four mutation operators. In "change order", two genes are randomly selected and exchanged. In "change concentrator", one gene is randomly selected and its value (concentrator) is replaced for a new random value (concentrator). In "change less distant concentrator", one gene is randomly selected and its value (concentrator) is replaced for a new value (less distant concentrator). In "multiple", a set of available mutation operators is established initially. The mutation set contains the three mutation operators applicable to the problem. In the initialization phase, each mutation operator has the same probability of being selected. From thereon and after every mutation is assigned a fitness value to the respective mutation operator based on its contribution to individual fitness. This fitness value is used for mutation selection. The mutation operators are selected using a tournament selection. The program chooses two random operators. The operator with the higher fitness will win. After a predefined number of mutations (NUM_PREV_OPS_MUT) the probabilities of each mutation operator are updated based on their contribution in the last mutations.

Multiple Mutation:

```
fitIni = fitness(individual)
if (first time)
      initialize numOperations
      initialize fitPreviousOperations
      initialize fitnessActual
      num_operations_total = 0
else
      If (num_operations_total = NUM_PREV_OPS_MUT)
         num_operations_total = 0
         For i=1 to NUM_OPERATORS do
             fitnessActual[i] = fitPreviousOperations[i] /
                                numOperations[i]
         initialize numOperations
         initialize fitPreviousOperations
num_operations_total = num_operations_total + 1
operator = random(NUM_MUT_OPERATORS)
for op = 1 to 2 do
      op = chooseRandomOperator()
      if (fitnessActual[op] > fitnessActual[operator])
           operator = op
switch(operator)
      case 1: mut = MutationChangeOrder()
      case 2: mut = MutationChangeConcentrator()
      case 3: mut = MutationChangeLDistantConcentrator()
run(mut, individual)
fitPreviousOperations[operator]=
      fitPreviousOperations[operator]+ fitIni -
      fitness(individual)
numOperations[operator] = numOperations[operator] + 1
```

5 Studied Examples

In order to test the performance of our approach, we use a collection of TA instances of different sizes. We use nine test instances of different difficulties. In general, the difficulty increases with the problem size.

The coordinates of terminals and concentrators were randomly obtained in a 100 x 100 grid, whereas the weights associated with each terminal (T) were randomly generated between 1 and 6 and the capacities associated with each concentrator (C) were randomly generate between 10 and 20.

Table 1 shows the main characteristics of these TA instances.

Table 1. TA instances

Problem	N	M	Total T capacity	Total C capacity
1	10	3	35	39
2	20	6	55	81
3	30	10	89	124
4	40	13	147	169
5	50	16	161	207
6	50	16	173	208
7	70	21	220	271
8	100	30	329	517
9	100	30	362	518

6   Results

Table 2 presents the best-obtained results with the GA. The first column represents the problem number and the remaining columns show the results obtained (Fitness, Time - Run Times, Gens – Number of Generations). The algorithm has been executed using a processor Intel Core Duo T2300. The algorithm was applied to populations of 200 individuals. The initial population was created using the Greedy Algorithm. The run time corresponds to the average time needed to obtain the best feasible solution. The values presented have been computed based on 100 different executions.

Table 2. Results for GA with multiple operators

Problem	GA		
	Fitness	Time	Gens
1	65,63	<1s	1
2	134.65	<1s	3
3	270,26	1s	42
4	286,89	1s	37
5	335.09	1s	52
6	371,12	1s	154
7	401,21	2s	168
8	563,19	8s	290
9	642.8	8s	960

In the tests carried out, the selection methods tournament and tournament with elitism are the methods that obtain solutions with the least cost (see Fig. 4).

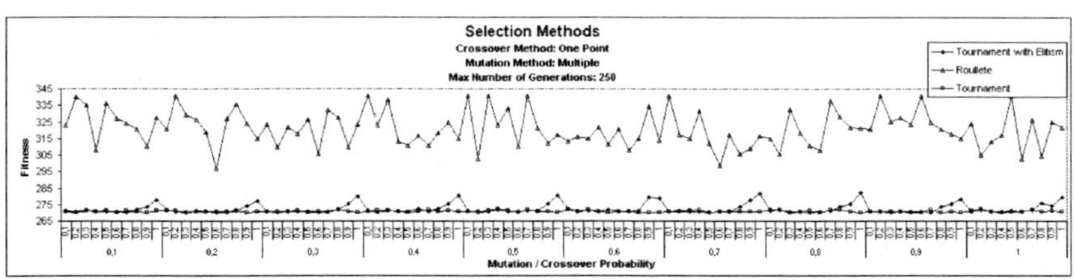

Fig. 4. Selection Methods – Problem 3

The best crossover methods used are "One Point", "2 Point", "4 Point", "Uniform" and "Multiple" (see Fig. 5, Fig. 6, Fig. 7, Fig. 8). The crossover probabilities depend on the initial population generated. The values presented in Fig. 5, Fig. 6, Fig. 7 and Fig. 8 have been computed based on 100 different executions of each method.

The multiple crossover method produces good results, but the difference from other good methods implemented is insignificant.

The best mutation strategy is "Multiple" (see Fig. 5, Fig. 9). The mutation probabilities depend on the initial population generated.

Table 3 presents the best-obtained results with the mutation methods implemented. The first column represents the problem number (Prob) and the remaining columns show the fitness obtained. The values presented have been computed based on 100 different executions of each method. The crossover method used was "Multiple".

It can easily be seen that multiple mutation method provides best solutions.

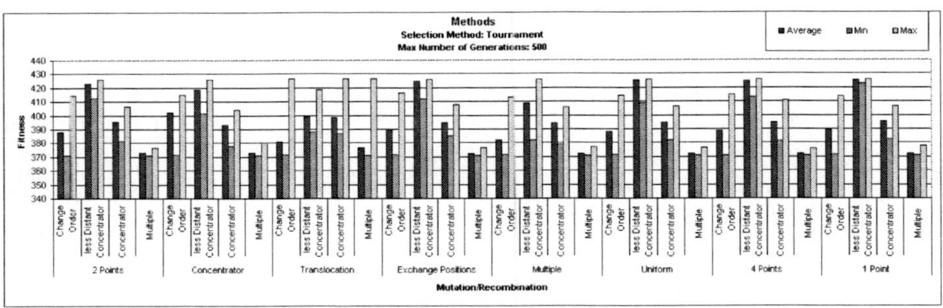

Fig. 5. Recombination and Mutation Methods – Problem 6

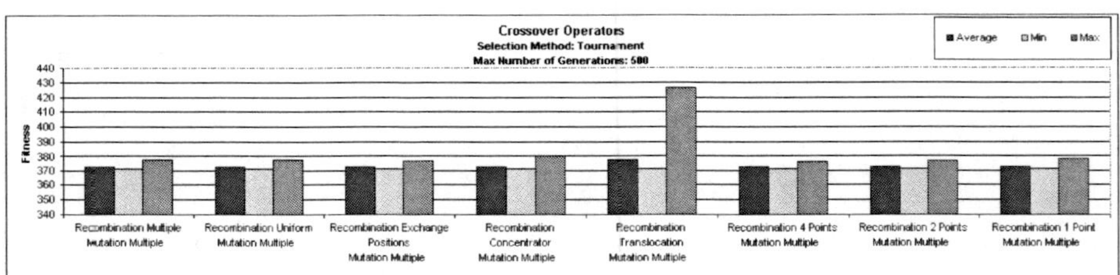

Fig. 6. Crossover Methods – Problem 6

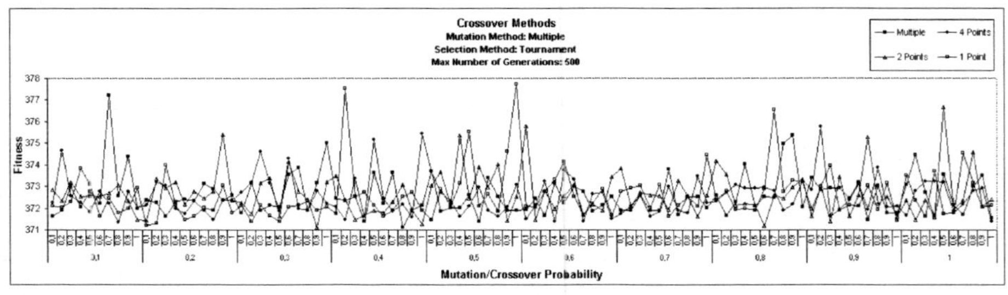

Fig. 7. Crossover Methods – Problem 6

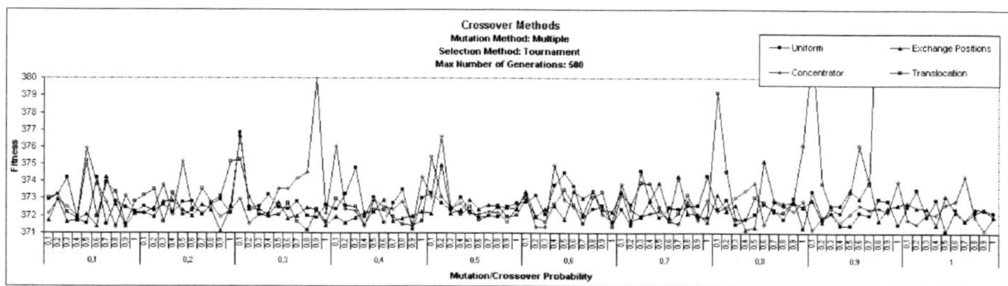

Fig. 8. Crossover Methods – Problem 6

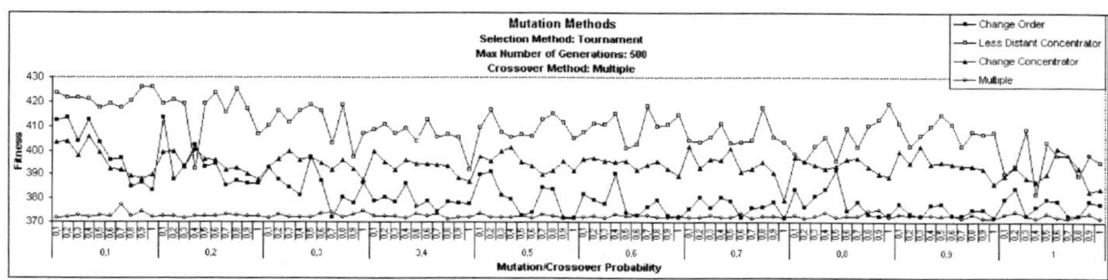

Fig. 9. Mutation Methods – Problem 6

Table 3. Comparison between mutation operators

Prob	Multiple	Change Order	Change Concentrator	Less Distant Concentrator
1	65,63	65,63	65,63	65,63
2	134.65	134,65	134,65	134,84
3	270,26	284,07	270,45	290,,48
4	286,89	286,89	288,53	291,24
5	335.09	335,09	338,23	338,59
6	371,12	371,48	379,2	381,87
7	401,21	401,45	407,54	409,83
8	563,19	563,75	575,43	610,14
9	642,83	703,78	660,78	752,09

7 Conclusions

In this paper we present a GA with multiple operators for crossover and mutation. We present a multiple mutation method and a multiple crossover method. In each step, the multiple mutation method selects one operator based on the amount fitness improvements achieved over a number of previous mutations. The multiple crossover method works similar to multiple mutation method. We present a novel operator selection method similar to tournament selection.

The implemented methods are used to solve a well-known problem, namely TA. The performance of the different methods implement are compared. Relatively to the problem studied the multiple mutation method presents better results. It can easily be seen that this method provides best solutions. The multiple crossover method in comparison with other methods doesn't produce significant best results. A great advantage of the proposed method is that it allows, through its execution, select crossover operators that are better adapted to the problem resolution.

The different methods implemented have driven to acceptable results. In any case, the implementation of new methods will permit to obtain better results. The implementation of parallel algorithms will speed up the optimization process.

References

1. Salcedo-Sanz, S., Yao, X.: A Hybrid Hopfield network-genetic algorithm approach for the terminal assignment problem. In: IEEE Transaction On Systems, Man and Cybernetics, vol. 34, pp. 2343–2353. IEEE Press, USA (2004)
2. Yao, X., Wang, F., Padmanabhan, K., Salcedo-Sanz, S.: Hybrid evolutionary approaches to terminal assignment in communications networks. In: Hart, W.E., Krasnogor, N., Smith, J.E. (eds.) Recent Advances in Memetic Algorithms and related search technologies, pp. 129–159. Springer, New York (2004)
3. Eiben, A.E., Smith, J.E.: Introduction to Evolutionary Computing. Springer, Berlin (2003)
4. Abuali, F., Schoenefeld, D., Wainwright, R.: Terminal assignment in a Communications Network Using Genetic Algorithms. In: 22nd Annual ACM Computer Science Conference, pp. 74–81. ACM Press, New York (2004)
5. Khuri, S., Chiu, T.: Heuristic Algorithms for the Terminal Assignment Problem. In: ACM Symposium on Applied Computing, pp. 247–251. ACM Press, New York (1997)
6. Atiqullah, M., Rao, S.: Reliability optimization of communication networks using simulated annealing. Microelectronics and Reliability 33, 1303–1319 (1993)
7. Pierre, S., Hyppolite, M.A., Bourjolly, J.M., Dioume, O.: Topological design of computer communication networks using simulated annealing. Engineering Applications of Artificial Intelligence 8, 61–69 (1995)
8. Glover, F., Lee, M., Ryan, J.: Least-cost network topology design for a new service: and application of a tabu search. Annals of Operations Research 33, 351–362 (1991)
9. Koh, S.J., Lee, C.Y.: A tabu search for the survivable fiber optic communication network design. Computers and Industrial Engineering 28, 689–700 (1995)
10. Xu, Y., Salcedo-Sanz, S., Yao, X.: Non-standard cost terminal assignment problems using tabu search approach. In: IEEE Conference in Evolutionary Computation, vol. 2, pp. 2302–2306. IEEE Press, Portland (2004)
11. Salcedo-Sanz, S., Portilla-Figueras, J.A., García-Vázquez, F., Jiménez-Fernández, S.: Solving terminal assignment problems with groups encoding: the wedding banquet problem. Engineering Applications of Artificial Intelligence 19, 569–578 (2006)
12. Holland, J.H.: Adaptation in Natural and Artificial Systems. The University of Michigan Press (1975)
13. Goldberg, D.E.: Genetic Algorithms in Search, Optimization and Machine Learning. Addison Wesley, Boston (1989)

An Efficient Hybrid Method for an Expected Maximal Covering Location Problem

R. Tavakkoli-Mogahddam[1], V.R. Ghezavati[2], A. Kaboli[2], and M. Rabbani[1]

[1] Department of Industrial Engineering, Faculty of Engineering,
  University of Tehran, P.O. Box: 11365/4563, Tehran, Iran
[2] Department of Industrial Engineering, Iran University of Science and Technology,
  P.C.: 16844, Tehran, Iran
  `tavakoli@ut.ac.ir`, `{ghezavati,kaboli}@iust.ac.ir`,
  `mrabani@ut.ac.ir`

Abstract. This paper presents a new mathematical model for an expected maximal covering location problem (EMCLP) that optimizes both location and allocation decisions. In real-world cases, traveling or lead times may change over the period of time. It is assumed that the traveling time between customers and emergency centers has an exponential distribution function. If this uncertain time is less than a critical time, the customer can be allocated to that emergency center (facility), which yields more flexibility for results and the proposed model. In this paper, we present a stochastic nonlinear model that selects the best location of emergency centers, and allocation decisions maximizing the total expected demand covered. To solve such a difficult model, an efficient hybrid method based on the simulation and genetic algorithm is proposed. Finally, some numerical examples are illustrated to show the effectiveness of this proposed method.

Keywords: Maximal covering, Location Problem, Simulation, Genetic algorithm, Uncertainty modeling.

1 Introduction

Location theory has attracted significant research effort since 1960s. Then, a number of problems have been proposed and the methodologies developed to solve these problems have been used to make decisions related to the location of facilities in many real-world problems. Brandeau and Chiu [1] presented a survey of location problems by determining more than 50 problems and indicating how those problems relate to one another (see Daskin [2] for detailed discussions on location models). One of these location problems is a maximal covering location problem (MCLP), which was initially developed to determine a set of facility locations. It maximizes a number of demand points covered or the total demands served by the emergency centers (facilities) within a specified critical (or maximal service) distance or time by a fixed number of facilities. This problem and its numerous extensions compose an important class of problems in the location literature. Traditionally, it is assumed that the traveling time between customers and facilities is certain. Customers' demands are also covered completely if it is located within the critical distance of the facility, and not covered outside of this distance. Normally, the associated model can be directly applied to most facility-location planning problems, such as location for warehouses, health-care centers, fire stations, recreation centers, libraries, and the like [3].

N.T. Nguyen, R. Katarzyniak (Eds.): New Chall. in Appl. Intel. Tech., SCI 134, pp. 289–298, 2008.
springerlink.com

Megiddo *et al.*, [4] defined the MCLP on a tree-network and presented a solution algorithm to solve the problem. Daskin *et al.*, [5] examined the developments, such as the concepts of multiple, excess, backup, and expected coverage to covering models. Schilling *et al.*, [6] provided a detailed review of the covering models in facility location problems. Araz *et al.*, [7] proposed a multi-objective MCLP based on an emergency vehicle location. They considered the maximization of the covered customers and minimization of the total travel distance from the emergency services. Karasakal [8] formulated the MCLP in the presence of partial coverage, and developed a solution procedure based on the Lagrangean relaxation. He showed the effect of the algorithm on the optimal solution by comparing it with the classical algorithms. Chan and Mahan [9] offered a variant of the MCLP to locate at most p signal-receiving stations. In their study, demands, named geo-locations, to be covered by these stations are distress signals and/or transmissions from any targets. Younies and George [10] introduced a zero–one mixed integer formulation for a maximal covering problem where points were covered by inclined parallelograms in a plane.

Daskin [11] proposed a version of a MCLP considering three assumptions: 1) Servers operate independently; 2) each server has the same busy probability; and 3) server busy probabilities are invariant with respect to their locations. This proposed model is as an extension of the MCLP that was first formulated by Church and ReVelle [12] in order to take into account for the possibility of the server unavailability due to a congested system. Church and Roberts [13] developed a model that addresses the relation between the quality of service and distance/service time to a limited extend. They used a piecewise linear step function instead of a more general form in order to have a tractable model. Berman and Krass [14] considered a generalization of the MCLP, which is allowed for partial coverage of customers, with the degree of coverage being a non-increasing step function of the distance to the nearest facility. Church and ReVelle [15] discussed the theoretical and computational links between the p-median, location set covering, and maximal covering location problems. The real distances in the p-median problem can be transformed into a binary form. The associated objective function minimizes the amount of population that is not covered within maximal service distance. Chung *et al.*, [16] added a new capacity constraint in order to define the capacitated MCLP. An upper limit is placed upon the total number of demand population that can be covered by an individual facility site.

The rest of this paper is as follows: Section 2 presents a new mathematical model for the given problem. Section 3 proposes the solution procedure based on simulation and genetic algorithms. The computational results are illustrated and discussed in Section 4. Finally, the remarking conclusion is given in Section 5.

2 Model Formulation

2.1 Parameters and Decision Variables

We present a novel, mathematical model for the probability maximal location problem. This model is defined as a function of the time of the demand point to the emergency centers. We assume that the demand point can be fully covered if travel time from emergency center to customer be less than coverage time. Because the travel time is probability (in this case, it has exponential distribution function), "full coverage" or "not coverage" will be probability. So, we maximize the expected demand that is covered. Suppose that there is a set of demand points (N), at which requests are generated, and a set of locations (M), where facilities may be opened. We also assume that requests at a demand point $i \in D$ are generated independent of the processes at other demand points in N. We develop a formulation based on the classical p-median formulation. However, instead of minimizing the total distance, this model maximizes the expected coverage of demand points by determining one of the selected facility sites, which ensures the maximum coverage level for each demand point.

d_i Demand for customer i.

λ_j Coverage radius for emergency center j.

C_j Capacity for emergency center j.

f_{ij} Random variable denoting traveling time between customer i and emergency
center j, which has exponential distribution function with the mean a_{ij}.

P Total number of emergency centers to be located.

r A probability coverage lower bound for the complete coverage of customers.

$X_{ij} = \begin{cases} 1 & \text{If demand of customer } i \text{ is satisfied by emergency center } j, \\ 0 & \text{Otherwise.} \end{cases}$

$U_j = \begin{cases} 1 & \text{If an emergency center located at site } j, \\ 0 & \text{Otherwise.} \end{cases}$

P_i Probability of coverage of customer i totally.

P_{ij} Probability of coverage of customer i by emergency j.

It is worth noting that in this model all of the emergency centers cannot serve all of the customers because any of the emergency centers has special coverage radius. If a customer ids not in this coverage, then the emergency center cannot serve that customer. This parameter makes this model more realistic. For example, an emergency center can cover the customers whose travel time between customers and emergency center is less than a given constant number. Now, if this time for the special customer is greater than this given constant, then the emergency center cannot cover the customer. We assume that this traveling time is probability and then "covering" or "not covering" a customer by an emergency center will be probabilistic. We measure the expected value of the objective function. (i.e., amount of demand $\times$ probability of coverage). Thus, this constraint must be added to the model in order to compute the coverage probability:

$$P_{ij} - \Pr(f_{ij} \leq \lambda_j) = 0 \Rightarrow P_{ij} - \left(\int_0^{\lambda_j} \frac{1}{a_{ij}} \cdot e^{-x\frac{1}{a_{ij}}} d_x \right) = 0 \tag{1}$$

$$P_i - \sum_j P_{ij} X_{ij} = 0 \Rightarrow P_i - \sum_j (1 - e^{-\frac{\lambda_j}{a_{ij}}}) X_{ij} = 0 \tag{2}$$

2.2 Proposed Model

$$\max \sum_{i}\sum_{j}[X_{ij}\, d_i\,]\, P_i \tag{3}$$

s.t.

$$X_{ij} \le U_j \qquad\qquad\qquad \forall i \in M, \forall j \in N \tag{4}$$

$$\sum_{j} X_{ij} \le 1 \qquad\qquad\qquad \forall i \in M \tag{5}$$

$$\sum_{i} X_{ij}\, d_i \le C_j\, U_j \qquad\qquad\qquad \forall j \in N \tag{6}$$

$$\sum_{j} U_j = p \tag{7}$$

$$P_i - \sum_{j}(1 - e^{-\frac{\lambda_j}{a_{ij}}}).X_{ij} = 0 \qquad\qquad \forall i \in M \tag{8}$$

$$X_{ij} = (0,1)\,, U_j = (0,1),\, P_i \ge 0 \tag{9}$$

The objective function (3) of the proposed model maximizes the total expected demands covered so far. Constraint (4) ensures that a customer can allocate to emergency center j when this emergency center is opened. Constraint (5) ensures that a customer cannot be allocated to more than one emergency center. Constraint (6) guarantees that the total demand assigned to emergency center j must be less that its capacity. Constraint (7) indicates that the total number of emergency centers to be located should be equal to p. Constraint (8) computes the probability of coverage customer i by all emergency centers, based on the exponential distribution. Constraint (9) determines the type of variables.

3 Solution Procedure

It is known that capacitated location problems belong to the class of NP-hard problems that cannot be solve in reasonable computational time. In this paper, we propose a nonlinear and stochastic programming model for the capacitated network design problem which is extremely hard to optimally solve in large sizes. Thus, we use an efficient meta-heuristic method based on simulation and genetic algorithm (GA). GA is a stochastic search and heuristic optimization technique, which has been widely adopted by many researchers to solve various problems. This algorithm was first developed by Holland [17]. It mimics the mechanism of genetic evolution in the biological nature and consists of a population of chromosomes (strings or individuals) that are composed of genes. These genes represent a number of values, called alleles. Each chromosome (genotype) represents one potential solution (phenotype). The process of genetic operators (i.e., crossover and mutation) is carried out in the pool; after that, an evolution is completed by creating new chromosomes (offspring). This offspring is expected to be stronger than the parents, but this may not always be true. GA does not rely on analytical properties of the function to be optimized [18]. In short, GA has two major processes: 1) GA is iteratively and randomly generating new solutions; and 2) these solutions are checked for the optimality according to predefined fitness functions. This becomes the most powerful principle of GA. It makes them widely suitable for finding an optimal solution in many complex problems, such as traveling salesman problem (TSP) and any forms of scheduling problems [19].

3.1 Simulation Process

Because of the complexity of the proposed model, we design a simulation model to calculate uncertain functions covering or not covering customers by emergency centers. First, we experiment the simulation process to generate a 0-1 matrix, namely Z_{ij}. Then, we define a lower bound for probability of coverage. If this probability is greater than the lower bound, we assume that the customer can be covered by emergency center completely. For instance, if probability of coverage customer 2 by Emergency Center 3 is 0.82 and we define the lower bound as 0.75, we assume that $Z_{23} = 1$ and it means that Customer 2 can be covered by Emergency Center 3 [20].

$$Z_{ij} = \begin{cases} 1 & \text{If Prob}(f_{ij} \leq \lambda_j) \geq r \\ 0 & \text{Otherwise} \end{cases} \tag{10}$$

3.2 Modified Model

By obtaining matrix Z_{ij}, the output of this simulation (i.e., the obtained matrix) is an input for GA. After running the simulation, we define a new binary decision variable imbedded in the proposed model as follows:

$$Z_{ij} = \begin{cases} 1 & \text{If the DC at site } j \text{ can cover customer site } i, \\ 0 & \text{Otherwise.} \end{cases}$$

$$\max \ Z = \sum_i \sum_j X_{ij} \, d_i \, Z_{ij} \tag{11}$$

s.t.

$$X_{ij} \leq U_j Z_{ij} \qquad \forall i \in M; \forall j \in N \tag{12}$$

Constraints 5 to 7.

$$X_{ij} = (0,\ 1) \ ; \ U_j = (0,\ 1) \tag{13}$$

Constraint (12) ensures that customer i can be allocated to emergency center j if this emergency center is opened and customer i can be covered by emergency center j.

3.3 Genetic Algorithm

Chromosome Structure. Each chromosome (string) representing a feasible solution consists of a 0-1 matrix. The dimension of this matrix is equal to the number of customers in the network by the number of emergency centers. The matrix represents the assignment of customers to emergency centers (ECs), and gets the value 1 in

$$
\text{Number of customers} \begin{cases} 1 \\ 2 \\ 3 \\ 4 \\ 5 \\ 6 \end{cases} \quad x_{ij} = \begin{array}{c} \text{Number of ECs} \\ \begin{array}{cccc} 1 & 2 & 3 & 4 \\ 0 & 1 & 0 & 0 \\ 0 & 0 & 1 & 0 \\ 0 & 0 & 0 & 0 \\ 0 & 1 & 0 & 0 \\ 0 & 0 & 1 & 0 \\ 0 & 1 & 0 & 0 \end{array} \end{array}
$$

Fig. 1. A sample of a chromosome structure

element [ij] if node i is assigned to node j. Fig. 1 shows a sample of the chromosome structure representing six demands, four ECs, and $p=2$.

The first generation is created by initializing the population of chromosomes, $X^k = [x_{ij}^k]$, ($k = 1,2,...,\text{pop\_size}$) from the potential region $(X) \mid g_i(x_i) \le 0, x_i = 1,2,...,n$ at random.

$$
[x_{ij}^k] = \begin{cases} 1 & \text{If emergency center } j \text{ covers customer } i \text{ in chromosome } k, \\ 0 & \text{Otherwise.} \end{cases}
$$

We first select p emergency centers that should be opened from all potential emergency centers randomly. In the second stage, a customer is selected randomly where this procedure is carried out for all customers. Then for this customer, a number of emergency centers are determined. Finally, a random emergency center from the determined emergency centers is selected randomly and customer in chromosome k is allocated to this emergency center. Because of some limited capacities on emergency centers, a number of customers cannot be allocated.

Fitness Function. The rank-based evaluation function is defined as the objective function for chromosome V_k, where $k=1, 2, ..., pop\_size$.

Selection Strategy. By generating a random real number, r, from the interval [0, 1], chromosome V_k is selected as a parent provided that $r < P$; where the parameter P is the probability of crossover or mutation operator.

Genetic Operators. It is very important to create new chromosomes (i.e., offspring) from the selected chromosomes (called parents) with the current population. This process is carried out by the use of genetic operators, namely crossover and mutation.

By using the crossover operator, chromosomes V_k are created, To determine which parents are selected for crossover operation, we repeat the following process from $k = 1$ to pop_size: generating a random number r from the interval [0, 1], chromosome V_k is selected as a parent, where $r < Pc$. Parameter Pc is the probability of crossover (or crossover rate). Then, we group the selected parents $V_1', V_2', V_3',...$ to pairs $(V_1', V_2'), (V_3', V_4'),....$ without loss of generality. Let us illustrate the crossover operator on each pair by (V_1', V_2'). First, we make a matrix with $i \times 2j$ dimension, where i is the number of customers and j is the number of emergency centers. In other words, we and make one matrix, where in each row if the number of X_{ij}, having value 1, is more than one, then select one of them randomly and set it to be 1 and set the other to be 0. By this procedure, a customer is not allocated to different emergency centers. In the new matrix, we have at most $2 \times p$ opened emergency centers. To create offspring, we select p emergency centers from opened emergency centers in this new matrix. In fact, we select p columns from the new matrix. Note that if column r $(r \leq j)$ is selected, then column $r + j$ must not be selected and if column r $(r \geq j)$ is selected, then column $r - j$ must not be selected. Finally, a number of offspring are created by crossover operator.

By the use of the mutation operator, chromosomes V_k are updated, where $k=1$, 2, ..., *pop_size*. Similar to the crossover operation of selecting parents, we repeat the following steps from $k=1$ to *pop_size*: generating a random number r from interval [0, 1], then chromosome V_k is selected as a parent provided that $r<Pm$; where the parameter *Pm* is the probability of mutation (or mutation rate). For each selected parent: $X^k = [x_{ij}^k]$, we mutate it with the following way: Select randomly from opened emergency centers and name it j, then select randomly from closed emergency centers and name it j'. Then with the following procedure, close emergency center j and open emergency center j'. First for those customers allocated to emergency center j, we should set $X^k = [x_{ij}^k] = 0$. Then, allocate unassigned customers to emergency center j' by considering the coverage radius and capacity constraint randomly. The selection process is based on selecting 50% from the best chromosomes and 50% randomly. Thus we obtain pop size copies of chromosomes, denoted also by V_k.

4   Computational Results

To measure the effectiveness of the proposed algorithm, we generate some random examples, solve them with the proposed algorithm, and then compare our GA solutions with global solutions. Further, to show the effectiveness of the proposed GA, some numerical examples are given and solved by a personal computer. All algorithms considered in this paper are coded in Visual Basic 6 and run on a Pentium IV PC with 1.5 GHz CPU and 256 MB RAM. The associated results are compared with global solutions obtained by the LINGO 8 software. In this section, we just check the validation of the proposed GA alone, without using the simulation. This GA is applied to solve the second model and it is assumed that matrix Z_{ij} is a deterministic parameter in order to measure just the efficiency of the proposed GA. Following measures the robustness of the hybrid genetic algorithm.

Suppose a company wants to locate new emergency centers, in which there are 20 customers. The decision maker needs to select emergency centers from six potential emergency centers to serve these customers. In Table 1, we compare solutions when different problems are taken with the same generations as a stopping rule. It appears that all the objective function and global optimum differ a little bit from each other. We present a parameter, called the percent error (i.e., (global optimum - objective function) / global optimum, where the global objective value is obtained by the

LINGO 8 software package). The last column, named 'error', in Table 1 shows the above errors. In this paper, we consider 20 population sizes and 150 generations for each test problem solved by the proposed algorithm. In Table 1, it is concluded that the percent error does not exceed 2.36% when different problems are selected. It implies that the proposed genetic algorithm is effective to solve the proposed model. Our aim of this part is to evaluate just GA operators. For this, we eliminated simulation process from the hybrid method and validate GA by solving some instances. To solve these problems by the proposed GA, each problem is solved with different GA parameters and the results based on the best solutions found are shown in Table 1.

To illustrate the effectiveness of the hybrid genetic algorithm, we solve a special problem for eight times with different and several runs for the simulation model. We show that this hybrid algorithm is also robust to the simulation settings. Also, parameter setting on GA and simulation are shown in Table 2. Consider a company wants to locate new emergency center, in which there are 20 customers. Suppose the decision maker needs to select fore emergency centers from 10, potential emergency centers to serve customers. In Table 2, we compare simulation solutions with global solutions when different simulation processes is taken with the same generations as a stopping rule. It appears that all the maximal covered demand differ a little bit from each other. We present a parameter, called the percent error (i.e., (best objective value - objective value) / best objective value). The best objective value is the maximum of

Table 1. Comparison between the GA and global solutions

	No. of Customers	P	No. of ECs	Pm	Pc	No. of Opened EC	Objective Function	Global Objective Function	Error (%)	CPU Time (Sec.)
1	10	3	5	1	0.5	1,2,3	196	196	0.0%	5
2	15	3	5	0.9	0.45	2,3,4	374	374	0.0%	5
3	20	3	6	1	0.5	1,4,6	496	500	0.8%	8
4	20	3	6	0.9	0.5	2,3,6	496	508	2.4%	8
5	20	2	5	0.95	0.45	1,4	360	360	0.0%	8
6	17	2	5	0.95	0.45	2,4	348	348	0.0%	8
7	20	3	6	1	0.5	2,3,4	503	503	0.0%	8.5

Table 2. Comparison between the simulation and global solutions

	No. of Customers	P	No. of ECs	Pop Size	Gen	No. of Opened EC	Objective Function	The Best Objective	Error (%)
1	30	4	10	20	300	4,6,8,9	682	682	0.0%
2	30	4	10	20	300	4,7,8,10	678	682	0.59%
3	30	4	10	20	300	2,3,4,10	668	682	2.05%
4	30	4	10	20	300	1,2,4,8	672	682	1.47%
5	30	4	10	20	300	2,3,6,8	678	682	0.59%
6	30	4	10	20	300	2,3,4,8	681	682	0.15%
7	30	4	10	20	300	1,3,8,10	672	682	1.47%
8	30	4	10	20	300	2,4,8,10	681	682	0.15%
9	30	4	10	20	300	2,4,8,10	675	682	1.03%
10	30	4	10	20	300	2,4,8,11	681	682	0.15%

all the ten maximal covered demand obtained so far. The last column in Table 2, namely 'error', shows this parameter. The percent error does not exceed 2.05% when different simulations are selected. It implies that the hybrid genetic algorithm is robust to the simulation settings and it is effective to solve the proposed model.

5 Conclusion

In this paper, we introduced the notions of expected coverage for the maximal covering location problem (MCLP). The proposed formulation provides a way of analyzing that covering or not covering for customers is probability. We formulated the problem as a non-linear integer programming model and proposed a hybrid genetic algorithm to solve it. We tested and verified the performance of the solution algorithm on randomly generated test problems. These experiments suggest that the proposed algorithm is efficient to generate high quality solutions in a short period of time. The main contributions of this research are to consider a set covering problem under uncertainty, and to propos a hybrid genetic algorithm to solve our new model. We suggest some new areas to develop mentioned model: by using this contribution in integrated logistic systems, it assumes that such inventory and transportation can be a suitable development as future research for the presented model. Also, aggregating this model with a routing problem and determining the order of services to the customers can be an interesting subject for future research. Because of applying the traveling time between emergency centers and customers in our model, considering time window constraint for customers to get service is another suitable development.

References

1. Brandeau, M.L., Chui, S.S.: An Overview of Representative Problems in Location Research. Management Science 35, 645–674 (1989)
2. Daskin, M.S.: Network and Discrete Location: Models, Algorithms, and Applications. John Wiley & Sons, New York (1995)
3. Chung, C.H.: Recent Application of the Maximal Covering Location Planning Model. J. of Operational Research Society 37, 735–746 (1986)
4. Megiddo, N., Zemel, E., Hakimi, S.L.: The Maximum Coverage Location Problem. SIAM Journal on Algebraic Discrete Methods 4, 253–261 (1983)
5. Daskin, M.S.: A Maximal Expected Covering Location Model: Formulation, Properties and Heuristic Solution. Transportation Science 17, 48–70 (1983)
6. Schilling, D., Jayaraman, V., Barkhi, R.: A Review of Covering Problems in Facility Location. Location Science 1, 25–55 (1993)
7. Araz, C., Selim, H., Ozkarahan, I.: A Fuzzy Multi-Objective Covering-Based Vehicle Location Model for Emergency Services. Computers & Operations Research 34, 705–726 (2007)
8. Karasakal, O., Esra, K.: A Maximal Covering Location Model in the Presence of Partial Coverage. Computers & Operations Research 31, 1515–1526 (2004)
9. Chan, Y., Mahan, J.M., Chrissis, J.W., Drake, D.A., Wang, D.: Hierarchical Maximal-Coverage Location–Allocation: Case of Generalized Search-and-Rescue. Computers & Operations Research 35, 1886–1904 (2008)

10. Younies, H., Wesolowsky, G.O.: A Mixed Integer Formulation for Maximal Covering by Inclined Parallelograms. Eur. J. of Operational Research 159, 83–94 (2004)
11. Daskin, M.S.: A Maximal Expected Covering Location Model: Formulation, Properties and Heuristic Solution. Transportation Science 17, 48–70 (1983)
12. Church, R.L., ReVelle, C.S.: The Maximal Covering Location Problem. Papers of the Regional Science Association 32, 101–118 (1974)
13. Church, R.L., Roberts, K.L.: Generalized Coverage Models and Public Facility Location. Papers of the Regional Science Association 53, 117–135 (1983)
14. Berman, O., Krass, D.: The Generalized Maximal Covering Location Problem. Computers & Operations Research 29, 563–581 (2002)
15. Church, R., ReVelle, C.: Theoretical and Computational Links between the p-Median, Location Set-Covering and Maximal Covering Location Problem. George. Anal. VIII, pp. 406–415 (1976)
16. Chung, C.H., Schilling, D.A., Carbone, R.: The Capacitated Maximal Covering Location Problem: A Heuristic Solution. Modeling and Simulation, Part 4 14, 1383–1388 (1983)
17. Holland, J.H.: Adaptation in Natural and Artificial Systems: An Introductory Analysis with Applications to Biology, Control, and Artificial Intelligence, 22nd edn. MIT Press, Cambridge (1992)
18. Goldberg, D.E.: Genetic Algorithms in Search, Optimization and Machine Learning. Addison-Wiley, MA (1989)
19. Chan Felix, T.S., Chung, S.H.: Multi-Criterion Genetic Optimization for Due Date Assigned Distribution Network Problems. Decision Support Systems 39, 661–675 (2005)
20. Hwang, H.S.: Design of Supply Chain Logistic System Considering Service Level. Computers & Industrial Engineering 43, 283–297 (2002)

Multi-sided Matching Lecture Allocation Mechanism

Yoshihito Saito, Takayuki Fujimoto, and Tokuro Matsuo

Department of Informatics, Graduate School of Engineering,
Yamagata University, 4-3-16, Jonan, Yonezawa, Yamagata, 992-8510, Japan
http://veritas.yz.yamagata-u.ac.jp/

Abstract. Elective Subject is one of important issues as education program in University. Students can declare their preferences directly by selecting it. In most of university, to allocate elective subjects to the students, university staffs poll students the lectures they want to take. However, due to the limitation of time and number of staffs, the hearing investigation includes the reason and the intention in which students select the lectures. Some students sometimes take a lecture for their career, for academical interest, and for assimilation of knowledge. However, some students might take the lecture following the crowd and take the lecture as Mickey Mouse. The latter case is undesirable for the higher education. To solve the problem, in this paper, we propose a new multi-step lecture allocation method based on students preferences and university intentions. Our protocol consists of the three steps of negotiations and three types of allocations. (1) The university warns the students who have never take a certain compulsory subject yet. The students can choose whether they attend the lecture or not. If the students answer they attend the lecture, the students are allocated to the lecture by priority. (2) The students inform the university of their reasons to take the lecture. The university allocates the lectures to the students based on their reasons. (3) They negotiate about the exchange of lectures to increase students' utilities with each student. Our protocol realizes the high performance of allocation compared with brute force algorithm and reducing computational costs compared with optimizations.

1 Introduction

In recent years, there are multiple course management systems to help university staffs' tasks. In this paper, we handle the lecture allocation method and its support system for elective subjects. University education program includes compulsory subjects and elective subjects. Students must take the compulsory subjects and elective subjects to meet minimum number of required credits. Elective subjects are opened in many universities and are selected by students based on their preferences and intentions.

In United States, many universities employ elective subject education program. If there are any constraints about number of seat in classroom, students can take lectures based on order of first arrival at the first course. In Japan, many universities employ the lecture allocation based on students' poll as the questionnaire before the term. University staffs forge a convergence of students' poll and conduct class organization by lecture allocation. Generally, there are some constraints of lecture allocation in university such as, number of classrooms, number of facilities, number of seats, and so on. Due to the constraints, university staffs determine the lecture allocation based on a certain rule, such as the order of first arrival, the order of student's grade and so on. When a student

N.T. Nguyen, R. Katarzyniak (Eds.): New Chall. in Appl. Intel. Tech., SCI 134, pp. 299–308, 2008.

cannot take a lecture in which he/she wants to take, he/she selects other lecture if there is vacancy seat in the lecture. When the student wants to take the same lecture, he/she must try polling again. The purpose of elective lecture is an education program where students can take the lecture based on their preferences and multidiscipline. However, in existing lecture allocation, it is difficult to reflect students' preferences since the determination method is quite simple without students' intentions. To solve the problem, in this paper, we propose a novel lecture allocation method based on students' multiple preferences. First, university informs students the number of required credit to graduate successfully. Second, students declare the lecture in which they want to take with their preferences. Third, the lectures are allocated based on students' multi-attribute preferences and university's educational intentions. Then, students negotiate with each other to increase their utilities and to realize the desirable allocation in sequence. As the reason why we employ the negotiation process, students' actual intentions are sometimes different from the preferences input to the computer. Our protocol includes the process to coordinate effectively.

The rest of this paper consists of the following six parts. In Section 2, we give preliminaries, explaining several terms and concepts of lecture allocation. In Section 3, we explain about our proposed class allocation mechanism. Then, in Section 4, example of our proposed method is shown. After that, in Section 5, discussions are presented to show the properties of our mechanism. Finally, we give our concluding remarks.

2   Lecture Allocation Problem

2.1   Elective Subject

Elective subject system is employed in many universities for basic course. In recent years, this system is employed in some high school. Elective subjects are lectures in which students can select. Students select lectures based on own preference. Credit-system high schools are high schools in which students can make curriculum. Elective subjects are center of education program in the high school and are also focused on as the effective education program.

Elective subjects have restrictions, which are spaces and times. As the result, many universities research lectures that students want to learn before the term open. Lecture allocations are decided based on university's method. Generally the methods are first-come-first-served, the order of student's choice, the order of grade. First-come-first-served is a method that allocates from the student applied. The order of student's choice is a method that allocates based on the order of high score from each student's poll. For example, students decide first selection, second selection, and third selection from lectures, which open same time. University allocates the first selected lecture by student. If its capacity is exceeded, university allocates the second selected lecture by student. If its capacity is exceeded, university allocates the third selected lecture by student. The order of grade is a method that allocates from high-grade students.

Existing methods do not allocate lecture based on student's purpose. As the result, student's preference is not reflected. Matsuo et al. proved in the simulation experiment that student's preference is not reflected in existing allocation methods [1]. They used utility to show quantify the preference. The satisfaction shown as utility is generated when you select something. For example, there are two selectable lectures. One is the management engineering, another one is the sociology of education. A student wants to get the teacher's license of information. In this case, he/she needs taking the lecture of the sociology of education. Thus, students attend a lecture based on many purpose and preference.

2.2 Student's Preference

When student selects lecture, he/she has a certain purpose and reason. The results of our investigate show that more than 80 percent students selects lecture based on interest and concern. We ask to students that they select three important attributes from five. Five attributes are (1) interest and concern, (2) human relationship, (3) time availability, (4) easy to get a credit, and (5) for the future. (1) is "interest and concern". Many students select this factor. When the student is interested in the lecture, he/she selects the lecture without the specialized field. (2) is "human relations". This factor is to attend a lecture based on teacher selection and friend. For example, the students who selected this factor think that they want to attend the lecture of the interesting teacher. The students who selected this factor think that they want to attend the lecture in which friends take. (3) is very simple factor. This factor is "time and schedule". The reason is that he had free time. (4) is a negative factor. The students who selected this factor think the unit is easily taken. (5) is factor for future. The students who selected this factor think about the job in the future. The students think that they want to attend the lecture for qualification authorization.

Matsuo et al. also suggested class allocation support system based on multi-attribute preference [2]. This method is based on educational consideration. However, this method has a certain problem regarding the situation in which some students have equal utility.

We show the example of this situation. There are two students that they have same preferences. They hope to take same lecture with each other. This lecture's capacity is assumed as only one person. The university should select either student for just one vacancy seat. In this paper, we propose the class allocation that clarifies such a situation. This allocation is based on student's many purposes and university's educational consideration.

3 Model

In this section, we show the definition and assumption for the lecture allocated to the student. These definitions are used in the class allocation algorithm. We use the concept of the utility in calculating allocation [3]. The utility is defined as satisfaction when the student took the lecture. When the satisfaction is high, the utility is also high.

- We define student as $S = \{s_1, ..., s_i, ..., s_n\}$. s_i is ith student in the set.
- We define lecture as $A = \{a_1, ..., a_k, ..., a_m\}$. a_k is kth lecture in the set. kth lecture's capacity is defined as r_k. t_k shows day and time in which kth lecture open. $t_k \in T$. T is a set of day and time that lectures can be opened.
- We define set of lecture that ith student wants to attend as G'_i and G_i. G'_i and G_i are sets of lectures which is important for ith student. G'_i is higher preference than G_i. $G'_i \succ G_i$.

- V_i^{ak} is ith student's evaluation value for kth lecture. $1 \leq i \leq n, 1 \leq k \leq m$. These evaluation values are shown as a set of multiple attributes and their values.

Assumption 1 (Uniqueness of lecture). *We assume that same multiple lectures are not held. Even though there are many students who want to take a lecture, the class is not divided/separated from multiple classrooms.*

Assumption 2 (Constraints on selecting a lecture). *We assume that the student cannot select two or more lectures for the same timetable.*

Assumption 3 (Students' data). *We assume that system has all students' learning results and history. Using this information, the system alerts the students who must take a certain lecture to graduate.*

In our method, students declare their multiple preferences in which they want to take. We employ multi-attribute utility theory (MAUT). This theory shows users' multiple preferences. Generally, MAUT handles problems for which outcomes are characterized by two or more attributes [4]. Student polls lectures in which he/she wants to select. We define as student's declaration G_i' is a lecture where the student s_i wants to take vary much. On the other hand, when the student declares G_i, he/she wants to take as much as possible. Otherwise, for the student allocated lecture as G_i, he/she can compromise losing the lecture after negotiation.

- For kth lecture, when the number of attribute is l, hth's attribute valuation is shown as $v_{i,h}^{ak}$. $1 \leq h \leq l$.
- ith student's evaluation value to kth lecture is shown as $V_i^{ak} = f\,(v_{i,1}^{ak}, ..., v_{i,h}^{ak}, ..., v_{i,l}^{ak})$. Function f shows multi-attribute utility function. In this paper, we define that f is multi-attribute linear utility function. We use this function, user's evaluation value is shown as $V_i^{ak} = \sum_{h=1}^{l} v_{i,h}^{ak}$.
- The university can decide controlled value c_h. This value can control utility function. The university sets high value for positive attribute and sets low value for negative attribute. $0 < c_h < 1, 1 \leq h \leq l$. c_h is controlled value which control hth attribute. We use this value, student's evaluation value is shown as $V_i'^{ak} = \sum_{h=1}^{l} c_h v_{i,h}^{ak}$.
- The students cannot know c_h.
- ith student's utility is defined as u_i^{ak} when the student take a kth lecture. u_i^{ak} can be described as satisfaction for lectures which is allocated.
- $V_i'^{ak}$ is defined as evaluation values of lectures that is allocated to the student and ith student's utility u_i^{ak} is defined as $V_i'^{ak}$.

$$u_i^{ak} = V_i'^{ak} \tag{1}$$

- ith student's utility is summation u_i^{ak}.

$$u_i = \sum u_i^{ak} \tag{2}$$

- All students' total utilities U is defined as sum of each student's utility.

$$U = \sum_{i=1}^{n} u_i \tag{3}$$

In this method, class allocation is three-steps. First step is class allocation for graduation. Second step is class allocation based on evaluation values. Last step is a negotiation process for the class allocation.

– We define the number of students allocated in each step as c_{ak}^{pre}, c_{ak}^{mid}, c_{ak}^{pos}.

$$c_{ak}^{pre} + c_{ak}^{mid} + c_{ak}^{pos} \leq r_k \tag{4}$$

4 Class Allocation Algorithm

The optimal resolution of combination of lecture allocation is calculated based on total utility U is maximum. Such allocation is shown as $argmax \sum_{i=1}^{n} u_i$ in combination of lectures. Generally, these problems about lecture allocation are NP hard problems. When number of students and number of lectures increase, solution spaces are extended. Some students have the compulsory subject that does not like. The student has the low evaluation value, and might not be able to get the lecture. The students have less incentive, because the students cannot graduate. The allocation is based on total utility U, however it is not good allocation since some students unsuccessfully cannot graduate. To solve the problem, we propose a new algorithm for lecture allocation using an empirical method. Advantages of our algorithm are that using negotiation agent. The negotiation agent prevents the situation that the compulsory subjects in elective subjects cannot be attended. This agent also prevents the situation that the mechanical allocation.

[STEP 1]
When the students start up the system, the system shows the compulsory subjects and necessary number of course credit.

[STEP 2]
Students take account of [STEP 1] and select lectures they want to take. ith student inputs evaluation values $v_{i,h}^{ak}$ of kth lecture. $V_i^{'ak}$ is decided based on the above multi-attribute linear utility function.

[STEP 3]
The system permits attending the lecture of the student who selects the lecture shown by [STEP 1]. The capacity of kth lecture is $r_k' = r_k - c_{ak}^{pre}$.

[STEP 4]
The system arranges the evaluation values of each lecture in descending order. The system permits attending a lecture of the student based on capacity R or R'. The system accords the students with high evaluation value priority over the students with low evaluation value. The method for kth lecture is,

(1) The system arranges the evaluation values of kth lecture in descending order.
(2) The highest evaluation value $V_i^{'ak}$ is stored in W.
(3) The system counts the number of students who have the evaluation value which equal W. The system stores the number in $temp$.

(4)**begin**

if $r'_k \geq temp$

    The system permits attending kth lecture to the student who is searched by (2).

    $r'_k = r'_k - temp$.

else

    The system suspends attending kth lecture of the student who is searched by (2).

    $r'_k = r'_k - temp$.

end.

(5)**begin**

if $r'_k > 0$

    $W = W - 1$.

    The system conducts (3), (4), and (5).

else

    $r'_k = r'_k + temp$

    The system finishes kth lecture's STEP4.

end.

[STEP 5]

After conducted [STEP 4] in all potential lectures, the mechanism allocates lectures based on negotiation among students with each lecture. However, the negotiation might reduce students' total utilities without any rules. To avoid such situation, compromising value q is employed by the university. The re-allocation is based on that student's utility is more than q. The re-allocation is also based on exchanging lecture among students through their negotiation. The order of negotiation is based on the minimum order of difference of student's utility and q. We show the concrete procedure of negotiation as follows.

(1) The number of student is searched based on $-a_k \in G'_i$. The information is recorded in $temp$.

(2)**begin**

if $r'_k \geq temp$

    Students searched in the above phase are given the permission of taking lecture a_k.

    For other students, the process goes to [STEP 6].

else

    As student who searched by the system in the above phase,

        student s_i is searched in which other lecture $a_{k*} \in G'_i$ is rejected.

    if There is student s_i who can not take other lecture $a_{k*} \in G'_i$ as

        the searched student in the above phase.

        As the students searched in the above phase, the system searches for student s_{i*}

            in which lecture $a_{k*} \in G_{i*}$ is allocated.

        if There are students s_{i*} who can take the lecture $a_{k*} \in G_{i*}$.

            Each valuation is compared and each student decides whether

                he/she exchanges other lecture or not.

            if $V'^{ak*}_{i*} - V'^{ak*}_i \geq q$.

                Student s_{i*}'s lecture a_{k*} is canceled.

                Student s_i is accepted to take the lecture a_{k*}.

The lecture a_k is allocated to the student s_{i*}.

$r'_k = r'_k - 1$.

else $V'^{ak*}_{i*} - V'^{ak*}_i < q$.

Students do not exchange their lectures.

if $r'_k = 0$ or there is no student to negotiate.

The process goes to [STEP 6].

else There is no student who is allocated a lecture or whose lecture is $a_{k*} \in G_{i*}$

for the searched lecture a_{k*} in previous step.

[STEP 6] is used for the searched student.

else all lectures included in G'_i are allocated to all searched students.

[STEP 6] is used for the searched student.

end

[STEP 6]

The system permits attending a lecture of the student based on capacity R'. The system accords the students with a little total course credits priority over the students with a lot of total course credits.

5   Example

In this section, we show an example that uses our algorithm.

Example.   There are 4 students and 3 lectures.

The lectures that s_1 wants to attend: $G'_1 = \{a_1\}$, $G_1 = \{a_2, a_3\}$.
The lectures that s_2 wants to attend: $G'_2 = \{a_2, a_3\}$, $G_2 = \{a_1\}$.
The lectures that s_3 wants to attend: $G'_3 = \{a_1, a_3\}$, $G_3 = \{a_2\}$.
The lectures that s_4 wants to attend: $G'_4 = \{a_2\}$, $G_4 = \{a_1\}$.

s_1's evaluation value: $\{V'^{a1}_1, V'^{a2}_1, V'^{a3}_1\}$ is (9, 8, 10).
s_2's evaluation value: $\{V'^{a1}_2, V'^{a2}_2, V'^{a3}_2\}$ is (10, 2, 8).
s_3's evaluation value: $\{V'^{a1}_3, V'^{a2}_3, V'^{a3}_3\}$ is (9, 9, 8).
s_4's evaluation value: $\{V'^{a1}_4, V'^{a2}_4, V'^{a3}_4\}$ is (8, 9, 0).

Each lecture's capacity: $\{r_1, r_2, r_3\}$ is (2, 2, 2).

The system shows lecture a_3 to the student s_2 in [STEP 1].

Fig. 1. The situation before allocation

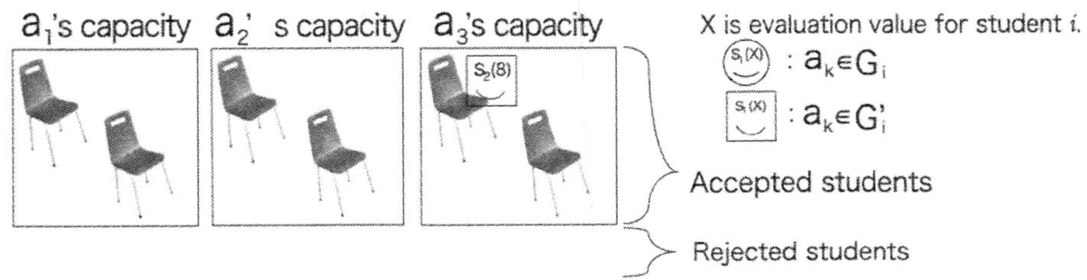

Fig. 2. [STEP3]

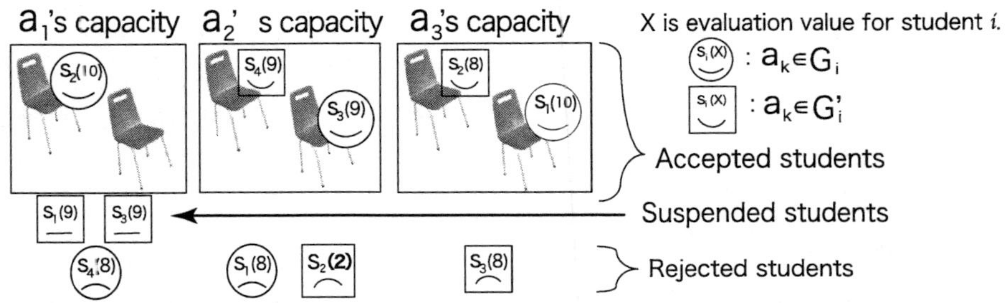

Fig. 3. [STEP4]

We allocate the lectures to the students in this situation. Fig1 is the situation before allocation. The system allocates these lectures to the students.

First, the system allocates the lecture a_3 to the student s_2. Fig2 is a result of the allocation. Student's valuation to a lecture is shown as parenthetical reference.

Next, the system allocates based on evaluation values. Fig3 is a result of the allocation.

a_1's capacity is 2. The students that want to attend the lecture a_1 is 4. The system arranges their evaluation values in descending order. The system allocates the lecture a_1 to the student s_2 because s_2's evaluation value is the highest. s_1's evaluation value and s_3's evaluation value is equal. These values are the second highest value. However, a_1's capacity is 1. As a result, the system suspends attending a_1. The system does not allocate s_4 to a_1 because s_4's evaluation value was lower than suspended students.

a_2's capacity is 2. The students that want to attend the lecture a_2 is 4. The system arranges their evaluation values in descending order. s_3's evaluation value and s_4's evaluation value is equal. These values are the highest value. The system allocates the lecture a_2 to the students s_3 and s_4. In this time, a_2's capacity is 0. As a result, the system does not allocate s_1 and s_2 to a_2.

a_3's capacity is 2. The students that want to attend the lecture a_3 is 3. However, student s_2 have already been allocated a_3 in First. The rest of the students is 2 person, and the rest of a_3's capacity is 1. The system arranges their evaluation values in descending order. The system allocates the lecture a_3 to the student s_1 because his/her evaluation value is the highest. In this time, a_3's capacity is 0. As a result, the system do not allocate s_3 to a_3.

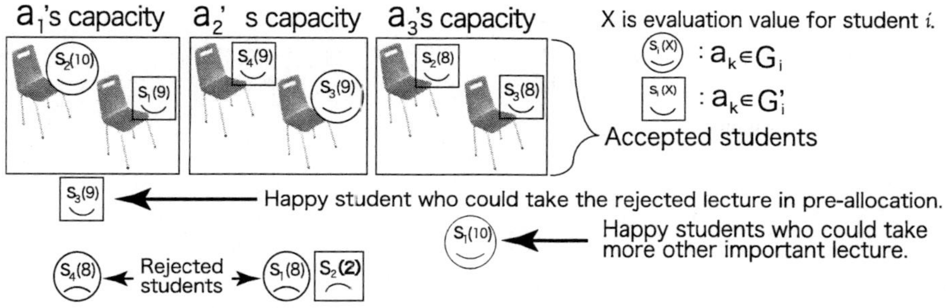

Fig. 4. The situation after allocation

Finally, the system allocates the suspended students based on negotiation. The system starts [STEP 5]. The lecture a_1 is included G'_1, and is included G'_3. $a_1 \in G'_1$, $a_1 \in G'_3$.

The student s_3 is not attended to the lecture a_3. The lecture a_3 is included G'_3. $a_3 \in G'_3$. The lecture a_3 is very important lecture for student s_3. The student s_1 is attended to the lecture a_3. a_3 is included G_1. $a_3 \in G_1$. This lecture is important lecture for student s_1. However, the lecture a_1 is more important for s_1.

The student s_1 negotiates with the student s_3. Their agents' aim is a better allocation. s_1's item for negotiation is permission for attending the lecture a_3. s_3's item for negotiation is surrender to the lecture a_1. The system allocates the lecture a_1 to the student s_1. The system cancels s_1's permission which attend the lecture a_3. The system allocates the lecture a_3 to the student s_3.

Result of negotiation, the system allocates two very important lectures. Fig4 is a result of the allocation. If the students does not negotiate, the system allocates one very important lecture and one important lecture. The negotiation is good for the student.

6   Discussion

In this paper, when the second choice is allocated, we use the method of allocating the first choice. Matsuo et al. suggested a method in which the student inputs multiple lectures in same timetable [5] [6]. In the method, when the lecture with the highest evaluation value is not allocated, the system communicates to the student with mail. The system allocates other lectures based on the result. We consider new method which mixed Matsuo's method and our method. We forecast that the new method is better allocation for the student.

In recent years, a lot of academic affairs information systems are operated. Kumamoto University in Japan operates the academic affairs information system. The system is called SOSEKI. SOSEKI is a system to store the score and the history of attending a lecture [7]. If the university uses both our algorithm and the academic affairs information systems which looks like SOSEKI, class allocation is more effectively.

We consider that we can decide the evaluation value from the factor other than the student's preference and purpose. The factor is student's profile and history of attending a lecture. If we can decide the evaluation value from the factor, the system is able to better class allocation for the students. In this situation, the students will not cancel allocated lectures. In this situation, the students will not attend the allocated lectures.

7   Conclusion

In this paper, we showed the method that the university allocates the lectures to the students. The method is allocation based on educational consideration and negotiation. The method allocates more important lectures for the students. This suggestion is one method, which discharges the problems. The problems appear when the system decides the evaluation value from the student's preferences and purposes. Although negotiation decreases total utility value, the system allocates more important lectures for the students. Our proposed method is better and efficient allocation for the students. Our future work includes an analysis of the situation where the multiple lectures are substitute and complementary.

References

1. Fujimoto, T., Matsuo, T.: A Lesson Selection and Class Organization Support System Based on Students' Utilities in Elective Subjects. In: CIEC, Computer and education, vol. 16, pp. 88–94 (2004)
2. Matsuo, T., Fujimoto, T.: A Class Organization Method based on Users Preferences. In: Proceedings of the IEICE General Conference (2005)
3. Varian, H.R.: Intermediate Microeconomics A Modern Approach, 7th edn., Keiso-Shobo (2007)
4. Konno, H.: Decision Making Method (in Japanese), Asakura-Shuppan (1992)
5. Fujimoto, T., Matsuo, T.: A Class Organization Support System based on Students' Utilities in Elective Subjects, Research reports of Kanagawa Institute of Technology. Part B. Science and technology 29, 69–74 (2005)
6. Matsuo, T., Fujimoto, T.: ClassCoordinator: Class Organization Support System, Human interface. The Transaction of Human Interface Society 7(3), 361–368 (2005)
7. Kita, H., Ida, M.: University Evaluation and University Information Database. Research in university evaluation (3) (2003)

AlineaGA: A Genetic Algorithm for Multiple Sequence Alignment

Fernando José Mateus da Silva[1], Juan Manuel Sánchez Pérez[2],
Juan Antonio Gómez Pulido[2], and Miguel A. Vega Rodríguez[2]

[1] Dept. of Informatics Engineering, School of Technology and Management,
  Polytechnic Institute of Leiria, Portugal
  `fernando.silva@estg.ipleiria.pt`
[2] Dept. Tecnologías Computadores y Comunicaciones,
  Escuela Politécnica, Universidad de Extremadura, Spain
  `{sanperez,jangomez,mavega}@unex.es`

Abstract. The alignment and comparison of DNA, RNA and Protein sequences is one of the most common and important tasks in Bioinformatics. However, due to the size and complexity of the search space involved, the search for the best possible alignment for a set of sequences is not trivial. Genetic Algorithms have a predisposition for optimizing general combinatorial problems and therefore are serious candidates for solving multiple sequence alignment tasks. We have designed a Genetic Algorithm for this purpose: AlineaGA. We have tested AlineaGA with representative sequence sets of the hemoglobin family. We also present the achieved results so as the comparisons performed with results provided by T-COFFEE.

Keywords: Multiple sequence alignments, genetic algorithms, bioinformatics.

1 Introduction

One of the most common tasks in bioinformatics is the alignment and comparison of DNA, RNA and Protein sequences. An alignment is a mutual placement of two or more sequences which exhibits where the sequences are similar and where they differ [1]. The problem of multiple sequence alignments is an optimization problem searching for the best alignment from large, complex search spaces [2].

Genetic Algorithms (GAs) are a biological inspired technology which conducts randomized search and optimization techniques guided by the principles of evolution and natural genetics. They are efficient, adaptive and robust search processes which produce near optimal solutions, having a large degree of implicit parallelism [1], making them appropriate for solving the problem of multiple sequence alignment.

We present a new application for multiple sequence alignment that uses a GA to fulfill its task: AlineaGA. This application was tested using representative sequence sets of the hemoglobin family. Then, the results provided by this application were compared with the ones provided by T-COFFEE [3], allowing to evaluate the quality of the implemented algorithm and to take conclusions regarding future developments.

2 Background

Since its introduction in the early seventies, multiple sequence alignment has become a cornerstone of modern molecular biology [4]. It may not be obvious, but multiple

N.T. Nguyen, R. Katarzyniak (Eds.): New Chall. in Appl. Intel. Tech., SCI 134, pp. 309–318, 2008.
springerlink.com   © Springer-Verlag Berlin Heidelberg 2008

alignments are present in most of the computational methods used in molecular biology. Multiple alignments are used to study molecular evolution, to help predict the secondary or tertiary structure of new sequences, RNA folding, gene regulation and polymerase chain reaction primer design [5]. They are also applied to different areas such as functional genomics, evolutionary studies, structure modelling, mutagenesis experiments and drug design.

2.1 Alignment

An alignment is a mutual placement of two or more sequences which exhibit where the sequences are similar, and where they differ. An optimal alignment is the one that exhibits the most correspondences and the fewest differences. It is the alignment with the highest score, but which may or may not be biologically meaningful [1].

Multiple sequence alignment can help compare the structure-relationship between sequences by simultaneously aligning multiple sequences to construct connections between the elements in different sequences [2]. Figure 1 shows an example of an alignment of four hypothetical protein sequences.

```
-TISCTGNIGAG-NHVKWYQQLPG
-RLSCSSIFSS--YAMYWVRQAPG
L-LTCTVSFDD--YYSTWVRQPPG
PEVTCVVSHEDPQVKFNWYVQ--G
```

Fig. 1. Example of a multiple sequence alignment

Each input sequence is represented in a different line and can have different lengths. Columns with different symbols in the alignment show that several mutation events take place. Columns with identical symbols, which are represented as bold characters, indicate that no mutation occurs. The symbol "–" is used to represent a space introduced in the sequence in order to improve the alignment result. This space is usually referred to as a gap. The introduction of gaps into sequence alignments allows the alignment to be extended into regions where one sequence may have lost or gained sequence characters not found in the other. A score measures the performance of the alignment. Lower mutation obtains higher scores. The best alignment is the one with the highest score.

2.2 Genetic Algorithms

GAs were introduced in 1975 by Holland [6]. They are stochastic algorithms whose search methods model some natural phenomena: genetic inheritance and Darwinian strife for survival [7]. Several differences distinguish them from classic optimization and search methods: (1) instead of working directly with the problem parameters, they use a codified version of them; (2) they use a population of candidate solutions in their search for a set of points; (3) they employ probabilistic transition rules [8]. The fundaments of GAs are based on three central elements: (1) a Darwinian concept of "fitness", which controls the degree of influence of an individual on future generations; (2) a "mating operator", which generates offspring for the next generation, and (3) "genetic operators", which establish the offspring genetic composition from the genetic material of its parents.

In GAs, adaptation proceeds by maintaining a population of structures from which new structures are created using genetic operators such as crossover and mutation [9]. Crossover combines the features of randomly chosen individuals (parents) to form two similar offspring by swapping corresponding segments of the parents. Mutation arbitrarily alters some values within the individual, by a random change with a probability defined by a mutation rate [5]. Each structure in the population has an associated fitness score, and these scores are used in a competition that determines which structures are used to form new ones [9].

A distinctive feature of GAs is their aptitude to take advantage of accumulating information about an initially unknown search space, in order to bias subsequent search into useful subspaces, making them suitable for problems with large, complex, and poorly understood search spaces [9].

3   AlineaGA Methods

This section presents a brief explanation about the methods of representation, selection, evaluation, crossover and mutation used by AlineaGA in order to perform the sequence alignment task.

3.1   Representation and Initialization

Our approach uses the same representation of SAGA [10] and RAGA [11]. This representation uses a real multiple sequence alignment as a data structure for each individual. This means that the chromosomes are represented by arrays of characters on which each line corresponds to a sequence in the alignment, and each column represents the residue present at a sequence position. Figure 2 presents an illustration of this representation.

W	G	K	V	N	–	–	V	D	E	V	–	G
W	D	K	V	N	E	E	E	–	–	–	V	G
W	G	K	V	G	–	–	A	H	A	G	E	Y
W	S	K	V	G	G	H	A	–	–	G	E	Y

Fig. 2. Example of the individual's representation

The possible values to each component of the individual (the alleles) are C, S, T, P, A, G, N, D, E, Q, H, R, K, M, I, L, V, F, Y, and W for protein sequences, and A, T, C and G for nucleotide sequences. In addiction the symbol "–" is used to represent a gap in the sequence.

In order to form the initial population, a desired number of individuals are generated by the following process: (1) load each sequence to each line of an array; (2) determine the size of the largest sequence; (3) determine a random number between 0 and 25% of the size of the largest loaded sequence; (4) Complete each sequence with the gap sign ("–") until it reaches the size of the largest sequence plus the random number of gaps calculated in (3). The gaps can be placed in the sequences accordingly to the preference of the user: at random positions within each sequence; at the beginning of each sequence; at the end of each sequence.

3.2 Evaluation

The sum-of-pairs score is used as a fitness measure to evaluate the population. This choice was made not only due to its relative speed and robustness [12], but also because of its simplicity. This score is assessed by scoring all of the pairwise comparisons between each residue in each column of an alignment and adding the scores together [13]. To do this calculus, it is necessary to use a scoring matrix which determines the cost of substituting a residue for another, and also a gap penalty value to determine the cost of aligning a residue with a gap.

The PAM 350 score matrix was implemented because it is used in reference programs such as CLUSTAL W [14] or T-COFFEE [3].

The gap penalty values are directly dependant on the scoring matrix. If gaps are introduced without a penalty, then they can be introduced at random and eventually all characters will be aligned in pairs of random sequences. There is not yet a theory for choosing a gap penalty for a given scoring matrix; therefore these values are set empirically. For PAM350 we use a gap opening penalty of -10 [15].

3.3 Selection

In order to perform selection two methods are available: Tournament selection and Roulette Wheel selection.

Tournament Selection. This type of selection runs a tournament between a specified number of individuals which are chosen randomly from the existing population. The winner of this tournament is the one that presents the higher fitness value. This operator allows adjusting the evolution pressure simply by changing the tournament size. A larger tournament size reduces the chance of selecting weaker individuals.

Roulette Wheel Selection. This type of selection associates a probability of selection with each individual in the population in proportion with its fitness value. Due to the use of a scoring matrix, it is possible that an individual has a negative fitness value. In order to assign the correct size in the wheel to every individual, a transformation is made to all fitness values by adding the worst fitness value in the population to all its individuals. With this transformation, the population will have greater than zero fitness values and consequently the correct corresponding size in the wheel.

3.4 Crossover

We opted for using one type of crossover: the one point crossover operator. This operator derives from Goldberg's standard one-point crossover operator [16] with an extension that treats the existing gaps in each sequence. Two parent alignments are combined through a single exchange. The first parent is cut straight at a randomly chosen position. The second one is tailored so that the right piece can be joined to the left piece of the first parent and vice versa. Any vacant space that appears at the junction point is filled with null signs [5]. Because of the specificity of this junction point, where rearrangements can occur, this operator combines both the traditional properties of a crossover and those of a local rearrangement mutation [10].

3.5 Mutation

We use two of the most popular and effective methods of mutation – the gap insertion and the gap shifting operator. In order to improve these methods some modifications were made which resulted in new mutation operators: the Smart Gap Insertion and Smart Gap Shifting. These operators act according to a user defined mutation probability. A mutation operator that removes columns of gaps was also implemented.

Gap Insertion. The gap insertion operator extends the alignments by inserting gaps. This operator produces a mutation identical to the GenAlignRefine [13] gap insertion operator which simply inserts a gap into every sequence of the alignment in a random fashion. However, in our implementation there's a small variation which allows the user to decide the number of gaps that he wishes to introduce on each mutation.

Smart Gap Insertion. This is a variation of the Gap Insertion operator, which only produces the mutation if the fitness of the mutated alignment is greater than the fitness of the original one. In order to do this mutation, a random position in the alignment matrix (line and column indexes) is chosen. Then a desired number of gaps are inserted on that position and all the other lines are filled with gaps until they all have the same size, as illustrated in Figure 3.

The insertion of these additional gaps is made at the beginning or at the end of the sequences lines according to a direction probability. The direction probability is initially set to 50% at the beginning of each generation, and according to the results will be increased (increasing the possibility of the gaps be inserted at the beginning of the remaining sequences), or decreased (increasing the possibility of the gaps be inserted at the end of the remaining sequences). Yet, the mutation and the change on the direction probability will only occur if the fitness of the generated alignment is greater than the original one. If the operator is unable to improve the alignment at the first attempt, it tries to choose another random position and repeats the whole process. The defined number of maximum attempts is set to 3.

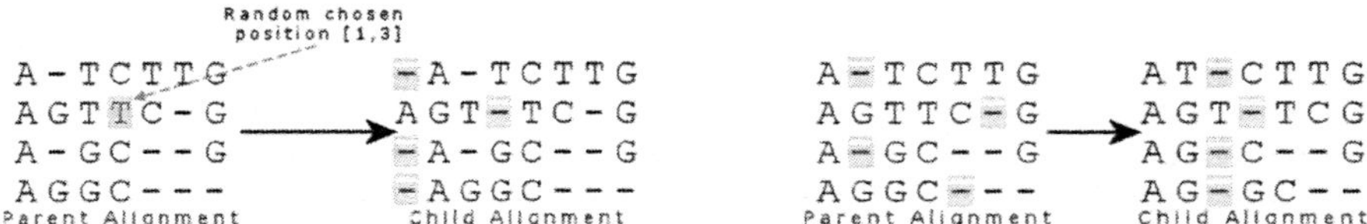

<table>
<tr><td>Fig. 3. Smart Gap Insertion mutation</td><td>Fig. 4. Gap Shifting mutation</td></tr>
</table>

Gap Shifting. Shifting gaps is another way to introduce new alignment configuretions. In RAGA's gap shifting operator, a gap is randomly chosen in an alignment and it is moved to another position [11]. We used the same concept, as it is demonstrated in Figure 4.

Smart Gap Shifting. The Smart Gap Shifting tries to move the gaps of an alignment until its fitness improves. As in the Smart Gap Insertion operator, the shift direction is determined by a direction probability which initially is set to 50% and that is updated when better alignments are found. The mutation only occurs if the fitness of the generated alignment is better than the original one.

Gap Column Remover. The function of this operator is to remove columns of gaps that can be generated by crossover and mutation operations. In order to "clean" the existing gap columns from alignments, this mutation occurs at the end of each generation. This operator isn't influenced by the mutation probability.

4 Testing and Results

This section presents the data sets used to test AlineaGA as well as the achieved results. In order to assess the quality of the resulting alignments, our results are compared with the ones achieved by T-COFFEE [3] Version 5.05, which is available online at http://www.ebi.ac.uk/t-coffee/.

4.1 Test Data Sets and Configurations

A representative set of sequences of the hemoglobin family was chosen as a data set to test the ability of aligning sequences. These includes α and β globins from mammals and birds. Subsets of 2, 4 and 6 (total) were used to assess the aptitude of the developed algorithm. This division was made because it is important to evaluate the effects of the number of sequences in the final results. The sequence test sets are: Subset 1: Human Hemoglobin α (HBA_HUMAN), Human Hemoglobin β (HBB_HUMAN); Subset 2: subset 1 sequences, Duckbill platypus Hemoglobin α (HBA_ORNAN), Duckbill platypus Hemoglobin β (HBB_ORNAN); Subset 3: subset 2 sequences, Anas platyrhynchos Hemoglobin α (HBA_ANAPL), Anas platyrhynchos Hemoglobin β (HBB_ANAPL). The test configurations which allowed AlineaGA to produce the best possible results are presented in Table 1.

Table 1. Test parameters configurations for AlineaGA

	Config. 1	Config. 2	Config. 3	Config. 4
Population Size	100	100	100	100
Number Generations	300	300	300	300
Initial Gap Insertion	Random	Random	Random	Random
Selection	Tourn. (2)	Tourn. (2)	Tourn. (2)	Tourn. (2)
Crossover Probability	0.2	0.2	0.2	0.2
Mx Prob. (Shift)	0.6	0.6	0.6	0.8
Gap Shifting	On	On	On	On
Smart Gap Shifting	On	On	On	On
Mx Prob. (Insert)	-	0.1	0.1	-
Gap Insertion	Off	Off	Off	Off
Smart Gap Insertion	Off	On (5 gaps)	On (10 gaps)	Off

Config., configuration; Tourn. (2), tournament size 2; Mx Prob. (Shift), mutation probability for shifting operators; Mx Prob. (Insert), mutation probability for insertion operators.

4.2 Results

The results obtained by AlineaGA for the test configurations and a brief comparison with T-COFFEE, are presented in Table 2.

Table 2. Results for AlineaGA and comparison with T-COFFEE

Data Set	Config.	Avg. Fitness Final Pop.	Fitness Best Individual	Id. Columns		Inserted Gaps	
				AlGA	T-COF	AlGA	T-COF
Subset 1	1	344.066(6)	356.466(6)	65	61	9	9
Subset 1	2	329.2	360.933(3)	63	61	9	9
Subset 1	3	327.033(3)	360.266(6)	65	61	9	9
Subset 1	4	339.733(3)	357.533(3)	64	61	9	9
Subset 2	1	1527.966(6)	1617.2	42	46	16	20
Subset 2	2	1446.4	1588.2	37	46	12	20
Subset 2	3	1400.9	1582.2	37	46	12	20
Subset 2	4	1531.66(6)	1623.166(6)	38	46	12	20
Subset 3	1	2675.4	2853.733(3)	30	39	17	29
Subset 3	2	2473.533(3)	2744.133(3)	29	39	17	29
Subset 3	3	2448.7	2720.466(6)	31	39	17	29
Subset 3	4	2789.233(3)	2914.9	29	39	17	29

Config., parameter configuration; Avg. Fitness Final Pop., average fitness of the final population; Id. Columns, fully identical columns; AlGA, AlineaGA; T-COF, T-COFFEE. Avg. Fitness Final Pop. and fitness of the best individual were obtained by averaging the results of 30 executions of each test.

Figures 5, 6 and 7 presents a comparison between the best individual found by AlineaGA and the results obtained by T-COFFEE. The identical regions between both alignments are filled in grey. Boxes mark interesting regions of alignment found by AlineaGA and not found by T-COFFEE. The symbol "*" identifies columns that are fully aligned.

While T-COFFEE fully aligns 61 residues, AlineaGA fully aligns 65. According to the optimal alignment definition, which states that it is the one that exhibits the most correspondences and the fewest differences, but which may or may not be biologically meaningful [1], AlineaGA presents a better result.

```
AlineaGA
HBA_HUMAN   MV-LSPADKTNVKAAWGKVGAHAGEYGAEALERMFLSFPTTKTYFPHF-DLS--H---GSAQVKGHGKKVADALT
HBB_HUMAN   MVHLTPEEKSAVTALWGKVN--VDEVGGEALGRLLVVYPWTQRFFESFGDLSTPDAVMGNPKVKAHGKKVLGAFS
            ** * *   *    * * ****    *  * *** *      *  *     *   * ***      *    ** ***** *

T-COFFEE
HBA_HUMAN   -MVLSPADKTNVKAAWGKVGAHAGEYGAEALERMFLSFPTTKTYFPHF------DLSHGSAQVKGHGKKVADALT
HBB_HUMAN   MVHLTPEEKSAVTALWGKV--NVDEVGGEALGRLLVVYPWTQRFFESFGDLSTPDAVMGNPKVKAHGKKVLGAFS
             *  *   *    * * ****    *  * *** *      *   *    *     *      *    ** ***** *

AlineaGA
HBA_HUMAN   NAVAHVDDMPNALSALSDLHAHKLRVDPVNFKLLSHCLLVTLAAHLPAEFTPAVHASLDKFLASVSTVLTSKYR
HBB_HUMAN   DGLAHLDNLKGTFATLSELHCDKLHVDPENFRLLGNVLVCVLAHHFGKEFTPPVQAAYQKVVAGVANALAHKYH
            ** *       ** **    ** ***  ** **      *    ** *    **** * *     *   * *     *   **

T-COFFEE
HBA_HUMAN   NAVAHVDDMPNALSALSDLHAHKLRVDPVNFKLLSHCLLVTLAAHLPAEFTPAVHASLDKFLASVSTVLTSKYR
HBB_HUMAN   DGLAHLDNLKGTFATLSELHCDKLHVDPENFRLLGNVLVCVLAHHFGKEFTPPVQAAYQKVVAGVANALAHKYH
            ** *       ** **    ** ***  ** **      *    ** *    **** * *     *   * *     *   **
```

Fig. 5. AlineaGA alignment (configuration 1) VS T-COFFEE alignment for subset 1

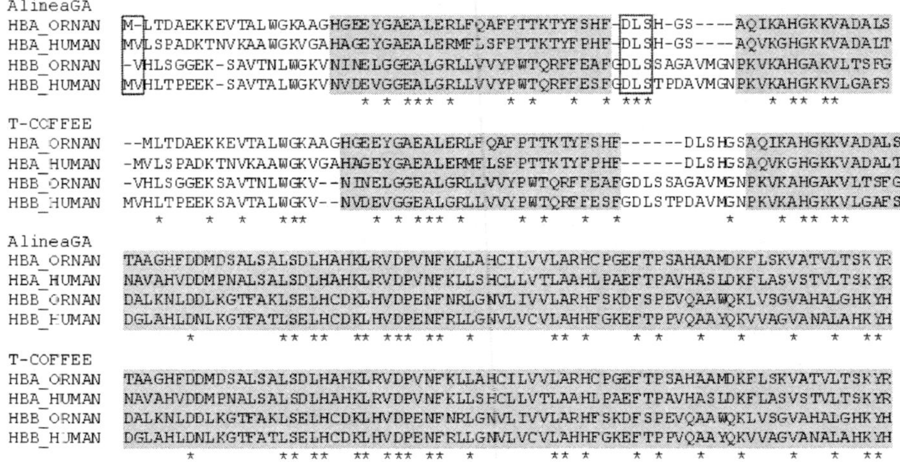

Fig. 6. AlineaGA alignment (configuration 1) VS T-COFFEE alignment for subset 2

T-COFFEE alignment presents 46 identical columns against the 42 aligned residues by AlineaGA. T-COFFEE achieves a better solution than AlineaGA however, there are interesting regions which are detected by AlineaGA but are missed by T-COFFEE.

Again, T-COFFEE performs better aligning 39 residues while AlineaGA aligns 31.

```
AlineaGA
HBB_ORNAN  -VHLSGGEKSAVTNLWGKVNINELGGEALGRLLVVYPWTQRFFEAFGDLSSAGAVMGNPKVKAHGAKVLTSFG
HBB_ANAPL  MVHWTAEEKQLITGLWGKVNVADCGAEALARLLIVYPWTQRFFASFGNLSSPTAILGNPMVRAHGKKVLTSFG
HBB_HUMAN  MVHLTPEEKSAVTALWGKVNVDEVGGEALGRLLVVYPWTQRFFESFGDLSTPDAVMGNPKVKAHGKKVLGAFS
HBA_ORNAN  -M-LTDAEKKEVTALWGKAAGHGEEYGAEA-LERL--FQ-AFPTTKTYFSHFDLSHGSAQIKAHGKKVADALS
HBA_HUMAN  -MVLSPADKTNVKAAWGKVGAHAGEYGAEA-LERM--FL-SFPTTKTYFPHFDLSHGSAQVKGHGKKVADALT
HBA_ANAPL  -MVLSAADKTNVKGVFSKIGGHAEEYGAET-LER---MFIAYPQTKTYFPHFDLSHGSAQIKAHGKKVAAALV
                  *          *          *    *                *     ** **

T-COFFEE
HBB_ORNAN  -VHLSGGEKSAVTNLWGKV--NINELGGEALGRLLVVYPWTQRFFEAFGDLSSAGAVMGNPKVKAHGAKVLTSFG
HBB_ANAPL  MVHWTAEEKQLITGLWGKV--NVADCGAEALARLLIVYPWTQRFFASFGNLSSPTAILGNPMVRAHGKKVLTSFG
HBB_HUMAN  MVHLTPEEKSAVTALWGKV--NVDEVGGEALGRLLVVYPWTQRFFESFGDLSTPDAVMGNPKVKAHGKKVLGAFS
HBA_ORNAN  --MLTDAEKKEVTALWGKAAGHGEEYGAEALERLFQAFPTTKTYFSHF-DLS-----HGSAQIKAHGKKVADALS
HBA_HUMAN  -MVLSPADKTNVKAAWGKVGAHAGEYGAEALERMFLSFPTTKTYFPHF-DLS-----HGSAQVKGHGKKVADALT
HBA_ANAPL  -MVLSAADKTNVKGVFSKIGGHAEEYGAETLERMFIAYPQTKTYFPHF-DLS-----HGSAQIKAHGKKVAAALV
                  *          *       * * *     *   *   *   *   **        *     ** **

AlineaGA
HBB_ORNAN  DALKNLDDLKGTFAKLSELHCDKLHVDPENFNRLGNVLIVVLARHFSKDFSPEVQAAWQKLVSGVAHALGHKYH
HBB_ANAPL  DAVKNLDNIKNTFAQLSELHCDKLHVDPENFRLLGDILIIVLAAHFTKDFTPECQAAWQKLVRVVAHALARKYH
HBB_HUMAN  DGLAHLDNLKGTFATLSELHCDKLHVDPENFRLLGNVLVCVLAHHFGKEFTPPVQAAYQKVVAGVANALAHKYH
HBA_ORNAN  TAAGHFDDMDSALSALSDLHAHKLRVDPVNFKLLAHCILVVLARHCPGEFTPSAHAAMDKFLSKVATVLTSKYR
HBA_HUMAN  NAVAHVDDMPNALSALSDLHAHKLRVDPVNFKLLSHCLLVTLAAHLPAEFTPAVHASLDKFLASVSTVLTSKYR
HBA_ANAPL  EAVNHVDDIAGALSKLSDLHAQKLRVDPVNFKFLGHCFLVVVAIHHPAALTPEVHASLDKFMCAVGAVLTAKYR
              *       ** **   ** *** **   *        * *     *   *   *      *    *   **

T-COFFEE
HBB_ORNAN  DALKNLDDLKGTFAKLSELHCDKLHVDPENFNRLGNVLIVVLARHFSKDFSPEVQAAWQKLVSGVAHALGHKYH
HBB_ANAPL  DAVKNLDNIKNTFAQLSELHCDKLHVDPENFRLLGDILIIVLAAHFTKDFTPECQAAWQKLVRVVAHALARKYH
HBB_HUMAN  DGLAHLDNLKGTFATLSELHCDKLHVDPENFRLLGNVLVCVLAHHFGKEFTPPVQAAYQKVVAGVANALAHKYH
HBA_ORNAN  TAAGHFDDMDSALSALSDLHAHKLRVDPVNFKLLAHCILVVLARHCPGEFTPSAHAAMDKFLSKVATVLTSKYR
HBA_HUMAN  NAVAHVDDMPNALSALSDLHAHKLRVDPVNFKLLSHCLLVTLAAHLPAEFTPAVHASLDKFLASVSTVLTSKYR
HBA_ANAPL  EAVNHVDDIAGALSKLSDLHAQKLRVDPVNFKFLGHCFLVVVAIHHPAALTPEVHASLDKFMCAVGAVLTAKYR
              *       ** **   ** *** **   *        * *     *   *   *      *    *   **
```

Fig. 7. AlineaGA alignment (configuration 3) VS T-COFFEE alignment for subset 3

5 Discussion

The implementation of this method is relatively simple and requires: coding the representation of the individuals – usually each individual is an alignment of sequences; three operations that allows evolving the existing population (selection, crossover and mutation) [17]; and an objective function which allows determining the fitness of the individuals.

AlineaGA uses several mutation operators. Some of these operators, namely the "Smart" ones, have improvements in its design relatively to existing solutions. Nevertheless, AlineaGA worked perfectly for two sequences, presenting better results than T-COFFEE [3], but didn't prove to be able to align more than a few sequences. This appears to be due to a lack of diversity in the population that is disabling the individuals to improve for better solutions.

There are still many enhancements that must be done to AlineaGA in order to achieve satisfying results. It is critical that a multiple sequence alignment tool succeeds in aligning more than just a few sequences. A better placement of the gaps along the sequences must be made and different mutation operators must be tested. Also, new fitness functions based on different scoring methods, such as travelling salesman or tree-based approaches, are possible straightforward developments.

In conclusion, one must not forget that biological reasoning is still mandatory in evaluating the results provided by multiple sequence alignment programs, because even evolution capable simulating machines don't do biology.

References

1. Pal, S.K., Bandyopadhyay, S., Ray, S.S.: Evolutionary computation in bioinformatics: A review. IEEE Transactions on Systems Man and Cybernetics Part C-Applications and Reviews 36, 601–615 (2006)
2. Horng, J., Wu, L., Lin, C., Yang, B.: A genetic algorithm for multiple sequence alignment. Soft Computing 9, 407–420 (2005)
3. Notredame, C., Higgins, D.G., Heringa, J.: T-Coffee: A novel method for fast and accurate multiple sequence alignment. Journal of Molecular Biology 302, 205–217 (2000)
4. Lecompte, O., Thompson, J.D., Plewniak, F., Thierry, J.-C., Poch, O.: Multiple alignment of complete sequences (MACS) in the post-genomic era, Gene, pp. 17–30 (2001)
5. Anbarasu, L.A., Narayanasamy, P., Sundararajan, V.: Multiple molecular sequence alignment by island parallel genetic algorithm. Current Science 78, 858–863 (2000)
6. Holland, J.H.: Adaptation in Natural and Artificial Systems. University of Michigan Press, Ann Arbor (1975)
7. Michalewicz, Z.: Genetic algorithms + data structures = evolution programs - Third, Revised and Extended Edition. Springer, Heidelberg (1996)
8. de Vega, F.F.: Parallel and Distributed Genetic Programming Models with applications to Logic Síntesis on FPGAs. Computer Science Department, vol. PhD., Universidad de Extremadura, Cáceres, p. 156 (2001)
9. De Jong, K.: Learning with genetic algorithms: An overview. Machine Learning 3, 121–138 (1988)
10. Notredame, C., Higgins, D.G.: SAGA: sequence alignment by genetic algorithm. Nucleic Acids Research 24, 1515–1524 (1996)

11. Notredame, C., O'Brien, E.A., Higgins, D.G.: RAGA: RNA sequence alignment by genetic algorithm. Nucleic Acids Research 25, 4570–4580 (1997)
12. Nicholas Jr., H.B., Ropelewski, A.J., Deerfield II, D.W.: Strategies for Multiple Sequence Alignment. BioTechniques 32, 572–591 (2002)
13. Wang, C., Lefkowitz, E.J.: Genomic multiple sequence alignments: refinement using a genetic algorithm. BMC Bioinformatics 6 (2005)
14. Thompson, J.D., Higgins, D.G., Gibson, T.J.: CLUSTAL W: improving the sensitivity of progressive multiple sequence alignment through sequence weighting, position-specific gap penalties and weight matrix choice. Nucleic Acids Research 22, 4673–4680 (1994)
15. L, J.: Calculate PAM Matrix. Wageningen Bioinformatics Webportal (2004),
    `http://www.bioinformatics.nl/tools/pam.html`
16. Goldberg, D.E.: Genetic Algorithms in Search, Optimization, and Machine Learning. Addison-Wesley Publishing Company, Reading (1989)
17. Zhang, C., Wong, A.K.C.: A genetic algorithm for multiple molecular sequence alignment. Comput. Appl. Biosci. 13, 565–581 (1997)

Investigation of Strategies for the Generation of Multiclass Support Vector Machines

Ana Carolina Lorena[1] and André C.P.L.F. de Carvalho[2]

[1] Centro de Matemática, Computação e Cognição,
Universidade Federal do ABC, 09090-400, Santo André, SP, Brazil
`ana.lorena@ufabc.edu.br`
[2] Instituto de Ciências Matemáticas e de Computação
Universidade de São Paulo, 13560-970, São Carlos, SP, Brazil
`andre@icmc.usp.br`

Abstract. Support Vector Machines constitute a Machine Learning technique originally designed for the solution of two-class problems. This paper investigates and proposes strategies for the generalization of SVMs to problems with more than two classes. The focus of this work is on strategies that decompose the original multiclass problem into binary subtasks, whose outputs are combined. The proposed strategies aim to investigate the adaptation of the decompositions for each multiclass application considered, using information of the performance obtained in its solution or extracted from its examples. The implemented algorithms were evaluated using benchmark datasets and real applications from the Bioinformatics domain. Among the benefits observed is the obtainment of simpler decompositions, which require less binary classifiers in the multiclass solution.

Keywords: Support Vector Machines, multiclass problems, Hybrid Intelligent Systems.

1 Introduction

Several problems involve the classification of data into categories or classes. From datasets whose data have known classification, Machine Learning (ML) [16] algorithms can be used in the induction of a classifier able to perform the desired discrimination for new data. Among the existing ML techniques are Support Vector Machines (SVMs) [5], which are originally formulated for the solution of two-class (binary) problems. This paper studies and proposes strategies for the generalization of SVMs to problems with more than two classes, known as multiclass problems.

There are two approaches for the generalization of SVMs to multiclass problems. The first, more frequent, is to decompose the problem in several binary subproblems. The second approach modifies the original training algorithm of the SVMs, creating multiclass versions, which in general yield to costly algorithms [8, 20]. Thus, the alternative of decomposing the multiclass problem into binary subproblems is more frequent.

N.T. Nguyen, R. Katarzyniak (Eds.): New Chall. in Appl. Intel. Tech., SCI 134, pp. 319–328, 2008.
springerlink.com

The focus of this work is on variations of the decomposition strategy. The objective was to investigate the adaptation of decomposition solutions, using binary SVMs, to each particular multiclass problem. Two approaches were followed in this process. The first uses information from the performance of the solutions to adapt them to the problems. The second uses information extracted from the datasets to define the decompositions performed.

This paper is structured as follows: Section 2 presents the main decomposition strategies from the literature. Section 3 discusses the motivations of this work and presents the research conducted. Section 4 presents a general discussion of experimental results obtained in the study. Section 5 concludes this paper.

2 Decomposition Strategies

Code-Matrix. The code-matrix approach unifies various decomposition strategies [1], which can be generally represented by a code-matrix $\mathbf{M}$. The lines of this matrix have codes attributed to each class. The columns of $\mathbf{M}$ define binary partitions of the k classes and correspond to the labels that these classes assume in the binary predictors generation. Thus, $\mathbf{M}$ has kxl dimension, in which l denotes the number of binary classifiers used. Each element of $\mathbf{M}$ assumes values in the set $\{-1, 0, +1\}$. An element m_{ij} with $+1$ value indicates that the class correspondent to line i assume positive label in classifier f_j induction. The value -1 designates a negative label and the value 0 indicates that data from class i do not participate in classifier f_j induction. Binary classifiers are trained to learn the labels represented in the columns of $\mathbf{M}$.

A new data $\mathbf{x}$ can be classified by evaluating the predictions of the l classifiers, which generate a vector $\mathbf{f}(\mathbf{x}) = (f_1(\mathbf{x}), \ldots, f_l(\mathbf{x}))$. This vector is then compared to the lines of $\mathbf{M}$. The data is attributed to the class whose line is closest according to some distance measure, like the margin-based for SVMs [1], used in this paper.

To decompose the multiclass problem, several strategies can be employed. Among the most common, one can cite the one-against-all (1AA), the one-against-one (1A1) [10] and the use of error correcting output codes [6].

In the 1AA strategy, given a problem with k classes, k binary classifiers $f_i(\mathbf{x})$ are generated. Each predictor is trained to distinguish a class i from the others. The representation of this technique is given by a matrix of dimension kxk, in which the diagonal elements have value $+1$ and the others, the value -1.

In the 1A1 decomposition, $\frac{k(k-1)}{2}$ binary classifiers are generated. Each one is responsible to differentiate a pair of classes (i, j), in which $i \neq j$. The code-matrix in this case has dimension kx$\frac{k(k-1)}{2}$ and each column corresponds to a binary classifier for a pair of classes. In a column representing the pair (i, j), the value of element in line i is $+1$ and the value of the member in line j is equal to -1. All other elements in the column have the value 0, indicating that examples from the other classes do not participate in this classifier induction.

In an alternative decomposition strategy, Dieterich and Bariki [6] proposed the use of error correcting output codes (ECOCs) to represent the k classes of the multiclass problem. One of the methods suggested to obtain ECOC codes is exhaustive, in which the obtained matrix has $2^{k-1}-1$ columns. The code of the first class is composed of $+1$ values. For each other class i, in which $i > 1$, it is composed of alternate runs of 2^{k-i} negative (-1) and positive $(+1)$ labels.

Hierarchical. An alternative way to solve multiclass problems with binary predictors can be performed by disposing them in a hierarchical structure, composed of nodes and ramifications. Internal nodes correspond to binary classifiers, the ramifications represent the outputs of these classifiers and the leaf nodes represent the problem classes. Figure 1 illustrates examples of the hierarchical structures employed in this work.

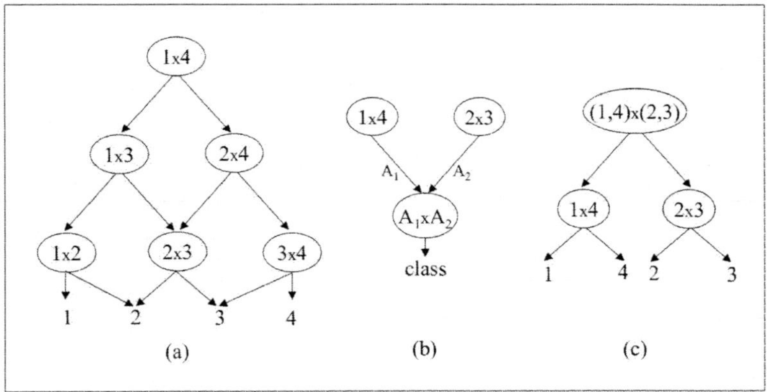

Fig. 1. Examples of hierarchical structures for a problem with four classes

To classify a new data, the nodes and ramifications are traversed accordingly until a leaf node is reached. To define the binary partitions of classes in the hierarchy, which is equivalent to decompose the multiclass problem, several strategies can be employed.

Platt et al. [18] suggested disposing the classifiers produced by the 1A1 decomposition in a Decision Directed Acyclic Graph (DDAG). Figure 1a shows an example of DDAG for a problem with four classes. A disadvantage of the DDAG, pointed in [9], is that different configurations of the binary classifiers in the graph may influence the results achieved. Besides that, depending of the correct class position in the graph, the number of evaluations with it may be unnecessarily high, resulting in a high cumulative error.

These facts motivated the development of an Adaptive DAG (ADAG) to combine the outputs of 1A1 classifiers [9], which corresponds to a DAG with reverse structure. Figure 1b presents an example of ADAG for a problem with four classes. The first layer of the ADAG has binary classifiers that distinguish pairs of all classes. The binary classifiers contained in the other layers are determined according to the binary predictions from previous levels, being adaptable for each data. In a prediction, the correct class is tested against others $\lceil \log_2 k \rceil$ times or less, while in the DDAG this number can reach $k - 1$. The occurrence of cumulative errors is thus minimized. Although the ADAGs have also demonstrated less dependency in relation to their structure, there can be still results disparities among different structures.

Another common type of hierarchical structure is of a tree, in which, apart from the root node, each node has one unique parent. Figure 1c illustrates an example of a tree for a problem with four classes. Trees require the training of $k - 1$ binary classifiers, the lower number among the decomposition strategies presented. The tree structure may influence its results, thus attention must be given to the definition of the binary partitions of the classes in each node of the tree. Generally, a specific criterion is applied recursively to subsets of classes, dividing them in two until one unique class remains [21, 22, 23].

3   Research Performed

This work is concerned with the investigation of the adaptation of decomposition solutions, using binary SVMs, to each particular multiclass problem.

3.1   Code-Matrix Design

A common criticism to the 1AA, 1A1 and ECOC strategies is that all of them decompose the problem a priori, without considering the characteristics of the applications [1]. In this work, Genetic Algorithms (GAs) [17] were used to design code-matrices related to each application, adapting them according to the performance obtained in the solution of the multiclass problems considered [14].

A matrix structure was adopted in the individuals' representation. Each individual corresponds to a possible code-matrix $\mathbf{M}$ of size kxl. Individuals of varying sizes (with different l values) were allowed in a same population. Herewith, the number of binary classifiers in the code-matrices is also determined by the GA. The maximum size of the matrices generated is limited by the user. The matrices must also have at least $\lceil \log_2 k \rceil$ binary classifiers, which corresponds to the minimum necessary to differentiate k classes.

The evaluation of the matrices was performed according to their performance in the multiclass problem solution. The GA searches matrices that minimize an error measure, estimated in a validation set. It also attempted to minimize the number of columns contained in the code-matrices. This criterion represents the search for simpler solutions and is in accordance to the Occam's razor, which affirms that, among several correct hypotheses, the simpler one must be chosen [16]. Equivalent columns were also avoided in the matrices, which would represent the use of identical binary classifiers in a same decomposition. Two columns are considered equivalent if all their elements are equal or complementary.

There are then three objectives: minimize the matrix error and its number of columns and avoid the presence of equivalent columns. Two multi-objectives GAs were applied to solve this task: a lexicographic and the algorithm SPEA2

(Strength Pareto Evolutionary Algorithm 2) [25]. Both GAs used the tournament selection method. The initial population was composed of random individuals with varying sizes. It is also possible to provide code-matrices of usual strategies, as the 1AA, to the initial population. Two cross-over operators and four types of mutation operators were defined. To opt for one particular type of cross-over and mutation, a probabilistic criterion was employed, so that types that produced better solutions in previous generations had more chance to be applied again.

3.2 DDAGs and ADAGs Structures

GAs were also used to search for permutations of nodes in a DAG (DDAG or ADAG), based on their performance in the multiclass problem solution. This problem can be reduced to a permutation problem, where a list of classes is ordered [15].

The individuals were represented as lists of classes. Every class has to be present in the list and no repetitions of classes are allowed. The individuals' evaluation is given by a multiclass error estimate in a validation set, which has to be minimized. Three cross-over operators, adapted from the literature related to the solution of the traveling salesman problem with GAs, were used. The codified GA also uses the elitism operator and the tournament selection method. The mutation is performed with the insertion operator.

3.3 Trees

Two algorithms were proposed to define the binary partitions of classes in a tree. Both of them make a hierarchical clustering of the problem classes according to their similarity. Each grouping defines directly a binary partition. A graph that weights the relation between the classes is used in this grouping. This graph has k vertices, which represent the k problem classes, and $\frac{k(k-1)}{2}$ edges between all pairs of vertices. The weights are calculated from information extracted from the dataset. Two weighting criterions were tested in this work.

The first considers the Euclidean distance between the centroids of the classes. Edges that connect classes with close centers are more similar and have lower weights than those that connect classes that have distant centers. The second criterion is based in the exam of confusion matrices obtained by Decision Trees (DTs) [19]. The weight of an edge is then calculated according to a measure that considers the confusion found by DTs in the distinction of these classes. Classes that show high confusion are considered more similar and their edges receive low weights, while higher weights are attributed to edges that show low confusion.

The Kruskal algorithm, originally used to find minimum spanning trees [2] in weighted graphs, was adapted in order to determine the tree partitions of classes. The generation of the tree operates in a bottom-up iterative way. A pseudo-code of this algorithm may be found in [11, 12].

The previous algorithm uses relations between pairs of classes to define the similarity among classes in the tree formation process. A modification was then tested, in which when two subsets of vertices of the weighted graph are joined, the graph structure is adapted. The vertices equivalent to these classes are joined in a unique vertex, as well as the classes' information extracted from data. In the centroid use, the centroid of the data of all joined classes is calculated. In the confusion classes case, a new confusion matrix is generated over a learning problem where the data from the grouped classes form a unique class. The connections with the other vertices are then recalculated.

4   Experimental Results

All strategies previously discussed were employed in the solution of Bioinformatics problems and of multiclass problems from benchmarks. The Bioinformatics datasets are from (the numbers of classes are indicated in parenthesis): protein structural class prediction (4), cellular localization of fungi proteins (9), diagnostic of lung cancers (5) and discrimination of leukocyte types (8). Eight datasets from the UCI repository [3] were also used: iris (3), car (4), glass (6), sat-image (6), segment (7), opt-digits (10), pen-digits (10) and vowel (11).

The experiments and analysis were divided in four groups. Next, the results obtained in these experiments are commented. Standard protocols from ML were followed and all accuracy results were statistically compared [7].

4.1   Code-Matrix Design

The lexicographic and SPEA2 GAs searched for code-matrices with performance similar or superior to the 1AA one, using less binary classifiers. The 1AA matrix was chosen as reference because it is the most used in practice and its results are generally good for SVMs [20].

The lexicographic GA was capable to determine solutions with accuracy rates similar and in some cases superior to the 1AA matrix, with the use of less binary predictors. However, in some datasets, the obtained matrices were similar or identical to the 1AA. The reduction of the number of binary classifiers in these datasets may thus be not adequate. This also confirms the good performance of the 1AA matrix for SVMs. The accuracy rates of the lexicographic GA solutions also compared to code-matrices that use more binary classifiers than 1AA: 1A1 and ECOC.

The SPEA2 GA minimized considerably the number of binary classifiers in the solutions, however in general it was not capable to maintain accuracy rates comparable to those of 1AA and also obtained some matrices with equivalent columns.

4.2   DDAGs and ADAGs Structures

Few variations were found among the accuracy rates of distinct DAGs structures. Due to this low variability, the application of GAs has lead to solutions with accuracy rates statistically similar to those of random structures. This result can be effect of the stability and good generalization ability of SVMs. It is also in accordance with the generalization bound in [18], which affirms that SVMs, which maximize the margin of separation between the classes in each node, are adequate to this type of hierarchical structure.

4.3 Trees

There were no significant differences among the accuracy rates of the trees determined by the two algorithms of Section 3.3. Thus, the performance of trees obtained through a grouping that considers the relation between pairs of classes was similar to that of trees in which this relation is calculated between subsets of grouped classes. The first algorithm has the advantage of being less costly.

Only in three of the twelve used datasets one of the criterions used to extract the classes information from the applications was prominent over the other. However, this fact indicates that, depending on the data characteristics, it can be more adequate to apply one of the criterions to weight the classes' relation.

Trees whose binary partitions were defined randomly in several cases had performance similar to those generated by the previous heuristics. They also demonstrated low accuracy variation, although it was higher than that found among distinct DAGs. The good generalization ability and stability of SVMs may have contributed for these results.

4.4 Different Strategies

Representatives of each of the previous groups of techniques and two types of direct algorithms, which reformulate the SVM training algorithm into multiclass versions, were compared. The code-matrix strategies are represented by the lexicographic GA solutions. Among the hierarchical strategies, the DDAGs and trees were analyzed. The DDAGs were obtained randomly. The trees are those generated by the first algorithm commented in Section 3.3, with weighting criterion of best performance for each dataset. The direct algorithms used were the ones of Weston and Watkins [24] (WW) and of Crammer and Singer [4] (CS).

Figure 2 plots the accuracy rates obtained in the twelve datasets considered in this work. Some datasets were divided with the stratified cross-validation, while for others the original train-test split was used. Thirty GA and DDAGs solutions were generated for each dataset partition and the accuracies plotted in these cases correspond to those of the mean performing solutions. For each dataset, horizontal lines connect accuracy results statistically similar at 95% of confidence level.

The lexicographic GA code-matrices and the CS algorithm were among the best performing techniques in all comparisons made. Both of them adapt code-matrices to the multiclass problems, although the methods used by each one of them in this process are different.

Although some techniques (WW, trees and DDAGs) presented an accuracy rate reduction in some datasets, in general the results of all multiclass strategies considered, which use different approaches in the generalization of SVMs to multiclass problems, were similar.

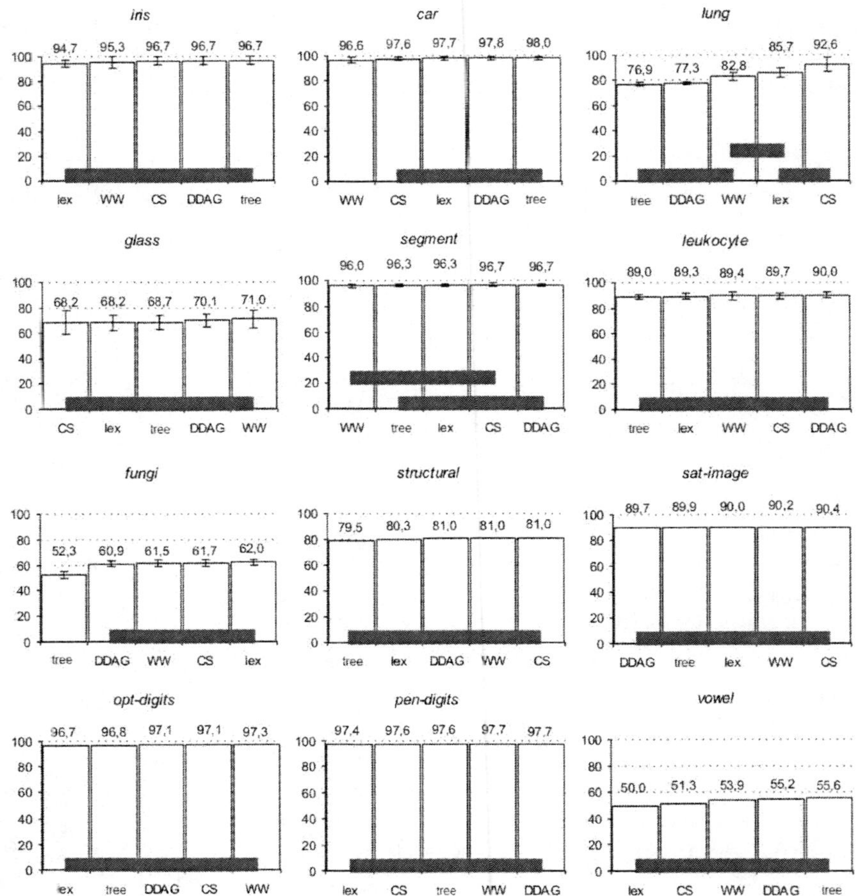

Fig. 2. Accuracy rate results of different multiclass approaches

Both trees and lexicographic GA solutions represent simpler decompositions, having less binary classifiers than the 1AA decomposition.

5   Conclusion

This paper presented a study in the solution of multiclass problems with binary SVMs. Algorithms for adapting the binary decompositions to each particular multiclass problem, using information about their performance or collected from the classes data, were proposed.

The studies with lexicographic GAs and trees demonstrated the viability of adapting the decompositions to the solution of each multiclass problem when SVMs are used as binary predictors. Both employed less binary classifiers in the multiclass solutions. In the GA case, there is a higher computational cost in this adaptation, mainly for problems with an elevate number of classes. As the GAs can be easily parallelized, this difficulty can be minimized.

Representatives of each of the groups of decomposition techniques and two types of direct algorithms, which reformulate the SVM training algorithm into multiclass versions, were also experimentally compared. In general, the accuracy results of all multiclass strategies considered, which use different approaches in the generalization of SVMs to multiclass problems, were similar.

In the experiments, several parameters of the GAs had common values in the different datasets. Variations of these values can produce better results. Similarly, the SVMs parameters were not varied. Adjusting these parameters for each binary classifier in a decomposition is a costly process. A modification of the individuals in the lexicographic GA was preliminarily investigated, demonstrating the viability of applying GAs in this search process [13].

The techniques proposed were also able to solve real practical problems from Bioinformatics, which present some challenges in multiclass prediction, as a class data unbalance.

As future work, it would be interesting to verify if the adaptation of the decompositions to the multiclass problems can benefit other ML techniques.

Acknowledgements

The authors would like to thank the financial support provided by the Brazilian research councils FAPESP and CNPq and Daniela Ushizima for the leukocyte dataset.

References

1. Allwein, E.L., Shapire, R.E., Singer, Y.: Reducing multiclass to binary: a unifying approach for margin classifiers. In: Proc 17th Int Conf on Machine Learning, pp. 9–16 (2000)
2. Aluha, R.K., Magnanti, T.L., Orlin, J.B.: Network flows: theory, algorithms and applications. Prentice Hall, Englewood Cliffs (1993)
3. Asuncion, A., Newman, D.J.: UCI repository of machine learning databases, http://www.ics.uci.edu/~mlearn/MLRepository.html
4. Crammer, K., Singer, Y.: On the learnability and design of output codes for multiclass problems. Machine Learning 47(2-3), 201–233 (2002)
5. Cristianini, N., Taylor, J.S.: An introduction to Support Vector Machines. Cambridge University Press, Cambridge (2000)
6. Dietterich, T.G., Bariki, G.: Solving multiclass learning problems via error-correcting output codes. J. Artificial Intelligence Research 2, 263–286 (1995)
7. Feelders, A., Verkooijen, W.: On the statistical comparison of inductive learning methods. In: Fisher, D., Lenz, H.-J. (eds.) Learning from data: artificial intelligence and statistics V, pp. 272–279. Springer, Heidelberg (1996)
8. Hsu, C.-W., Lin, C.-J.: A comparison of methods for multi-class support vector machines. IEEE Transactions on Neural Networks 13, 415–425 (2002)

9. Kijsirikul, B., Ussivakul, N.: Multiclass Support Vector Machines using adaptive directed acyclic graph. In: Proc. Int. Joint Conf. on Neural Networks, pp. 980–985 (2002)
10. Kreβel, U.: Pairwise classification and Support Vector Machines. In: Advances in Kernel Methods - Support Vector Learning, pp. 185–208 (1999)
11. Lorena, A.C., Carvalho, A.C.P.L.F.: Minimum spanning trees in hierarchical multiclass Support Vector Machines generation. In: Ali, M., Esposito, F. (eds.) IEA/AIE 2005. LNCS (LNAI), vol. 3533, pp. 422–431. Springer, Heidelberg (2005)
12. Lorena, A.C., Carvalho, A.C.P.L.F.: Protein cellular localization prediction with multiclass Support Vector Machines and Decision Trees. Computers in Biology and Medicine 37, 115–125 (2007)
13. Lorena, A.C., Carvalho, A.C.P.L.F.: Multiclass SVM design and parameter selection with genetic algorithms. In: IEEE Digital Proc. IX Brazilian Symp. on Neural Networks (2006)
14. Lorena, A.C., Carvalho, A.C.P.L.F.: Evolutionary design of multiclass support vector machines. J. Intelligent and Fuzzy Systems 18, 445–454 (2007)
15. Lorena, A.C., Carvalho, A.C.P.L.F.: Design of Directed Acyclic Graph Multiclass Structures. Neural Network World 17, 657–674 (2007)
16. Mitchell, T.: Machine learning. McGraw Hill, New York (1997)
17. Mitchell, M.: An introduction to Genetic Algorithms. MIT Press, Cambridge (1999)
18. Platt, J.C., Cristianini, N., Shawe-Taylor, J.: Large margin DAGs for multiclass classification. In: Advances in Neural Information Processing Systems 12, 547–553 (2000)
19. Quilan, J.R.: Induction of Decision Trees. Machine Learning 1(1), 81–106 (1986)
20. Rifkin, R., Klautau, A.: In defense of one-vs-all classification. J. Machine Learning Research 5, 1533–7928 (2004)
21. Schwenker, F.: Hierarquical support vector machines for multi-class pattern recognition. In: Proc. 4th Int. Conf. on Knowledge-Based Intelligent Systems and Allied Technologies, pp. 561–565 (2000)
22. Takahashi, F., Abe, S.: Decision-tree-based multiclass support vector machines. In: Proc. 9th Int. Conf. on Neural Information Processing, vol. 3, pp. 1418–1422 (2002)
23. Vural, V., Dy, J.G.: A hierarchical method for multi-class Support Vector Machines. In: Proc. 21st Int. Conf. on Machine Learning, pp. 831–838 (2004)
24. Weston, J., Watkins, V.: Multi-class Support Vector Machines. Tech Rep CSD-TR-98-04, Dep. Computer Science, University of London (1998)
25. Zitzler, E., Laumanns, M., Thiele, L.: SPEA2: improving the strength pareto evolutionary algorithm. In: Evolutionary methods for design, optimization, and control, pp. 95–100 (2002)

Part 5
Intelliengence Technologies in Computer and Telecommunication Systems

Multiagent Monitoring System for Complex Network Infrastructure

Marcin M. Michalski and Tomasz Walkowiak

Institute of Computer Engineering, Control and Robotics, Wroclaw University of Technology, ul. Janiszewskiego 11/17, 50-372 Wroclaw, Poland
marcin.mich@gmail.com, tomasz.walkowiak@pwr.wroc.pl

Summary. This paper presents specification of the agent monitoring system, responsible for supervision of the network elements (switches, workstations). Mechanism described below, handles passive role in aspects of the network administration. It is dedicated for monitoring predefined system statistics of the network members (hosts, switches, routers, etc.), like CPU usage, memory usage, network traffic.

1 Introduction

In times of rapid IT development, computer network became irreplaceable in all modern companies. Not only giant corporation, but small firms as well, depend on computer networks as on a medium for exchanging information. According Polish Central Statistical Office[1], in 2006 percentage of Polish companies, that used computers, reached almost 93% and almost 90% of them had access to Internet. During last 20 years digital technology became common good and demand for it constantly increasing. Nowadays computer networks are very complex systems, consist of hundreds, thousand or even million devices, very often spread across huge distance. Administration of these networks is very difficult and network reliability is critical for the companies interests. Moreover, most of networks contains many different devices with different operating systems. This causes additional problems, like common interface for all devices, so that i.e. device with Little-Endian architecture was able to "communicate" with Big-Endiad one.

With improvement of administration in mind, many software companies offers commercial tools for network monitoring. One of the most advanced monitoring tool is IBM Tivioli Monitoring. This is very complex platform that provides professional tools for gathering data and presents it to administrator in very convenient graphical diagrams. This feature helps to have a quick view of what is happening on the network. This software is also supplied with special feature that may inform administrator about any threats like i.e. bottlenecks. Additionally Tivioli makes possible to use gathered data in data mining researches because all data are stored in data bases. Tivioli architecture is divided in three layer with top level manager, through middle layer to low level agent layer. It provides

---

[1] *http://www.stat.gov.pl/gus/index_ENG_HTML.htm*

N.T. Nguyen, R. Katarzyniak (Eds.): New Chall. in Appl. Intel. Tech., SCI 134, pp. 331–340, 2008.
springerlink.com

support for Windows and UNIX like platforms and has very professional help desk support from IBM.

Other solutions are not as advanced as IBM Tivioli software, however there is some interesting software like OpManager[2], that provides interface for monitoring network traffic, CPU usage, memory and disk usage. It can also generates daily (weekly) statistics. Producer assures that it can cooperate with SNMPTRAP mechanism[3]. According to data from product web site, there is free version of this software. However it does not contain any interesting features like mentioned SNMPTRAP handling, report generation or CPU usage monitoring.

AdRem NetCrunch[4] is another monitoring platform. This system is dedicated for network data analysing basing on SNMP protocol. It makes possible to gather system statistics characteristic for monitored workstation (i.e. memory usage). It available many interesting features like physical and logical topology recognition. This is achieved through ICMP scanning and SNMPGET queries. It is also supplied with alert mechanism, a dedicated event manager, that handles with events like network interface state monitoring or filtering `syslog`[5]. Moreover it has very well developed network monitoring interface, among other things capacity monitoring. Still this product has its disadvantages. First of all it does not provide multiagent handling. System statistics are gathered by enquiring other devices, except SNMPTRAPS that are triggered from SNMP servers.

From free software there are usually local applications like `top` [6] for Linux/Unix or Task Manager for Windows. There is also network version of the `top` application called RPCTOP [2], however like original `top` it can only read CPU and memory usage. The main player on open-source market of network management software (which has network monitpring capability) is OpenNMS [7]. OpenNMS is a truly distributed, scalable platform for all aspects of the FCAPS network management model [8]. Currently, OpenNMS focuses on three main areas:

- Service Polling - determining service availability and reporting on same.
- Data Collection - collecting, storing and reporting on network information as well as generating thresholds.
- Event and Notification Management - receiving events, both internal and external, and using those events to feed a robust notification system, including escalation.

The main motivation for designing a new solution described in this paper was to derive the best features from commercial products, add more flexibility in monitoring configuration and delivers product as an open source, under GPL

---

[2] http://manageengine.adventnet.com/products/opmanager

[3] *SNMPTRAP* - mechanism that makes possible to supervise system, by receiving information from SNMP server.

[4] http://www.adrem.com.pl/netcrunch

[5] *Syslog* - logs exchange standard in IP networks.

[6] `Top` command line tool that presents system statistics of the local workstation.

[7] http://www.openms.org/

[8] FCAPS is a network management functional model defined by ITU-T and ISO in specification M.3400 - *http://www.itu.int/*

licence [9]. Therefore, we target simlar group like OpenNMS. However, we focus mainly on monitoring system statistics, a multiagent flexible architecture (to avoid a problem of enlarging the network traffic due too sending monitoring information) and designing of an extensible XML based language for setting up event generation rules.

2 Assumptions

The main goal was to design and implement Open Source multiagent monitoring platform. This system, in assumptions, was capable of processing data from remote sites of the network and presenting it to the network administrator in easy to absorb way. Very important assumption was, that system had to be muliti-agent. It was dedicated to manage large number of remote agents[10], dispersed across all network. Their job was to provide management unit with interested data (network traffic, CPU usage). Besides gathering data, system had to be easily to configure and works against provided rules (this rules will be called *Statistic Rules* an they will concern *System Statistics*). This system additionally had to be equipped with report generator which will inform administrator, by sending report (Alert or Warning), in case of any rules violation. Another very important feature of that system was its hierarchical and modular architecture. Architecture, derived from IBM Tivioli, was meant to be easy to integrate with other security environment and easy to extend with new functionalities.

This system should serve passive role in aspect of network monitoring, however it should provide communication interface for external system that would be responsible for active actions, like security reconfiguration. Thanks to this solution system could be easily combined with many security platforms.

As it was mentioned before, because of different types of devices that are in typical computer network, this system had to be platform independent.

3 System Overview

Basing on assumptions (section 2), system architecture has been made and presented on the figure 1. Designed system was called NESMA (**NE**twork **S**tatistics **MA**nagement).

Arrows on the figure defines transmission flow direction (in NESMA system it is always bidirectional transmission and envelopes symbolize adequately reports (alerts and warnings) and inquiries (actualization demands). During figure 1 analysis it is clearly to notice that system has hierarchical architecture. Application of this solution increases system scalability and modularity. Moreover, presented model reflects responsibilities division characteristic of single agent, local network segment and global system.

---

[9] *http://www.gnu.org/copyleft/gpl.html*

[10] Agent - intelligent application capable of communicating, monitoring its environment and making an autonomous decisions.

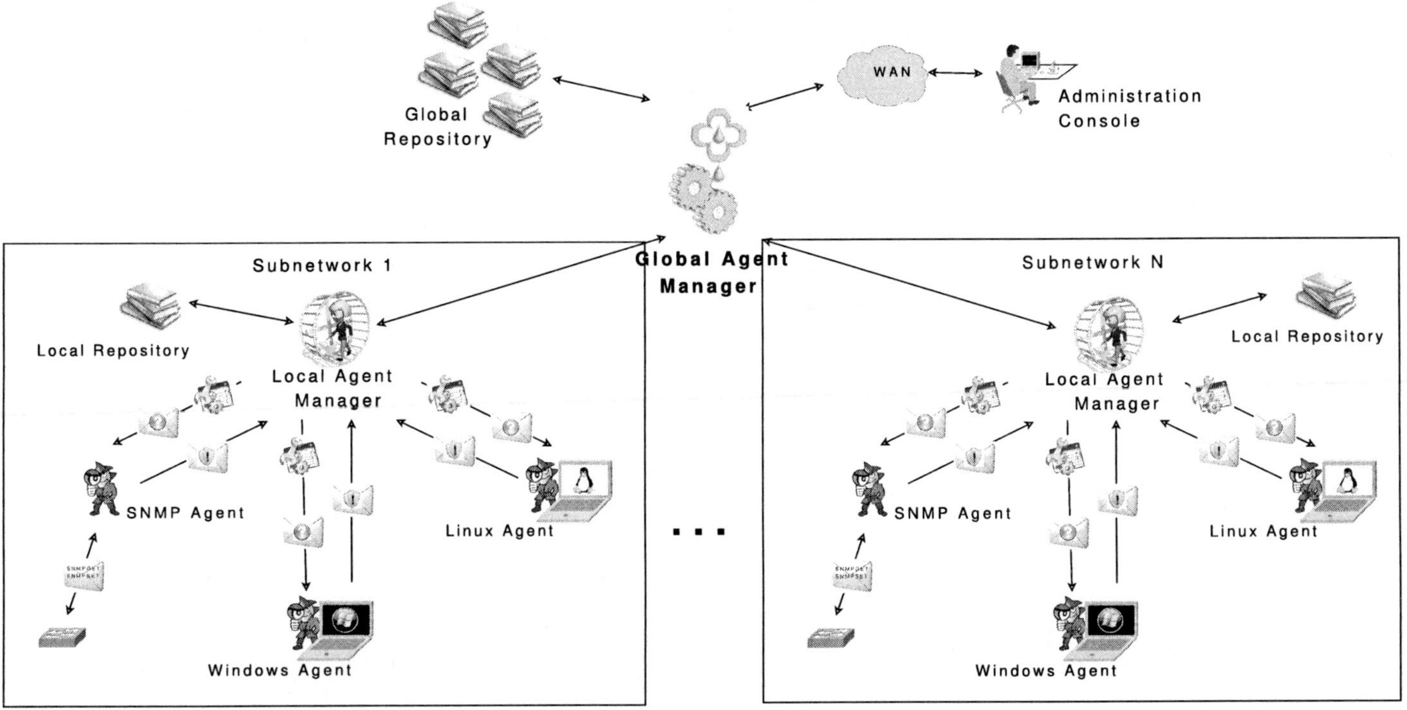

Fig. 1. Conceptual schema of the NESMA system

Hierarchical structure is also very important in aspect of transfer limitation. If there was no middle layer between GAM (*Global Agent Manager*) application and agents, each actualization of the system statistics would have involved transfer of many bytes from hundreds of agents. This could be very critical for the network bandwidth. With middle layer LAM (*Local Agent Manager*) which is responsible for the data aggregation, data transfer is limited.

In the NESMA system 4 hierarchical layers are distinguished:

- Agent - the lowest layer, responsible for gathering system statistics from the local OS[11] or, in case of SNMP agent[1], from SNMP server,
- Local Agent Manager - network segment management level, where agent manager aggregate data from subordinate agents, and sends it to the GAM application,
- Global Agent Manager - global network management level, where application receives aggregated data from subordinate LAM applications and presents it to the network administrator,
- Administration Console - administrator interface for system management.

Agents receives from LAM managers set of statistics rules called SSR (System Statistic Rules, the rules are stored in XML format). Basing on these rules the know what kind of data they should gather and when they should inform administration about rules violation.

In NESMA system 3 types of agents are distinguished: Linux, Windows and SNMP Agent. First two tapes of agent works on local workstation, third one can be located on different workstation and monitor more than single device (by means of SNMP protocol - SNMPGET and SNMPWALK). Because NESMA in assumption is a passive system it does not interfere in the network configuration, that is why no SNMPSET command is used.

Data transfer can be initialized by the agent and by the LAM as well (in case of quicker actualization demand). Except SSR, agent needs another configuration data called ASCONF (Agent Startup CONFiguration). ASCONF contains data needed for connection with LAM managers and may additionally contain, in case of SNMP agent, SNMP servers connections parameters and MIB to OID mapping. After initiation, basing on ASCONF data, agent have to register itself in the LAM manager.

LAM layer mediate between agent layer and GAM layer. It is responsible for:

- processing configuration, received from GAM layer,
- supervising subordinate agents (send them configurations),
- informing GAM about new registered agents (agent amount is dynamic),
- aggregating data provided by subordinate agents,
- transferring aggregated data to the GAM application,
- forwarding agent reports to the GAM layer,
- generating reports, that consider network segments, and sending them to GAM.

---

[11] OS - Operating system.

LAM receives SSR data from GAM. Some part of it are dedicated to the LAM application and some other to agents. Agent specific data are forwarded to the registered agents and LAM specific data remains on the LAM in its repository called Local Repository. When new agent is registered, it receives default SSR configuration specific for its type (Linux,Windows,SNMP). This configuration was previously received by LAM from GAM application. LAM informs GAM about new agent and GAM can send SSR reconfiguration request if the default settings are not satisfying. Data received by the LAM application are stored in the Local Repository and periodically are sent to the GAM application in aggregated form. In case of any SSR rules violation report is generated. Similar to agent, LAM also requires start up configuration data called LSCONF (LAM Start up CONFiguration).

Global Agent Manager is LAM applications management system. It can influence on local network segments monitoring process by providing configuration to the LAM application.

Its main responsibilities are:

- providing subordinate LAM application monitoring configuration (SSR rules),
- processing, received from subordinate managers, system statistics,
- storing received information in to global repository,
- presenting data to the network administrator,
- gathering reports and informing administrator about any violations.

GAM is a mediator between NESMA system and Administration Console. It receives instructions from administrator and deploy them to the system. Except LAM specific configuration, network administrator also provides default configuration that GAM application provides to the new registered Local Agent Manager.

Provided rules are stored in the global repository, so that in case of shutting down all system, after power up system will be full operational.

After receiving SSR rules LAM applications send periodic statistic updates to GAM, in an aggregated form. These data are stored to the global repository and can be presented to the network administrator any time he requests.

Very important feature is informing administrator about any rules violation. It provides administrator with reports from subordinate LAM managers. After informing administrator each report is marked as read.

GAM may also generate alerts i.e. after 3 bad logins (bad password). Moreover similar to agents and LAMs it uses startup configuration that contains data necessary for connection with repository and for setting listening ports that are used for communication with LAMs and administrator. Administration console provides communication interface with NESMA system. It makes possible to retrieve system statistics from LAMs application (aggregated data) and from concrete agents as well (non aggregated, agent specific data). Moreover it may send request for to update data immediately.

4   Technology Solutions

Presented system architecture requires some mechanism for cooperation between agents. These mechanism had to be platform independent. Moreover it had to provide safety data exchange, and transfer limitation to save bandwidth. This section describe proposed technology solutions its advantages and disadvantages.

4.1   Linux Agent

Three solution was proposed:

- *libstatgrab* [6] library interfaced,
- texttt/proc file system interface combined with Java application,
- JNI[12] [5] combined with *libstatgrab* library.

libstatgrab library is dedicated interface for retrieving system statistics from Unix like operating systems. It is written in C language It provides acces to such data like CPU usage, memory usage, disks usage, process counters, network traffic, etc. It is supported by *Solaris 2.x*, *Linux 2.2/2.4/2.6*, *FreeBSD 4.x/5.x*, *NetBSD 1.6.x*, *OpenBSD 3.x*, *DragonFly BSD 1.0*, *HP-UX*, and *Cygwin*. This solution does requires any additionally tools or middleware platforms.

`procfs` is a file system that provides special files from which it is possible to read any necessary statistics. Most of Unix applications like `top`, use this interface to gather necessary data. Compare to *libstatgrab*, it is much more complicated solution, which requires implementation of the `procfs` parsers. Additionally this approach my by very problematic in term of platform independence, because on different architecture this file system may have platform specific features.

JNI makes possible to combine Java application with C code. So it is possible to use Java platform independence features with *libstatgrab* library, that is written in C.

First solution is very effective. It does not requires any additional effort, like JNI and *libstatgrab* integration. However it has very major disadvantage which is complicated communication interface with LAM manager that is written in Java. Second solution may case many problems in term of platform independence. That is why third solutions was adapted.

4.2   Other Agents

Decision for SNMP agent technology solution was not very hard. SNMP Java Package is unrivalled free software for managing SNMP enquiries. It provides mechanisms for `SNMPGET` and `SNMPWALK` requests and it is platform independent as well.

Windows agent type is not considered in this paper it is planned as an future extension for the NESMA system.

---

[12] JNI - *Java Native Interface.*

LAM and GAM Managers had to be platform independent. That is why, to provide this feature, they were implemented in Java with JDBC as an interface to repository that was based on MySQL database engine.

4.3 Communication

Communication was one of the major difficulty during designing NESMA system. Because system is dedicated for many different platforms it had to provided platform independent communication interface. Many solutions was like i.e. socket communication, ONC RPC[13] [11] (RemoteTea[14], Netbula JRPC[8]), BEEP [9], XML based protocols (SOAP[15] [4] [10], XML-RPC [3]), was taken under consideration.

Finally two communication technology ware applied in the system. CORBA for communication between LAM and agents and XML-RPC for communication between LAMs and GAM applications.

CORBA was used from two main reasons. Because transfer data are not aggregated, it was decided not to use XML based protocols, that sends all data as strings. As a result it causes data redundancy. Another important issue was that in spite of common Java platform on LAM and agents sides, it was decided to use platform independent communication protocol (instead of RMI), because it makes possible to add new enhancements for the system that do not depend on specific platform, i.e. add Windows Agent that are based on .NET Framework instead of Java.

For inter-segment communication (LAM to GAM), XML-RPC was applied. This decision was made from two reasons. First of all XML based protocols, like SOAP or XML-RPC, transfer string data by means of http protocol. Thanks to that it is easier to transfer data between network segments, because http protocol is usually granted in firewall rules. As for the data redundancy, because data are aggregated it is rather minor problem. Still to reduce data size, XML-RPC was chosen instead of SOAP because it offered all needed features and was lighter than SOAP.

5 Summary

These paper describes distributes, multiagent monitoring system prototype. This system is not ready to be a product but a start for future development. However it is already very complex system that applies such technologies like JNI, CORBA, XML-RPC, JDBC, XML, XSD, JAXB, JAXP, XPath, DOM, MySQL, Java, C.

During development, major effort was dedicated to multiagent layer handling. Thanks to that NESMA system is capable of managing many agents, dispersed

---

[13] *Open Network Computing Remote Procedure Call.*
[14] http://remotetea.sourceforge.net/
[15] SOAP - *Simple Object Access Protocol.*

across network, that provide interesting data. These agents may monitor single workstations (Linux Agents) and multiple network devices as well (SNMP agents). This makes possible to automate process of gathering data and transfer it to GAM. System is also equipped with mechanism that may enforce system statistic enquiries (administrator may ask for specific data update, i.e. CPU usage from specific network segment or even from specific host).

Specially of multiagent handling flexible configuration languages was designed (ASCONF,LSCONF,GSCONF and SSR, all based on XML) that are describes in detailed in [7]. Additionally languages validation mechanism was implemented.

Distinguished to lower layer, top levels of the NESMA system need enhancement. Alert and warning reporting mechanism should be based on the IODEF[16] standard. Moreover administration console, that currently is a simple CLI application, should be developed in more user friendly form (GUI application that is capable of presenting not only data but diagram summaries as well) and Windows Agent should be implemented.

NESMA system test were performed on Intel Pentium M 1.73MHz (cache 2048KB) CPU with 512MB memory. Agent application cased less than 30% of the CPU user mode usage. Although after initiating all services of the NESMA system on single workstation CPU usage reached almost 90%, however it is assumed that each of the system element will be located on separate devices and managers application (LAM and GAM) will be running on more advanced equipment than average workstations, so it should not exceeded 30% of the CPU usage.

NESMA system currently is a proof of concept (prototype) rather than ready for usage software, however major parts of its source code was implemented, so that it could be reused during development of the monitoring system product.

During NESMA system development we realized that working on prototype is very complex process in which designing phase is the most important one. During this phase team of architects should perform research process to indicate most suitable option if there is any. Every so often it is very difficult to decide which solution is the best, because each of them has its own advantages and disadvantages (i.e. communication protocol in NESMA system). It is very important to have more than one architect, so that system could reflect common requirements not a personal point of view . Very intensive researches may save lots of work, that would be spent on inventing functionalities, that has already been implemented by someone else (i.e. using JAXB instead of writing XML parser from scratch). Designing process should also consider future system enhancement, by creating very flexible framework, easy to customized according to everyone's needs (flexible XML configuration languages, enhancement for new types of agents).

Acknowledgment. We acknowledge the support of the Polish Ministry of Science and Higher Education (grant SPUB 208/6.PR UE/2006/7).

---

[16] http://www.ietf.org/rfc/rfc3067.txt

References

1. Case, J.: RFC 1157 - Simple Network Management Protocol (SNMP), Network Working Group, MIT Laboratory for Computer Science (1990)
2. Egorytchev, V., Rybnikov, V.: PRCTOP - CPU and process statistics monitor on a remote node, Hamburg (2001)
3. Kidd, E.: XML-RPC HOWTO, Source Builders (2001)
4. Koftikian, J.: Simple Object Access Protocol (SOAP), Technische Universitat Hamburg (2000)
5. Liang, S.: The Java Native Interface - Programmer's Guide and Specification. Sun Microsystems Inc., Palo Alto (1999)
6. libstatgrab Home Page: `http://www.i-scream.org/libstatgrab/` [Accessed November 1, 2007]
7. Michalski, M.: Multiagent Monitoring System for Complex Network Architecture, Master's thesis, Wroclaw University of Technology (2007)
8. Netbula JRPC Home Page: `http://netbula.com/javarpc` [Accessed November 1, 2007]
9. Rose, M.: RFC3080: The Blocks Extensible Exchange Protocol Core, Invisible Worlds Inc. (2001)
10. Shin, S.: SOAP 1.2, Sun Microsystems Inc. (2004)
11. Srinivasan, R.: RFC 1831: Remote Procedure Call Protocol Specification Version 2. Sun Microsystems Inc., Palo Alto (1995)

Multiagent Approach to Autonomous Reorganization of Monitoring Task Delegation in Distributed IDS

Karolina Jeleń, Piotr Kalinowski, Wojciech Lorkiewicz, and Grzegorz Popek

Institute of Information Science and Engineering
Wrocław University of Technology
50-370 Wrocław, Janiszewskiego 11/17, Poland
{Karolina.Jelen,Piotr.Kalinowski}@student.pwr.wroc.pl,
{Grzegorz.Popek,Wojciech.Lorkiewicz}@pwr.wroc.pl

Abstract. Intrusion detection systems (IDSs) are exposed to highly dynamic and demanding environments. Moreover, analysis engines based on either signature recognition or anomaly detection most often are large modules which consume a significant amount of resources. In distributed IDSs individual monitoring entities are capable of detecting local intrusions based on performed observations. In this paper we propose that agents are assigned some collection of monitoring tasks, which need varying amount of computational resources. These needs vary over time causing occasional overload of single monitoring entities and resulting in a need for tasks' reassignment. An analysis performed on distant objects brings an additional load over the system. Therefore to reduce an additional network traffic invoked by multi-object reassignment a method of single object's delegation is proposed, based on neighbours' lists and distances between monitoring agents and observed objects. Designed solution has been implemented with JADE and compared with a random delegation solution.

Keywords: intrusion detection system, reorganization, JADE, monitoring.

1 Introduction

Many strategies and methods of IDSs have been developed and studied (see [2,3]). Over last years many of these systems tend towards a distributed manner (DIDS [8], GrIDS [12], EMERALD [7], AAFID [9,10]), as they introduce highly extended monitoring entities performing local observations. The results of these observations are passed to a higher level entities in order to form a global system perspective. Such systems are referred to as distributed IDSs, as they are based on local – and in that sense on distributed – data analysis (see [11]). Purely local character of performed observations' analysis is arguable as there exists a certain need for hierarchical (centralized) approach. Therefore, such systems are not completely distributed in terms of data analysis as the part of it is performed at higher levels of the hierarchy.

Nevertheless the aforementioned systems and frameworks neglect the problem of optimization of monitoring task distribution which is a very important aspect of monitoring processes carried out in the IDS. As stated in [1] the "analysis engines based on either signature recognition or anomaly detection are often large modules analyzing system audit logs, user activities and system state. This consumes a significant amount of resources in terms of CPU usage, disk I/O and memory usage". It was assumed that every individual monitoring entity is capable of performing any

N.T. Nguyen, R. Katarzyniak (Eds.): New Chall. in Appl. Intel. Tech., SCI 134, pp. 341–350, 2008.
springerlink.com

observation and that for every agent, its set of observed objects is static. However in real situations there exist several physical limitations of each monitoring unit and therefore the redistribution of monitoring assignment is necessary.

Throughout this paper we address the problem of monitoring task redistribution among distributed individual monitoring entities. Each monitoring entity is a part of a pre-assumed IDS which focuses only on the Anomaly Detection. As proposed in [5,6], local anomalies can be further analyzed on a higher layer of the system.

1.1 Decentralized Approach

We assume a multiagent system that manages the distribution of an observation task upon a distributed set of objects and performs this monitoring task. Each agent is both a monitoring entity (i.e. is capable of performing objects analysis) and a part of managing system (i.e. is capable of performing reorganization tasks). As we adopt a strict decentralized approach[1] every agent has equal computational capabilities and is equally important. As a monitoring entity, each individual agent is capable of performing assumed task of observation upon a delegated set of objects and of evaluating its current load status. In particular, every agent is able to identify its overload, i.e. when assigned tasks require more computational power and resources than it is equipped with. If an agents is overloaded, it needs to be assigned less tasks (reorganization is needed), and if it is not overloaded, it is competent of acquiring more tasks. In the former situation there exists a need for the reorganization, whereas in the latter the reorganization (e.g. to balance loads) is optional. As a managing entity the agent is capable of performing planning and reorganization of monitoring tasks assignment. In order to sustain individual and autonomous character of each agent and to enable reorganization, mutual communication mechanisms are needed.

1.2 General Idea of Monitoring Task Redistribution

We assume that there exists a set of objects that need to be observed and are located on distributed servers – at least one on each server. There exists a set of agents dispersed among servers – at most one at each server. Every observation enforces varying load over the monitoring unit (an agent). The load is defined upon a given relation between objects and monitoring entities, and modeled as a function (see Def. 3). Each object implies a direct load demand dependent on current object status and an additional load demand dependant on agents' server locations in relation to objects' servers location. Every agent observes an assigned subset of objects. It should be underlined that the monitoring task must be performed continuously during the system activity. The agent analyzes its current load status during every time step, and compares it to a given load threshold which describes the amount of observational work the agent is capable of performing. At each discrete step the agent refreshes its current load status, by calculating the load value (see Def. 3). If it is above the individual load threshold then a reorganization is needed and agent has to perform a local optimization. A single object is delegated to a certain known agent.

---

[1] Decentralized approach naturally arises from the network character and as stated in the introduction such approach becomes more and more popular in IDS systems.

There are four following issues in the introduced approach that need to be described and analyzed: an initial status of the system, a mechanism for a determination of a load status, a mechanism for a choice of objects to be delegated and a mechanism for a choice of an agent to receive the delegated object.

2   MAS Delegation Model Approach

We apply the idea described in [11] and [4] in which it is assumed that only one agent exists per server. We modify this approach in such a way, that there are less agents than servers and that agents located on different servers transmit information about all detected anomalies to agents from the higher layer of the IDS using ACL Messages. Agents are a group of processes which run independently and are designed to detect anomalous events and violations of security policy. They consist of three major modules: a communication interface, monitoring mechanism and delegation mechanism.

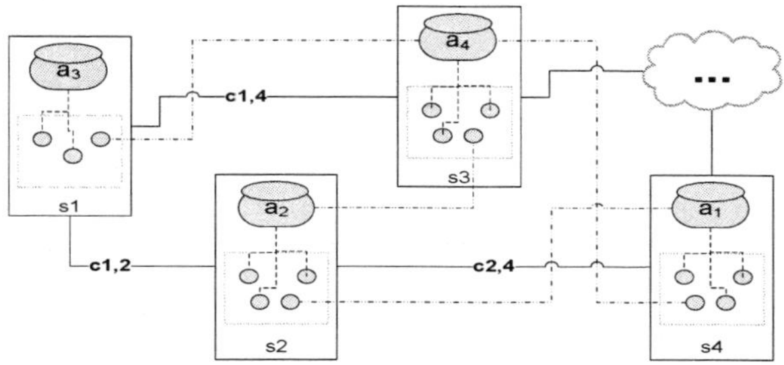

Fig. 1. Decentralized monitoring system scheme consisting of: 4 servers (rectangles), 15 objects (circles), 4 agents (pots), local (dashed line) and distributed (dotted-dashed lines) assignments of objects to agents and inter-server connections (solid lines)

2.1   Formal Definition and System Scheme

We introduce formal notation used in further chapters. Let the monitoring system be denoted by relational system *Sys* defined as follows:

Definition 1. $Sys=<G,O,A,LocO,LocA>$, where $G=<S,Con>$ represents the network of servers $S=\{s_1,s_2,...,s_K\}$ physically connected according to relation $Con \subseteq S \times S$; $O=\{o_1,o_2,...,o_N\}$ denotes the set of objects to be monitored; $A=\{a_1,a_2,...,a_M\}$ denotes the set of agents; $LocO \subseteq O \times S$ denotes the location of objects on servers such that if $(o^*,s^*) \in LocO$ then $\forall s \in S$ if $s \neq s^*$ then $(o^*,s) \notin LocO$ where $N \geq K$ and $K >> M$; $LocA \subseteq A \times S$ such that if $[(a^*,s^*) \in LocA] \Rightarrow \forall s \in S$ $[s \neq s^* \Rightarrow (a^*,s) \notin LocA] \wedge \forall a \in A [a \neq a^* \Rightarrow (a,s^*) \notin LocA]$.

It is assumed, that the system is static, as $G,O,A,LocO,$ and $LocA$ do not vary over time. Private knowledge of every single agent is represented in form of the following dynamic relational system:

Definition 2. Agent's a_j knowledge at time point $t \in [t_0, \infty)$ is denoted as by $KA_j(t) = <L_j, AS_j, C_j(t), O_j(t), Ob_j(t), N_j, LocN>$, where L_j denotes the load threshold; $AS_j \in S$ denotes the agent's location; $C_j(t) = <c_{j,1}(t), c_{j,2}(t), ..., c_{j,K}(t)>$, where $C_j(t) \in [0,1]^K$, is the cost vector of observing objects located on other servers and if $AS_j = s_m$ then $c_{j,m} = 0$; $O_j(t) \subseteq O$ denotes the set of objects in the scope of observation; $Ob_j(t) = \{<o, w(t), c(t)>: o \in O_j, (o,s) \in LocO\}$ denotes the current observational state, where $w(t)$ denotes the current load demands of object o and $c(t)$ denotes the load resulting from the fact that object o is located on server s (can be interpreted as a transmission load); $N_j \subseteq A$ denotes the neighborhood agents and $LocN_j \subseteq N_j \times S$ represents an agent's knowledge about neighborhood agents location.

It is assumed that every agent is capable of communicating only with neighborhood agents and therefore, the reorganization is limited only to the overloaded agent's neighborhood.

Agents' current load is based on the state of observed objects and on the cost of observing objects located on servers other than the one the agent resides at.

Definition 3. The load of an agent a_j at time point t is represented with a function $F(Ob_j(t), C_j(t))$, such that

$$F: 2^O \times [0,1]^K \to R.$$

Function F has to fulfill following claims:

1. $F(Ob_j(t), C_j(t)) > F(Ob_j(t) - \{o_i\}, C_j(t))$
2. **If** $(o, s_m) \in LocO$, $C_j(t) = <c_{j,1}, ..., c_{j,m}, ..., c_{j,K}>$, $C'_j(t) = <c_{j,1}, ..., c'_{j,m}, ..., c_{j,K}>$, where $c_m < c'_m$ and $o \in O_j$ **then** $F(Ob_j(t), C'_j(t)) > F(Ob_j(t), C_j(t))$
3. $F(\emptyset, C_j(t)) = 0$

2.2 Problem Description

As it was mentioned above, agents' load partially depends on the location of observed objects. It is assumed, that significant portion of the load is not related to the analysis itself but to the mentioned location of objects. Therefore, it is possible to construct a global assignment of objects in such a way, that the total load generated by objects is minimal. Assumed goal of the reorganization task is to locally minimize a part of the load which is correlated with objects' placement.

Due to the actual overhead induced by objects' delegation upon the network activity, the number of invoked reassignments should be minimized. Reassignment should be an unusual operation performed in order to restore a normal state of the system. The reassignment of objects requires a break in the process of their observation and a relocation of already collected data correlated with them. Therefore, the lower number of reassignments is better in terms of reduction of data transfers.

Let a system Sys, an agent a_j and its individual knowledge $KA_j(t)$ at time point t be given. The problem of reorganization of monitoring task can be stated as a selection of delegation object $o*$ from the set $O_j(t)$ and determination of agent $a*$ from the neighborhood, such that $o*$ is the farthest (in terms of the cost $C_j(t)$) object from a_j and $a*$ is an agent closest (in the exact terms) to $o*$ capable of observing it in terms of load. Such claim results in following definition.

Definition 4. By reorganization target realized by agent a_j we understand a pair (o^*,a^*) such that $o^* \in O_j(t)$ and $a^* \in N_j$, and following conditions hold

$$\left(o^*, c^*\right) = \underset{\left\{(o,c_m)\in LocO: \quad o\in Ob_j(t)\right\}}{\arg\max} c_{j,m}(t) \tag{1}$$

$$a^* = \underset{\left\{a\in N_j(t): \quad (a,s_w)\in LocN ,(o^*,s_m)\in LocO\right\}}{\arg\min} c_{w,m}(t) \tag{2}$$

2.3 System Cycle

Cycle of the management task is described in a following discrete step manner:

Step 1. Periodically every agent verifies its load status by calculating its current load and compares it with assumed load threshold.

Step 2. For every agent, if it is overloaded, it performs local optimization task and determines the object that should be delegated based on Def. 4.

Step 3. Every overloaded agent determines an agent from its neighborhood according to Def. 4 and delegates chosen object to that agent.

Step 4. Each chosen agent receives respective object and adds it to the set of observed objects. Each delegating agent removes respective object from its set of observed objects.

Step 5. Each target agent checks its load status and performs Step. 2.

Steps of a given above cycle are precisely described in further parts of the paper.

3 Decentralized Solution — Delegation Approach

Let the system $Sys=\langle G,O,A,LocO,LocA\rangle$ be given (see Def. 1) and let for every agent a_j its knowledge KA_j be known. Let us assume that in the beginning state the system as a whole is not overloaded, which means that although some agents might be overloaded, there exists at least one distribution of monitoring tasks in which no agent is overloaded. Then, there exists such a chain of delegations resulting in none of monitoring agents being overloaded.

3.1 Initial State

Dynamic state of the system at the time moment t is described from each agent point of view by its neighborhood N_j and set of objects $O_j(t)$ it is monitoring. The knowledge about the initial state of the system is complete in such a way, that each agent is initialized with a local part of this knowledge represented as $O_j(t_0)$ and N_j.

Let two most intuitive approaches to initial state generation be considered. Firstly, the initial state of the system can be a direct consequence of a structure of an environment or a result of performed pre-organization for given initial load values. For example, agents may be physically connected by a network cable, may be geographically close to each other, etc. In case of agent-object relation this approach can be understood as certain preferences of the designers or a result of a certain clustering of the network. Secondly, the initial state can be generated at random, where objects are randomly assigned to agents and have a randomly defined neighborhood.

3.2 Overload

At a given discrete time point t each agent a_j calculates at a its load value v_j as defined by its load determination function F (see Def. 4). The load is compared with agent's load threshold L_j. If the load value $v_j{\geq}L_j$ then we assume that the agent is overloaded and it should initiate the reorganization task. Otherwise, from the agent's point of view there is no need for the reorganization.

It should be noted that the load threshold does not denote a real physical limitation of agent's observation task capabilities, like appropriate memory space, CPU usage, etc. Rather, it denotes an alarm value, which suggest the need for reorganization, as the potential raise of load would cause agent's inability to perform assumed observation tasks. In consequence the load threshold should be defined below the real physical limitation level and should provide a reasonable amount of computational reserves to assure that the agent will not be physically overloaded before reorganization ends.

3.3 Object Delegation

To locally minimize the distance between agents and objects, an overloaded agent needs to determine an object to be delegated to another agent (see Step 2) based on a distance of its objects from the target agent. There are various possible approaches to choice of an object for delegation. Probably the simplest one is to choose one random object from the objects being observed by an agent. It shall be used hereafter as a reference choice algorithm. Nevertheless such approach introduces many undesirable properties, as for example it may delegate the closest object to another monitoring entity and as such even worsen the situation. Whilst many drawbacks the random approach seems to be still better than the lack of reorganization method.

We will incorporate a method that does not use the information about the load of observed objects, but the information about the distance to the object source. This information is used to determine objects which sources (servers they are located on) are the most distant. Such objects are observed quite inefficiently, meaning that there is high overhead related to the mentioned distance. Perhaps the neighbors of the agent could be able to observe such object in a more efficient way. Formally we define the delegation object $o*$ for agent a_j in time point t as given in equation (1). Assuming an additive character of load determination function F and small impose of single object's load in comparison to the load imposed by the network, the object $o*$ is a result of maximization of the load gain (See Equation 3). Load gain has to be understood as a difference between load imposed by observing the object o* by agent a_j to the load imposed by observing the object o* by agent a_i.

$$o* \leftarrow \max_{\left\{o\in O_j:\ (o,c)\in LocO,\ a_i\in N_j(t)\right\}} F(\{o\},C_j(t)) - F(\{o\},C_i(t)) \tag{3}$$

3.4 Delegation Target

After a choice of an object for delegation it is necessary to select an agent from the neighborhood which the object will be delegated to. The target agent can be chosen on the basis of its distance from the source of the delegated object. The resulting agent has to be as close to the source as possible, trying to locally bring the system to a state, in which every object is assigned to the closest monitoring agent. Of course, such the state may be unreachable due to the load generated by monitoring tasks, but in general an idea of stabilization is reflected and realized by this solution. It can be assumed, that when there is no abnormal activity in the network, then a state in which no object can be reassigned closer to its source is stable and destabilizations are caused only by overloads generated by certain attacks and are as small as possible. Formally we define the delegation target $a*$ for agent a_j in time point t and for delegation object $o*$ as given in equation (2).

4 Decentralized Approach Implementation

Implementation has been done using JADE agent platform where observation agent was implemented along with a set of objects that were registered for its observation. There are two crucial sets of information stored in an agent: a set of information about sources from which each respective objects has been delegated to it and a set that contains information about which agents are its neighbors.

During each observation cycle the current load is calculated by an agent and if it exceeds a particular predefined threshold, the agent begins reassignment. After a choice of an object to delegate, the agent asks its neighbors about their distance to the source of the selected object. If any of the neighbors is located closer to the object source than the overloaded agent, selected object is delegated to that agent. If none of neighbors is closer to the object source, the agent delegating an object chooses the next farthest object and asks for distance for this object. An agent repeats asking until it finds an agent closer to the object's source. If none of the neighbors is located closer to the source of any of the objects being observed by the delegating agent, the agent chooses an object and target agent so that the distance to the object source is increased the least. The exception is that the agent cannot delegate the object back to the agent from which it received this object.

The process of object delegation involves communication between two agents constituting source agent and target agent of delegation and is outlined at Figure 2.

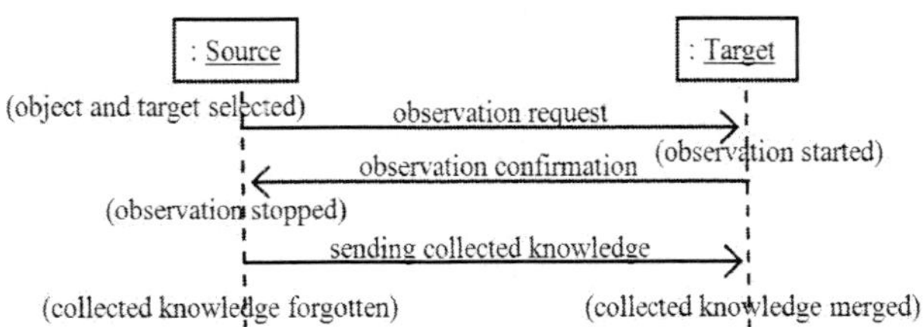

Fig. 2. A sequence diagram of the object delegation operation

After choice of an object, the agent sends a query to its neighbors excluding the previous source of the object (the communication scheme is outlined in Figure 3). All the properly functioning queried neighbors reply with information about their distance from the object. This information is stored and compared with the distance of the delegating agent. If any of neighbors claims to be closer to the object, the neighbor which replied with the smallest distance is chosen. Otherwise, the process is repeated for the next object from the list of candidates. If none of objects is appropriate for delegation according to above claims, stored information about distances between observed objects and neighbors is used to select an object-neighbor pair with the smallest distance gain in comparison to a distance between this object and the delegating agent. Chosen object is then delegated to a chosen agent. The source agent sends a message to the chosen agent indicating that an observation process has to be started and waits for confirmation. Target agent starts the observation and sends confirmation. Once it is received, source agent stops observing the object and sends the knowledge about its past states to the target agent, which is merged with data already gathered by the target agent. One delegation may not be enough to unload overloaded agent. Therefore, such delegations are initialized until success.

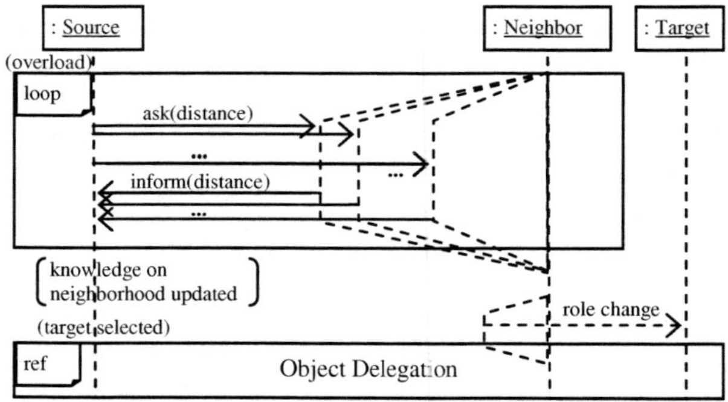

Fig. 3. A sequence diagram of communication with neighbors during object reassignment

To perform a comparison test and evaluate proposed solution, random solution was also implemented. In this case overloaded agent chooses randomly both an object and a target neighbor agent and performs a delegation. Further we have prepared a simple environment consisting of 5 agents observing 20 virtual objects. Each object was assigned to a separate source. Initially objects generated relative loads from 0 to 0.2 chosen randomly and increased by the distance penalty. Those loads were periodically modified using a random disturbance to simulate dynamic environment. Each agent initially was assigned a set of objects with loads summing up to at most 0.8. Object reassignment was performed if current load of an agent exceeded 0.9. The simulation lasted for 100 cycles of observation and was run 10 times for each algorithm.

Above experiment is not a methodologically complete verification of the proposed solution. Rather than that, it is a test of performed implementation, which in addition shows some tendencies of the solution. It is easy to observe that in modeled case

Table 1. Results comparison for proposed algorithm and random selection algorithm

	Random	Proposed
Load average	0.88	0.83
Load std deviation	0.04	0.07
Distance average	60.05	55.58
Distance std deviation	13.86	23.97
Average delegation count	140.8	5.8
Std deviation of delegation count	66.93	0.63

proposed algorithm reduced average load by locally reducing distance between agents and observed objects and that the reduction reaches 5.68% of the value resulting from an application of the random selection algorithm. Indeed, average distance between agents and assigned objects was reduced by about 7.4%.

The most important thing in above table of results is a value of an average delegation count. Number of agents and objects were chosen in such a way, that agents were highly loaded and in such critical situation, proposed delegation algorithm resulted with wise object delegations, which greatly reduced a number of delegations in comparison to random algorithm.

5 Summary and Further Research

A decentralized approach towards multiagent monitoring system reorganization was presented and analyzed. Reasons for introduction of the proposed presented were pointed out and a procedure of the task's redistribution was proposed. A system was implemented using JADE to perform tests and compare different algorithms. Proposed algorithm was initially tested and compared with a random choice algorithm. Usage of the proposed algorithm resulted in great reduction of an average delegation count, therefore reducing system stress correlated with transfers of monitoring tasks.

Future works are correlated with various parts of the proposed solution. Firstly, other models of decentralized approach can be described and tested. Secondly, a more advanced testing environment should be developed in order to gather better and more precise results before starting tests in a large-scale network environment. Also, an approach with emergency load value should be compared to other solutions, e.g. load levels, probabilistic delegation, load prediction mechanisms.

References

1. Gopalakrishna, R., Spafford, E.H.: A Framework for Distributed Intrusion Detection using Interest Driven Cooperating Agents
2. Intrusion Detection & Prevention. McGraw-Hill, New York (2003)
3. Axelsson, S.: IDS: A Taxonomy and Survey. Tech. rep., Department of Computer Engineering, Chalmers University of Technology, Goteborg, Sweden (2000)
4. Jansen, W., Mell, P., Karygiannis, T., Marks, D.: Applying mobile agents to intrusion detection and response. Tech. rep., NIST Interim Report - 6416 (October 1999)

5. Juszczyszyn, K., Kolaczek, G.: Attack pattern ontology for network traffic anomalies and intrusion detection. In: 16th International Conference on Systems Science, Wrocław, Poland, September 4-6 (2007)
6. Juszczyszyn, K., Nguyen, N.T., Kolaczek, G., Grzech, A., Pieczynska, A., Katarzyniak, R.: Agent-based Approach for Distributed Intrusion Detection System Design. In: Alexandrov, V.N., van Albada, G.D., Sloot, P.M.A., Dongarra, J. (eds.) ICCS 2006. LNCS, vol. 3993, Springer, Heidelberg (2006)
7. Porras, P.A., Neumann, P.G.: EMERALD: event monitoring enabling responses to anomalous live disturbances. In: 1997 National Information Systems Security Conference (October 1997)
8. Snapp, S., Brentano, J., Dias, G., et al.: DIDS (Distributed Intrusion Detection System) – motivation, architecture, and an early prototype. In: Proceedings of the 14th National Computer Security Conference (October 1991)
9. Balasubramaniyan, J.S., Garcia-Fernandez, J.O., Isacoff, D., Spafford, E., Zamboni, D.: An architecture for intrusion detection using autonomous agents. In: Proceedings of the 14th Annual Computer Security App. Conference, IEEE Computer Society, Los Alamitos (1998)
10. Spafford, E.H., Zamboni, D.: Intrusion detection using autonomous agents. Computer Networks 34(4), 547–570 (2000)
11. Barrus, J., Rowe, N.C.: A distributed autonomous-agent network-intrusion detection and response system. In: Proceedings of Command and Control Research and Technology Symposium, Monterey, CA, June 1998, pp. 577–586 (1998)
12. Staniford-Chen, S., Cheung, S., Crawford, R., Dilger, M., Frank, J., Hoagland, J., Levitt, K., Wee, C., Yip, R., Zerkle, D.: GrIDS-a graph based intrusion detection system for large networks. In: Proceedings of the 19th National Inf. Sys. Security Conf. (1996)

An Intelligent Service Strategy in Linked Networks with Blocking and Feedback

Walenty Oniszczuk

Bialystok University of Technology, Faculty of Computer Science, ul. Wiejska 45A,
15-351 Bialystok, Poland,
`walenty@ii.pb.bialystok.pl`

Abstract. An intelligent service strategy is one of the key elements in ensuring quality of service in a computer network. This paper presents a new analytical model for investigating a linked computer network with blocking and feedback (service according to the HOL priority scheme.) It describes behaviour of a computer network exposed to an open Markovian queuing model with blocking. The model illustrated below is very accurate, derived directly from a two-dimensional state graph and a set of steady-state equations, followed by calculations of Quality of Service (QoS) parameters. Collected numerical results indicate that the proposed open queuing network model with blocking and feedback can provide accurate performance estimates of a network. In our examples, the performance is calculated and numerically illustrated by varying buffer capacities, regulating intensity of the input flow and altering feedback probability. In addition, blocking probabilities in such network are calculated.

Keywords: intelligent service strategy, Markov chain, network performance analysis, feedback and blocking.

1 Introduction

This paper is aimed at designing an analytical model of intelligent service procedures in computer networks; seeking to obtain high network utilization, acceptable delay time, and some degree of fairness among users. In the past couple of years, there has been a strong demand for computer networks that can provide adequate quality of service (QoS) among users. Such demand, initiated a need for a solution where the servers play an active role in congestion control and collision avoidance. A series of intelligent service procedures were proposed to control queue lengths and to promote fairness among task generating sources.

Most of computer networks are connection oriented, which are also known as linked in series. There are many blocking models of linked in series networks that can be used to provide insight into the performance of those networks. Blocking models, if they can be solved efficiently, are often used in network planning and dimensioning. Due to obvious resource constrains, realistic models have finite capacity buffers, where the queue length cannot exceed its arbitrary maximum threshold. When the queue length sreaches its capacity, the buffer and the server are said to be full (blocking factors). Queuing network models (QNMs) with finite capacity queues and blocking provide powerful and practical tools for performance evaluation and predication of discrete flow systems in computer systems and

N.T. Nguyen, R. Katarzyniak (Eds.): New Chall. in Appl. Intel. Tech., SCI 134, pp. 351–361, 2008.
springerlink.com

networks. In recent years, extensive research in this field produced many results that are well explained in the literature. An excellent study may be found in the well-known series of books by Perros [12] and Balsamo [1]. In addition, many interesting theories and models appeared in a variety of journals and at worldwide conferences in the field of computer science, traffic engineering and communication engineering [2–5, 8, 10-11, 13].

Despite all the research done so far, there are still many important and interesting models to be studied. For example, finite capacity queues under various blocking mechanisms and synchronization constraints, such as those involving feedback service or priority scheduling, where in a feedback queue, a task with a fixed probability can return to the previous node immediately after its service at the current node. Although feedback queues have already been extensively studied in literature; see [6, 7, and 9], series queues with feedback are more complex object for research than the queues without feedback.

The rest of the paper is organized into 5 sections. In section 2, the analytical model is presented and explained. Section 3, analyzes a three-node network with blocking and feedback. Additionally, several procedures for calculating the quality of service (QoS) parameters and other measures of network effectiveness are presented. Model implementation and numerical examples are described in Section 4.   And finally, conclusions are drawn in Section 5.

2 Model

We consider an open queuing model with a single task class and three stations: a source, station A and B (see Fig. 1).   Tasks arrive from the source at station A according to the Poisson process with rate λ. The service rates at each station are $\mu_1{}^A$, $\mu_2{}^A$ (for feedback tasks) and μ^B, respectively. After service completion at station A, the task proceeds to station B. Once it finishes at station B, it gets sent back to station A for re-processing with probability σ. We are also assuming that tasks are leaving the network with $1- \sigma$ probability. Service at each station is provided by a single exponential server.

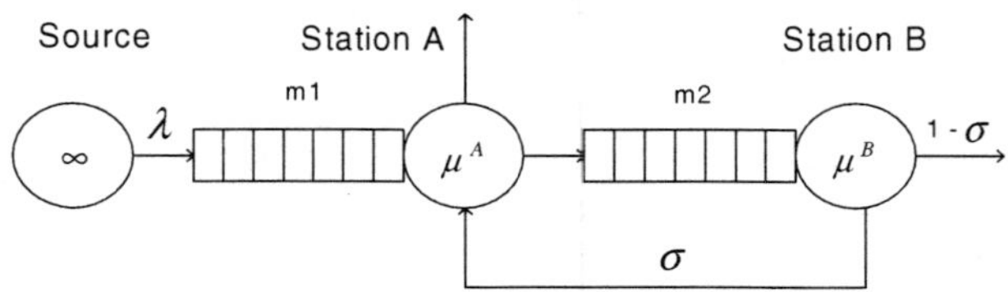

Fig. 1. Illustration of the three-station network model with feedback

A feedback task is served at station A according to a non-preemptive priority scheme (head-of-line (HOL) priority discipline) independently of all other events, where a task cannot get into service at station A (it waits at station B – blocking factor) until the task currently in service is completed. Once finished, each re-processed task departs from the network. The successive service times at both stations are assumed to be mutually independent and they are independent of the state of the network.

Blocking in such networks occurs when a station reaches its maximum capacity, which, in turn, may momentarily stop the traffic of all incoming tasks to that station. Let's say that between station A and station B, there is a common waiting buffer with finite capacity $m2$. When the buffer fills up completely, the accumulation of new tasks from the first station is temporarily suspended. Similarly, if the first buffer (with capacity $m1$) in front of the first station gets full, then that source station is momentarily blocked. This is a classical mechanism for controlling the intensity of an arriving task streams from a source station.

According to our model specification, when either of the buffers is full, any task upon completion of service at the source station or at station A, is forced to wait at its station. The task flows from the source station to station A or from the first station-to-station B depends only on service process in stations A or B respectively. Physically, blocked tasks stay on the source station or on station A, but the nature of the service process in stations A and B, allows one to treat them as located in additional places in the buffers and they belong to stations A or B. If the server A is busy, any task which needs a repeated service in this station, after completed its service in the station B, is forced to wait in this station (blocking factor). The nature of the service process in this case depends only of the service rates in station A. It allows one to treat this task as located in additional places in the buffer A.

Because of the blocking mechanism, with finite capacity buffers in a multistage network with feedback - as depicted above, it is possible that deadlocks can occur. For example, assume that station A is blocked by station B (the second buffer is full). In such situation, it is possible that a task in station B, upon its completion may get send back to station A, which in turn, will cause a deadlock. In this paper, we assume that deadlocks are detected and resolved instantaneously by an intelligent service controller without any delay, simply by exchanging the blocked tasks.

Intelligent service strategy principles (set of control laws and procedures) in a computer network include:

1. Procedure for feedback tasks **"no two priority services in succession"** (preventing a possible congestion in the first buffer),
2. Mechanisms for checking the current buffer occupancy (resource allocation policy by blocking operations),
3. Procedures for detecting and resolving a possible deadlock.

3 Analysis

The queuing network model described in Section 2 is a multistage queuing system with recycling (or feedback as it is sometimes called) that also allows blocking. All external tasks that arrive from the source station are assumed to be dispersed according to the Poisson process, whereas the service time at each station is defined as a random variable with an exponential distribution. We assume that each successive service at either station and the inter-arrival times are independent of each other. Under such assumptions, the queuing system, we are describing, can be represented by a continuous-time Markov chain, in which the underlying Markov process can analyze the stationary and transient behavior of the network. We consider this network in its stationary conditions. As such, the queuing network model reaches a steady-state condition and the underlying Markov chain has a stationary state distribution, where if each queue has finite capacity, the underlying process yields finite state space.

The state of the queuing network with blocking and feedback can be described by random variables (i,j,k), where i indicates the number of tasks at the first station, j indicates the number of tasks at second server and k represents the state of each server (see Fig. 2 and Fig. 3). Here, the index k may have the following values: $0, 1, 2, 3, 4$. If $k = 0$ - idle network, $k = 1$ – regular task service, $k = 2$ - priority task service, $k = 3$ - blocking one station and regular task service at the other one, $k = 4$ - blocking one station and priority task service at the other one.

Based on our analysis of the state space diagrams, the process of constructing the steady-state equations in the Markov model can be divided into several independent steps, which illustrate some similar, repeatable schemas (see Fig. 2 and Fig. 3). These steady-state equations for states without blocking are:

$$\lambda \cdot p_{0,0,0} = \mu^B (1-\sigma) \cdot p_{0,1,1} + \mu_2^A \cdot p_{1,0,2}$$

$$(\lambda + \mu^B \sigma + \mu^B (1-\sigma)) \cdot p_{0,j,1} = \mu_1^A \cdot p_{1,j-1,1} + \mu_2^A \cdot p_{1,j,2} + \mu^B(1-\sigma) \cdot p_{0,j+1,1}$$
$$\text{for } j = 1, \dots, m2$$

$$(\lambda + \mu^B \sigma + \mu^B (1-\sigma)) \cdot p_{0,m2+1,1} = \mu_1^A \cdot p_{1,m2,1} + \mu_2^A \cdot p_{1,m2+1,2} + \mu^B(1-\sigma) \cdot p_{0,m2+2,3}$$

$$(\lambda + \mu_1^A) \cdot p_{1,0,1} = \lambda \cdot p_{0,0,0} + \mu^B(1-\sigma) \cdot p_{1,1,1} + \mu_2^A \cdot p_{2,0,2}$$

$$(\lambda + \mu_1^A) \cdot p_{i,0,1} = \lambda \cdot p_{i-1,0,1} + \mu^B(1-\sigma) \cdot p_{i,1,1} + \mu_2^A \cdot p_{i+1,0,2}$$
$$\text{for } i = 2, \dots, m1+1$$

$$(\lambda + \mu_2^A) \cdot p_{1,0,2} = \mu^B \sigma \cdot p_{0,1,1} + \mu^B(1-\sigma) \cdot p_{1,1,2} + \mu_2^A \cdot p_{1,1,4}$$

$$(\lambda + \mu_2^A) \cdot p_{i,0,2} = \lambda \cdot p_{i-1,0,2} + \mu^B(1-\sigma) \cdot p_{i,1,2} \qquad \text{for } i = 2, \dots, m1+1$$

$$(\lambda + \mu^B(1-\sigma) + \mu_1^A + \mu^B \sigma) \cdot p_{i,j,1} = \lambda \cdot p_{i-1,j,1} + \mu_1^A \cdot p_{i+1,j-1,1} + \mu^B(1-\sigma) \cdot p_{i,j+1,1} + \mu_2^A \cdot p_{i+1,j,2} \qquad \text{for } i = 1, \dots, m1+1, \ j = 1, \dots, m2$$ (1)

$$(\lambda + \mu^B(1-\sigma) + \mu_1^A + \mu^B \sigma) \cdot p_{i,m2+1,1} = \lambda \cdot p_{i-1,m2+1,1} + \mu_1^A \cdot p_{i+1,m2,1} + \mu^B(1-\sigma) \cdot p_{i,m2+2,3} + \mu_2^A \cdot p_{i+1,m2+1,2} \qquad \text{for } i = 1, \dots, m1+1$$

$$(\lambda + \mu^B(1-\sigma) + \mu_2^A + \mu^B \sigma) \cdot p_{1,j,2} = \mu_1^A \cdot p_{1,j,3} + \mu^B(1-\sigma) \cdot p_{1,j+1,2} + \mu^B \sigma \cdot p_{0,j+1,1} + \mu_2^A \cdot p_{1,j+1,4} \qquad \text{for } j = 1, \dots, m2$$

$$(\lambda + \mu^B(1-\sigma) + \mu_2^A + \mu^B \sigma) \cdot p_{i,j,2} = \lambda \cdot p_{i-1,j,2} + \mu_1^A \cdot p_{i,j,3} + \mu^B(1-\sigma) \cdot p_{i,j+1,2} \qquad \text{for } i = 2, \dots, m1+1, \ j = 1, \dots, m2$$

$$(\lambda + \mu^B(1-\sigma) + \mu_2^A + \mu^B \sigma) \cdot p_{i,m2+1,2} = \lambda \cdot p_{i-1,m2+1,2} + \mu_1^A \cdot p_{i,m2+1,3} + \mu^B \sigma \cdot p_{i-1,m2+2,3} \qquad \text{for } i = 2, \dots, m1+1, \ j = m2+1$$

$$(\lambda + \mu^B(1-\sigma) + \mu_2^A + \mu^B \sigma) \cdot p_{1,m2+1,2} = \mu_1^A \cdot p_{1,m2+1,3} + \mu^B \sigma \cdot p_{0,m2+2,3}$$

For states with blocking the equations are:

$$(\lambda + \mu^B(1-\sigma) + \mu^B \sigma) \cdot p_{0,m2+2,3} = \mu_1^A \cdot p_{1,m2+1,1}$$

$$(\lambda + \mu^B(1-\sigma) + \mu^B \sigma) \cdot p_{i,m2+2,3} = \lambda \cdot p_{i-1,m2+2,3} + \mu_1^A \cdot p_{i+1,m2+1,1}$$
$$\text{for } i = 1, \dots, m1$$

$$(\mu^B(1-\sigma) + \mu^B \sigma) \cdot p_{m1+1,m2+2,3} = \lambda \cdot p_{m1,m2+2,3} + \mu_1^A \cdot p_{m1+2,m2+1,1}$$

$$\mu_1^A \cdot p_{m1+2,0,1} = \lambda \cdot p_{m1+1,0,1} + \mu^B(1-\sigma) \cdot p_{m1+2,1,1}$$

$$(\mu_1^A + \mu^B(1-\sigma) + \mu^B \sigma) \cdot p_{m1+2,j,1} = \lambda \cdot p_{m1+1,j,1} + \mu^B(1-\sigma) \cdot p_{m1+2,j+1,1}$$
$$\text{for } j = 1, \dots, m2$$

$$(\mu_1{}^A + \mu^B(1\text{-}\sigma) + \mu^B\sigma) \cdot p_{m1+2,m2+1,1} = \lambda \cdot p_{m1+1,m2+1,1}$$
$$\mu_2{}^A \cdot p_{m1+2,0,2} = \lambda \cdot p_{m1+1,0,2} + \mu^B(1\text{-}\sigma) \cdot p_{m1+2,1,2}$$
$$(\mu_2{}^A + \mu^B(1\text{-}\sigma) + \mu^B\sigma) \cdot p_{m1+2,j,2} = \lambda \cdot p_{m1+1,j,2} + \mu^B(1\text{-}\sigma) \cdot p_{m1+2,j+1,2} +$$
$$\mu_1{}^A \cdot p_{m1+2,j,3} \qquad \text{for } j = 1, \dots, m2$$
$$(\mu_2{}^A + \mu^B(1\text{-}\sigma) + \mu^B\sigma) \cdot p_{m1+2,m2+1,2} = \lambda \cdot p_{m1+1,m2+1,2} + \mu^B\sigma \cdot p_{m1+1,m2+2,3} +$$
$$\mu_1{}^A \cdot p_{m1+2,m2+1,3}$$
$$(\lambda + \mu_1{}^A) \cdot p_{1,j,3} = \mu^B\sigma \cdot p_{1,j,1} + \mu_2{}^A \cdot p_{2,j,4} \qquad \text{for } j = 1, \dots, m2+1$$
$$(\lambda + \mu_1{}^A) \cdot p_{i,j,3} = \mu^B\sigma \cdot p_{i,j,1} + \lambda \cdot p_{i-1,j,3} + \mu_2{}^A \cdot p_{i+1,j,4}$$
$$\text{for } i = 2, \dots, m1+1, \; j = 1, \dots, m2+1$$
$$\mu_1{}^A \cdot p_{m1+2,j,3} = \mu^B\sigma \cdot p_{m1+2,j,1} + \lambda \cdot p_{m1+1,j,3} \qquad \text{for } j = 1, \dots, m2+1$$
$$(\lambda + \mu_2{}^A) \cdot p_{1,j,4} = \mu^B\sigma \cdot p_{1,j,2} \qquad \text{for } j = 1, \dots, m2+1$$
$$(\lambda + \mu_2{}^A) \cdot p_{i,j,4} = \mu^B\sigma \cdot p_{i,j,2} + \lambda \cdot p_{i-1,j,4} \quad \text{for } i = 2, \dots, m1+1, \; j = 1, \dots, m2+1$$
$$\mu_2{}^A \cdot p_{m1+2,j,4} = \mu^B\sigma \cdot p_{m1+2,j,2} + \lambda \cdot p_{m1+1,j,4} \qquad \text{for } j = 1, \dots, m2+1$$

$$(2)$$

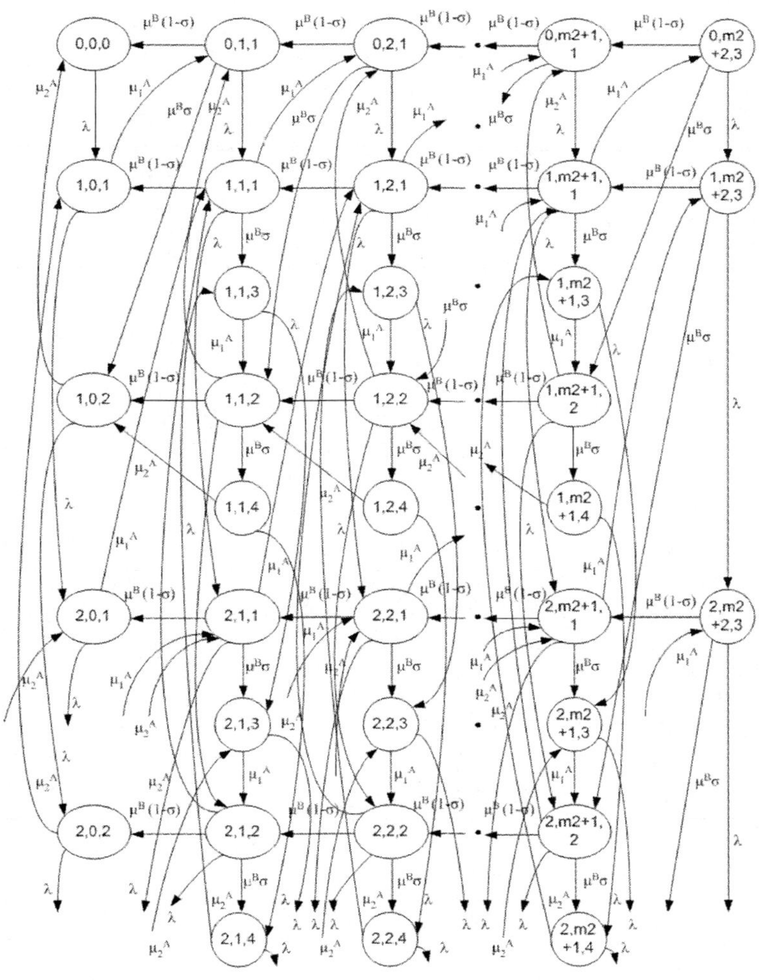

Fig. 2. Two dimensional network state diagram (first part)

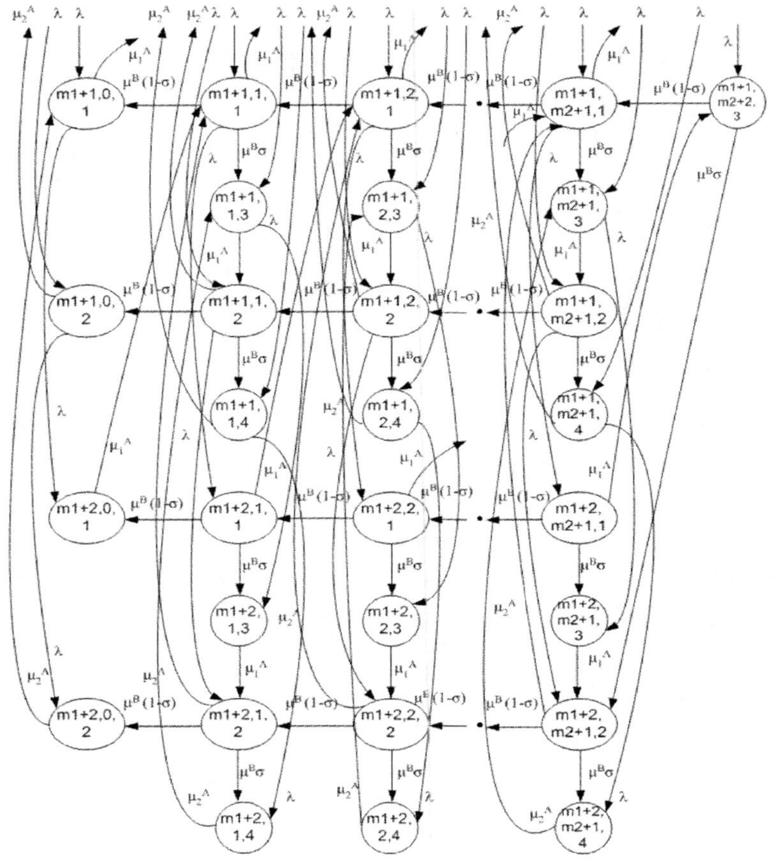

Fig. 3. Two dimensional network state diagram (second part)

A queuing network with blocking and feedback, under appropriate assumptions, is formulated here as a Markov process. The stationary probability vector can be obtained from (1) and (2) using numerical methods for linear systems of equations.

When steady-state probabilities are known, one can easily obtain various performance measures such as:

1. Station A blocking probability p_{blA}:

$$p_{blA} = \sum_{i=0}^{m1+1} p_{i,m2+2,3} \tag{3}$$

2. Source station blocking probability p_{blS}:

$$p_{blS} = \sum_{j=0}^{m2+1}(p_{m1+2,j,1} + p_{m1+2,j,2}) + \sum_{j=1}^{m2+1}(p_{m1+2,j,3} + p_{m1+2,j,4}) + p_{m1+1,m2+2,3} \tag{4}$$

3. Station B blocking probability p_{blB}:

$$p_{blB} = \sum_{i=1}^{m1+2} \sum_{j=1}^{m2+1} (p_{i,j,3} + p_{i,j,4}) \tag{5}$$

4. Both stations (source and station A) simultaneous blocking probability p_{blAS}:

$$p_{blAS} = p_{m1+1,m2+2,3} \tag{6}$$

5. Both stations (source and station B) simultaneous blocking probability p_{blBS}:

$$p_{blBS} = \sum_{j=1}^{m2+1} (p_{m1+2,j,3} + p_{m1+2,j,4}) \tag{7}$$

6. The mean blocking time at station A:

$$t_{blA} = \frac{n_{blA}}{\mu^B} = \frac{1}{\mu^B} \cdot \sum_{i=0}^{m1+1} (1 \cdot p_{i,m2+2,3}) \tag{8}$$

7. The mean blocking time at source station:

$$t_{blS} = \frac{n_{blS}^1}{\mu_1^A} + \frac{n_{blS}^2}{\mu_2^A} + \frac{n_{blS}^3}{\mu^B} = \frac{1}{\mu_1^A} \cdot [\sum_{j=0}^{m2+1} (1 \cdot p_{m1+2,j,1}) + \sum_{j=1}^{m2+1} (1 \cdot p_{m1+2,j,3})] +$$

$$+ \frac{1}{\mu_2^A} \cdot [\sum_{j=0}^{m2+1} (1 \cdot p_{m1+2,j,2}) + \sum_{j=1}^{m2+1} (1 \cdot p_{m1+2,j,4})] + \frac{1}{\mu^B} \cdot p_{m1+1,m2+2,3} \tag{9}$$

8. The mean blocking time at station B:

$$t_{blB} = \frac{n_{blB}^1}{\mu_1^A} + \frac{n_{blB}^2}{\mu_2^A} = \frac{1}{\mu_1^A} \cdot \sum_{i=1}^{m1+2} \sum_{j=1}^{m2+1} (1 \cdot p_{i,j,3}) + \frac{1}{\mu_2^A} \cdot \sum_{i=1}^{m1+2} \sum_{j=1}^{m2+1} (1 \cdot p_{i,j,4}) \tag{10}$$

9. The mean waiting time in the buffer B:

$$w_B = \frac{(v_{B1} + v_{B2})}{\mu^B} + \frac{v_{B3}}{\mu_i^A} + \frac{v_{B4}}{\mu_2^A} = \frac{1}{\mu^B} \cdot [\sum_{j=2}^{m2+1} (j-1) \cdot p_{0,j,1} +$$

$$+ \sum_{i=1}^{m1+2} \sum_{j=2}^{m2+1} ((j-1) \cdot (p_{i,j,1} + p_{i,j,2})) + \sum_{i=0}^{m1+1} (m2 \cdot p_{i,m2+2,3})] +$$

$$+ \frac{1}{\mu_1^A} \cdot \sum_{i=1}^{m1+2} \sum_{j=2}^{m2+1} (j-1) \cdot p_{i,j,3} + \frac{1}{\mu_2^A} \cdot \sum_{i=1}^{m1+2} \sum_{j=2}^{m2+1} (j-1) \cdot p_{i,j,4} \tag{11}$$

10. The mean response time at station B:

$$q_B = w_B + \frac{1}{\mu^B} + t_{blB} \tag{12}$$

358   W. Oniszczuk

11. The mean response time at station A:

$$q_A = \frac{1}{\mu_1^A} + \frac{1}{\mu_2^A} \cdot \sigma + t_{blA} + w_A \qquad (13)$$

12. The average network response time:

$$t_{thr} = \frac{1}{\lambda} + t_{blS} + q_A + q_B \qquad (14)$$

13. The effective input stream rate (intensity):

$$\lambda_l = \frac{1}{\frac{1}{\lambda} + t_{blS}} \qquad (15)$$

14. The mean waiting time in the buffer A:

$$w_A = \frac{(v_{A1} + v_{A3})}{\mu_1^A} + \frac{(v_{A2} + v_{A4})}{\mu_2^A} =$$

$$= \frac{1}{\mu_1^A} \cdot [\sum_{i=2}^{m1+1} \sum_{j=0}^{m2+1} (i-1) \cdot p_{i,j,1} + \sum_{j=0}^{m2+1} m1 \cdot p_{m1+2,j,1} + \sum_{i=2}^{m1+1} \sum_{j=1}^{m2+1} (i-1) \cdot p_{i,j,3} +$$

$$+ \sum_{i=1}^{m1} (i \cdot p_{i,m2+2,3}) + \sum_{j=1}^{m2+1} m1 \cdot p_{m1+2,j,3} + m1 \cdot p_{m1+1,m2+2,3}] + \qquad (16)$$

$$- \frac{1}{\mu_2^A} \cdot [\sum_{i=2}^{m1+1} \sum_{j=0}^{m2+1} (i-1) \cdot p_{i,j,2} + \sum_{j=0}^{m2+1} m1 \cdot p_{m1+2,j,2} + \sum_{i=2}^{m1+1} \sum_{j=1}^{m2+1} (i-1) \cdot p_{i,j,4} +$$

$$+ \sum_{j=1}^{m2+1} m1 \cdot p_{m1+2,j,4}]$$

4 Numerical Results

To demonstrate our analysis procedures of a three-station network with feedback and blocking proposed in Section 3, we have performed numerous calculations. These calculations were realized for many parameters combinations by varying the arrival rate (λ), the mean service rates at both stations (μ_1^A, μ_2^A, μ^B), and the value of the feedback probability σ. For the first group of calculations the following parameters were chosen: the service rates in station A and station B are equal to $\mu_1^A = 0.7$, $\mu_2^A = 0.6$, $\mu^B = 1.0$. The inter-arrival rate λ from the source station to station A is 1.0 and $\sigma = 0.5$. The buffers capacities $m1$ and $m2$ are set to fluctuate within a range from 0 to 10.

Based on such parameters the following results were obtained and the majority of them are presented in Table 1 and Figure 4.

For the second group of experiments the following parameters were chosen: the service rates in station A and station B are equal to $\mu_1^A = 3.0$, $\mu_2^A = 4.0$, $\mu^B = 2.0$.

The inter-arrival rate λ from the source station to station A is *2.0*. The feedback probability σ is changed from *0.0* to *1.0* with step of *0.1*. Buffer capacities are changed within the range from *0* to *10* but *m1+m2* is constant and it is equal to *10*. For this model the following results were obtained and the majority of them are presented in Figure 5.

The results of the experiment clearly show that the effect of the feedback phenomena must be taken into account when analyzing performance of a computer network. As noted above, feedback probability σ and blocking factor considerably change the performance measures in such networks. Figs. 4-5 illustrate dependencies of effective input and service rates with blocking probabilities on the feedback probability and buffer capacities.

Table 1. The series queues measures of effectiveness

$m1=m2$	w_A	w_B	t_{blA}	t_{blS}	t_{blB}	t_{thr}
0	0.000	0.000	0.093	0.938	0.369	5.662
1	1.310	0.415	0.053	0.902	0.530	7.472
2	2.702	0.878	0.036	0.886	0.601	9.364
3	4.140	1.332	0.026	0.878	0.638	11.276
4	5.606	1.756	0.019	0.874	0.659	13.177
5	7.090	2.142	0.015	0.871	0.672	15.053
6	8.583	2.491	0.012	0.870	0.681	16.899
7	10.084	2.804	0.009	0.869	0.688	18.715
8	11.589	3.083	0.007	0.868	0.692	20.501
9	13.097	3.332	0.006	0.867	0.696	22.259
10	14.607	3.552	0.004	0.867	0.698	23.990

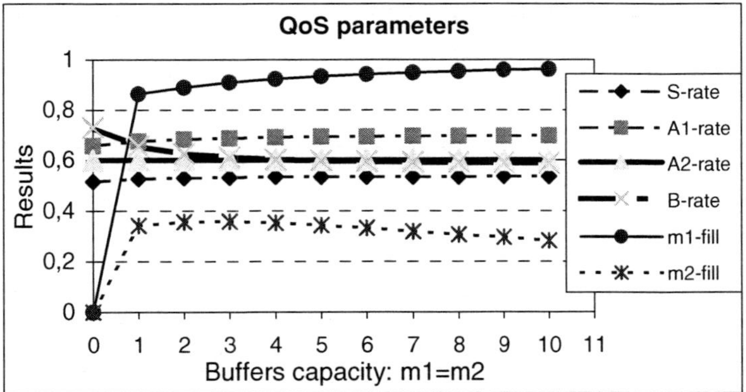

Fig. 4. Graphs of QoS parameters, where *S-rate* is the effective input rate (see (17)), *A1-rate* is the effective service rate of regular tasks in station A (blocking factor), *A2-rate* is the service rate of priority tasks in station A (no blocking), *B-rate* is the effective service rate in station B (blocking factor), *m1-fill* and *m2-fill* are buffers filling coefficients

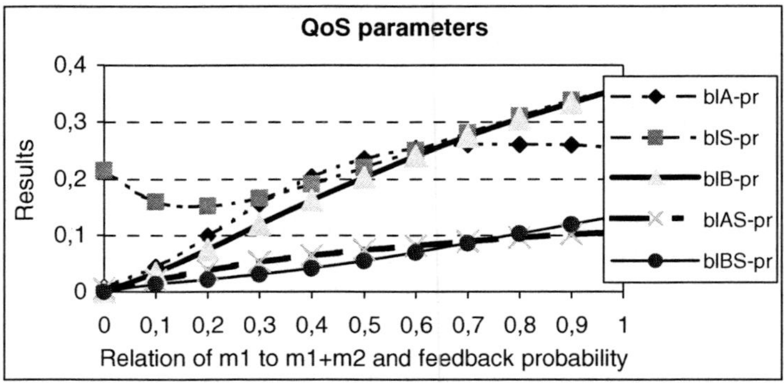

Fig. 5. Graphs of QoS parameters, where, *blA-pr* is the station's *A* blocking probability, *blS-pr* is the source station's blocking probability, *blB-pr* is the station's *B* blocking probability, *blAS-pr* is the simultaneous blocking probability of the source station and station *A*, *blBS-pr* is the simultaneous blocking probability of the source station and station *B*

5   Conclusions

In this paper, the problem of analytical (mathematical) modelling and calculation of the stationary state probabilities for a multistage network with recycling and task blocking is investigated. Tasks blocking probabilities and some other fundamental performance characteristics of such network are derived, followed by numerical examples. The results confirm importance of a special treatment for the models with blocking and with HOL feedback service (**no two priority services in succession principle**), which justifies my research. The results can be used for capacity planning and performance evaluation of real-time computer networks where blocking and feedback are present (intelligent service strategy).

Acknowledgements. This work is supported by the Bialystok University of Technology S/WI/5/03 grant.

References

1. Balsamo, S., de Nito Persone, V., Onvural, R.: Analysis of Queueing Networks with Blocking. Kluwer Academic Publishers, Boston (2001)
2. Balsamo, S., de Nito Persone, V., Inverardi, P.: A review on queueing network models with finite capacity queues for software architectures performance predication. Performance Evaluation 51(2-4), 269–288 (2003)
3. Badrah, A., et al.: Performance evaluation of multistage interconnection networks with blocking – discrete and continuous time Markov models. Archiwum Informatyki Teoretycznej i Stosowanej 14(2), 145–162 (2002)
4. Boucherie, R.J., van Dijk, N.M.: On the arrival theorem for product form queueing networks with blocking. Performance Evaluation 29(3), 155–176 (1997)
5. Economou, A., Fakinos, D.: Product form stationary distributions for queueing networks with blocking and rerouting. Queueing Systems 30(3/4), 251–260 (1998)

6. Gomez-Corral, A., Martos, M.E.: Performance of two-stage tandem queues with blocking: The impact of several flows of signals. Performance Evaluation 63, 910–938 (2006)
7. Kim, C.S., et al.: The BMAP/G/1-> ./PH/1/M tandem queue with feedback and losses. Performance Evaluation 64, 802–818 (2007)
8. Kouvatsos, D., et al.: A cost-effective approximation for SRD traffic in arbitrary multi-buffered networks. Computer Networks 34, 97–113 (2000)
9. Mei van der, R.D., et al.: Response times in a two-node queueing network with feedback. Performance Evaluation 49, 99–110 (2002)
10. Oniszczuk, W.: Analysis of an Open Linked Series Three-Station Network with Blocking. In: Pejaś, J., Saeed, K. (eds.) Advances in Information Processing and Protection, Springer Science+Business Media, LLC, New York, pp. 419–429 (2007)
11. Onvural, R.: Survey of closed queuing networks with blocking. Computer Survey 22(2), 83–121 (1990)
12. Perros, H.G.: Queuing Networks with Blocking. In: Exact and Approximate Solution, Oxford University Press, New York (1994)
13. Tolio, T., Gershwin, S.B.: Throughput estimation in cyclic queueing networks with blocking. Annals of Operations Research 79, 207–229 (1998)

Advances in Automated Source-Level Debugging of Verilog Designs

Bernhard Peischl, Naveed Riaz, and Franz Wotawa*

Technische Universität Graz
Institute for Software Technology (IST)
Inffeldgasse 16b/2, A-8010 Graz, Austria
{peischl,nriaz,wotawa}@ist.tugraz.at

Abstract. Developing models for fault localization in HDL designs has been an active research area in recent years. Whereas research on circuit verification is typically conducted on Verilog programs, research on fault localization has recently focused on the VHDL domain. The research presented herein focuses on fault localization models for Verilog designs and thus promotes the investigation of the relationships between models for property verification and fault localization. Primarily we focus on two novel contributions. First, this article points out notable semantic differences between VHDL and Verilog models and discusses its implications for fault localizations. Secondly, we advance existing work by incorporating multiple testcases and provide first empirical results obtained from the the ISCAS 89 benchmarks indicating our novel technique's applicability for real world designs.

Keywords: model-based diagnosis, software debugging, debugging of hardware designs, multiple testcases.

1 Introduction

Tool support for mastering today's ever increasing complexity of hardware designs has been an active research area during the last decade. Most of the research prototypes particularly address fault detection in terms of automated test generation or verification of properties. Whereas these kind of tools provide considerable aid in detecting a fault, there is almost no support for locating a fault once it has been detected. Shrinking design windows, ever increasing design complexity, but particularly expensive prototyping demand focusing the debugging task, in particular fault localization, on the source code level [4].

Most recent research efforts concentrate on bringing together verification techniques, such as model checking, and fault localization. For example, the authors of [6] point out how to employ a model checker for localizing faults in C programs. Particularly for the Verilog HDL numerous tools for model checking exist. Moreover, and in contrast to VHDL [8], Verilog [9] has a formalized and thus

---

* Authors are listed in alphabetical order.

N.T. Nguyen, R. Katarzyniak (Eds.): New Chall. in Appl. Intel. Tech., SCI 134, pp. 363–372, 2008.
springerlink.com   © Springer-Verlag Berlin Heidelberg 2008

well-defined synthesis and simulation semantics even for the register-transfer level. Thus to intermingle research on fault detection and model checking, Verilog offers an excellent test bed. It is thus of uttermost importance to provide diagnosis models for the Verilog language.

The research reported herein reports on the development of an automated source-level debugger for Verilog designs and extends previous research by two major contributions. First, it points out notable semantic differences between Verilog and VHDL and discusses the implications on debugging models and fault localization capabilities. Secondly, we advance previous work on fault localization in terms of employing multiple testcases rather than only a single testcase. First empirical results on the well-known ISCAS 89 benchmark suite indicate that our novel approach is applicable in a practical setting.

In section 2 we introduce model-based software debugging of HDL designs (a methodology which originates from Artificial Intelligence) refer to previous research, and introduce some basic definitions. Section 3 points out the challenges in providing a debugging model for Verilog semantics and illustrates our novel model in terms of a simple example. In Section 4 we show how to exploit multiple testcases for software debugging. In Section 5 we employed both, our novel model for the Verilog semantics alongside with multiple testcases for our approach's evaluation. After discussing the empirical results obtained we conclude our article.

2 Model-Based Software Debugging of HDL Designs

Model-based software debugging (MBSD) is an application of Model-based Diagnosis (MBD) [13, 2] to locate errors in computer programs.

The basic idea behind MBD is to compare a model - essentially capturing a system's correct behavior, to an observed behavior of this system. Typically the model is supplied by the system's designer and the observed behavior is given in terms of discrete measurements. MBD employs the behavior anticipated by the model and the actual observed behavior to reason about possible defects in the system.

The key idea of adapting MBD for software debugging is to exchange the roles of the model and the actual observations: the model reflects the behavior of the incorrect program, while the testcase specifies the correct and expected result. Again, the differences between the values computed by the program and the anticipated results allow to pinpoint at model elements that - when assumed to behave incorrectly - allow to get rid of the misbehavior.

Each of the components can operate in a normal mode - the component C functions as specified - classically denoted by $\neg AB(C)$. In the abnormal mode the component C exhibits any faulty behavior, which is denoted $AB(C)$ in the following. In terms of the component-connection model, the overall model captures both, the program's syntax as well as semantics. For example, Figure 2 illustrates the component-connection model for the Verilog code snippet given in Figure 1.

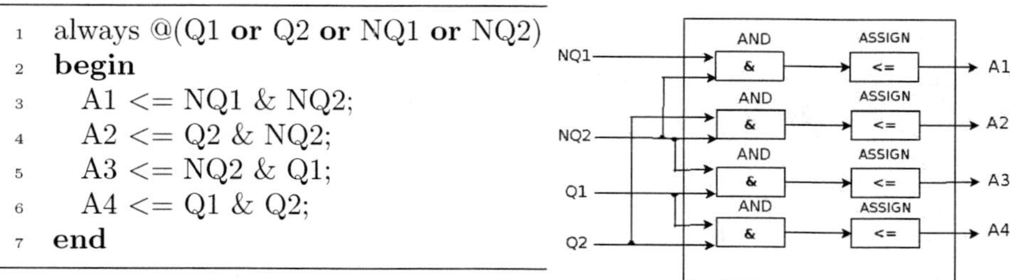

Fig. 1. A simple Verilog example   **Fig. 2.** The corresponding component-connection model

Using the consistency-based view on diagnosis as defined by [14], a diagnosis system can be formally stated as a tuple $(SD, COMP)$ where SD (system description) is a logical theory modeling the behavior of the given program to be debugged, and $COMP$ is a set of components, i.e., statements or expressions. A diagnosis system together with a set of observations OBS – that is, a testcase, forms a diagnosis problem. A diagnosis Δ is a subset of $COMP$, with the property that the assumption that all components in Δ are abnormal (they do not behave as expected), and the rest of the components is behaving normal (they behave as expected), is required to be consistent with SD and OBS. Formally, Δ is a diagnosis if and only if $SD \cup OBS \cup \{\neg AB(C) | C \in COMP \setminus \Delta\}$ is consistent. Mutation testing can also be significantly enhanced by using MBD. The author of [15] describe the relationship between MBD and mutation testing.

3 Modeling Verilog Semantics

For the VHDL language, we have developed models at different levels of abstraction. Recently developed value-level models allow for localizing the cause of misbehavior at the level of statements or expressions. Whereas capturing the simulation semantics of VHDL is intractable for larger designs [11], debugging models employing VHDL's synthesis semantics allow for debugging synthesizeable real-world designs [12].

The most notable difference between VHDL and Verilog on the register-transfer level is the semantics of the assignment statement. Whereas assigning a blocking and non-blocking assignment to a single variable is disallowed in VHDL, Verilog allows for this [1]. Although this is not a common practice in everyday development work, the Verilog language provides this freedom and the underlying diagnosis model must be able to capture this execution semantics appropriately. Figure 3 illustrates the main intricacies in terms of a small Verilog

---

[1] A blocking assignment assigns the most recent value of the right hand side to the left hand variable, whereas a non-blocking assignment assigns the values before evaluating the always' blocks (or process in VHDL) statements. Details can be found in [8, 9].

snippet: According to the Verilog semantics, depending on the execution order of non-blocking assignments, we obtain different values for the variables A and B. Regarding our example, the values of *cond*1 and *cond*2 determine this specific execution order.

Our novel model (see Figure 5) captures this semantics by introducing independent variables, in the following denoted A^* and A' respectively B^* and B'. Whereas A^* and B^* record scheduled transactions regarding the non-blocking assignments, A' and B' record transactions with respect to blocking assignments. Both variables are fed into a specific component that, depending on sequence of executed assignments, allows for assigning the value corresponding to the Verilog semantics. Note that this specific component (referred as evaluation component in the following) is not equipped with abnormality assumptions, as there is no correspondence with any element in the source. Depending on the Boolean value at its condition port, this component propagates either the first or the second input to its output. For all target variables x on which both types of assignments are applied, we introduce independent identifiers x^* and $x\prime$, and apply these variables to the corresponding inputs respectively. Given the specific port *cond* for the condition of the component, the semantic of this specific evaluation component (denoted with EC in the following) is mapped to the following rules:

$$(cond(EC) = T) \rightarrow (out(EC) = inp_1(EC))$$
$$(cond(EC) = F) \rightarrow (out(EC) = inp_2(EC))$$

Note, that in our concrete implementation, all rules are converted to Horn clause rules before deploying them on the diagnosis engine. For the remainder of the components we can employ almost the same model for the VHDL language [12, 11].

<table>
<tr><td>

```
 1
 2   always @(posedge clk)
 3   begin
 4     if (cond1)   //if1
 5     begin
 6        A = x;
 7        B = A;
 8        if (cond2)     //if2
 9           A <= e;
10        else
11           B <= d;
12     end
13   end
14   ....
```

</td><td>

```
 1   ....
 2   always @(posedge clk)
 3   begin
 4     if (cond1)   //if1
 5        A <= x;
 6     else if (cond2) //if2
 7        A = y;
 8     else
 9        A <= z;
10   end
11   ....
```

</td></tr>
</table>

Fig. 3. Blocking and non-blocking assignments

Fig. 4. Nested conditional statements

Consider code given in Figure 3. If *cond*1 is *false* then A and B will remain unchanged. If *cond*1 and *cond*2 are *true* then, the new value of A will be be value assigned using nonblocking assignment at line 9 (then part of if2) and the new value of B will be the value assigned using blocking assignment at line 7 (then part of if1). If *cond1* is *true* and *cond2* is *false* then the new value of B will be the value assigned using nonblocking assignment at line 11 and the new value of A will be the value assigned by blocking assignment at line 6. Our model captures this whole scenario as depicted in Figure 5. Evaluation components for each variable (Depicted as EC1 for B and EC2 for A), based on the condition applied to them decide whether they have to pass on the value assigned using nonblocking assignment or blocking assignment. If condition is *true* then the value assigned using nonblocking assignment is passed on, otherwise the value assigned using blocking assignment is passed. The condition to be applied is computed as follows.

For each variable v the condition to be applied on its evaluation component is the conjunction of all the conditions in the if/nested if structure which if *true*, leads to a nonblocking assignment for v getting executed. Consider variable A in Figure 3, if both *cond1* and *cond2* are *true* then it will lead to the nonblocking statement A <= e; getting executed. So the condition to be applied on evaluation component of A will be *cond1* $\wedge$ *cond2*. Similarly for B, if *cond1* is *true* and *cond2* is *false* then it will lead to the nonblocking statement $B <= d$; getting executed. So the condition to be applied on evaluation component of B will be *cond1* $\wedge$ ($\sim$ *cond2*) .

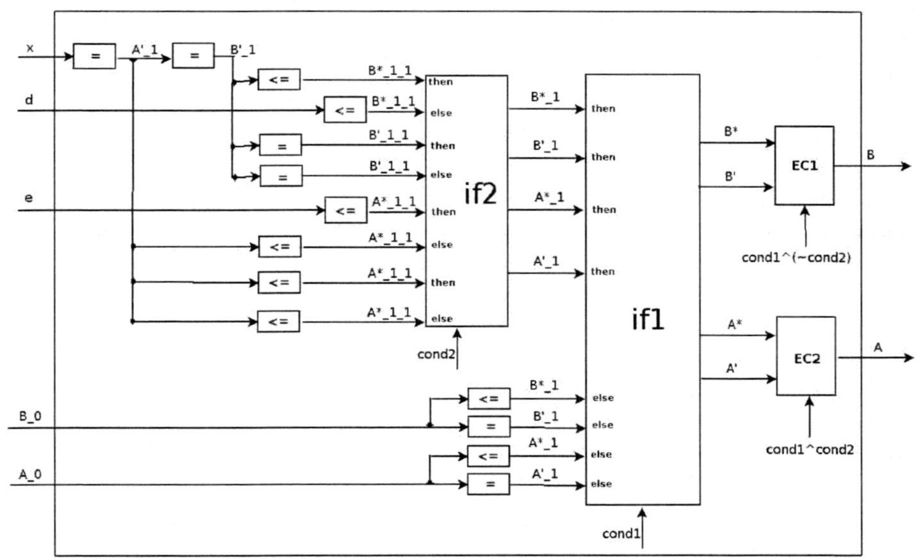

Fig. 5. Component Connection Model for program depicted in Figure 3

Now consider Figure 4. A nonblocking assignment on A will get executed if either *cond1* is *true* or both *cond1* and *cond2* are *false*. So the condition to be applied on evaluation component of A in this case will be *cond1* $\vee$ ($\sim$ *cond1* $\wedge$ $\sim$ *cond2*).

4   Multiple Testcases

These days a large number of tests are carried out on HDL designs for verification purpose and usually each test is carried out independent of others. In model based diagnosis each testcase has different discrimination capabilities for finding diagnosis candidates, so we can receive different information from each testcase. There is a growing need that the information received from one testcase should in some way act as additional information for the other testcases. This way the fault localization process can be improved.

Meerwijk and Priest [10] have described an approach for using multiple testcases. Reiter [14] also described the concept of usage of additional measurements for efficient diagnosis. His concept was further explored by Hou [7]. However all these approaches have some underlying assumptions which can become very difficult (or even impossible) to fulfill if we consider sequential HDL designs. All these approaches assume that primary inputs do not change in the testcase to be run.

The authors of [7, 14] take an approach where the diagnosis candidates received from previous measurement are able to predict the value of the next measurement and then the predicted values are compared with actual measurement to add new or remove old diagnosis candidates. This is only possible when primary inputs do not change. The solution described by Meerwijk and Priest [10] is also workable but for only those designs where there is a simple one to one relationship between primary inputs and primary outputs, e.g, if we have received an output 101 for input 001, for all next inputs 001 the output should be constant.

In sequential circuits we can have different outputs for same inputs in different clock cycles. It has been described by authors of [11] that we should consider the signal changes that occur within a certain period of time for computing diagnosis for HDL designs. These signal changes are normally strongly dependent on changes in the primary inputs. Finding proper testcases can also become very difficult for HDL designs with small number of primary inputs, if primary inputs are to be kept static. It is clear that these approaches do not suffice to temporal nature of HDLs (specially if we consider sequential circuits) where a fault might only be visible after a number of clock cycles and with changing primary inputs.

One approach can be to run all testcases independently and get all the diagnosis resulting from these testcases and then either take union or intersection of these diagnosis . This approach also has some flaws.

In case of union, the number of diagnosis resulting from multiple testcases will always be increasing so it will not suit us because at the end we will receive more diagnosis from multiple testcases than a single testcase. Using this approach, the fault localization process will become more difficult. Intersection can work if we consider only single fault diagnosis, but this method cannot be used for locating multiple faults. To illustrate this we consider a simple scenario: Suppose we have two testcases and our design has two errors. One testcase catches one error and the second testcase the second one. Then their intersection will be {}. So we will not be able to locate any error using multiple testcases although the individual testcases were able to catch a subset of actual faults.

Our approach relies on reusing Reiter's well known Minimal Hitting Set Tree Algorithm [14] for computing diagnosis. This algorithm was later revised by Greiner [5]. At the beginning we use Greiner's algorithm to get a Minimal Hitting Set directed acyclic graph. The nodes labeled by $\sqrt{}$ represent the minimal diagnosis for current testcase. Further testcases reuse the same directed acyclic graph. This directed acyclic graph is traversed using breadth first strategy and further testcases modify this acyclic directed graph by adding new nodes or closing/pruning/reusing old nodes. The main idea is that, first an acyclic directed graph D is created, the nodes which represent minimal diagnosis and are labeled by $\sqrt{}$ are replaced with conflict sets CS returned from next testcases, if such conflict sets exist . If no such conflict set exists then this node also represents the minimal diagnosis for next testcases as well.

5   Empirical Results

In this section we evaluate our approach using circuits from ISCAS 89 benchmark suite. We introduced a functional fault by substituting a randomly selected statement with another one, e.g., changing an *and* operator with an *or* operator. We obtained the error revealing input in terms of circuit equivalence checking using the VIS [1] model checker and recorded the number of components that made up our debugging model and the number of single fault candidates. We recorded the number of single fault candidates returned after the application of 1 to 5 testcases together for 4 as well as for 8 cycles. Table 1 outlines the results obtained using multiple testcases. Table 2 provides a reference to the maximum and minimum number of diagnosis obtained for a circuit using a single testcase independently. Figure 6(a) and 6(b) are graphical representation of these results for cycle no. 4 and cycle no. 8 respectively.

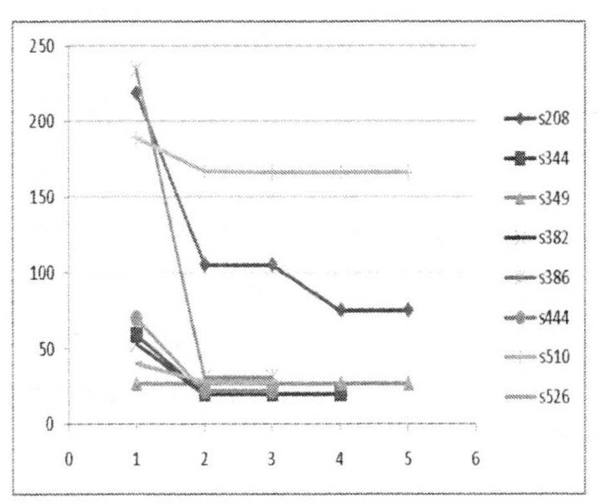

(a) multiple testcases, cycle no. 4

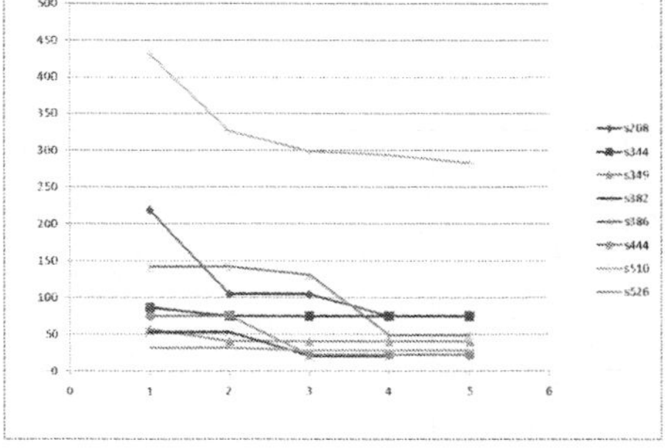

(b) multiple testcases, cycle no. 8

Fig. 6. Empirical Results for Multiple Testcases

Table 1. Empirical Results Using Multiple Testcases for the ISCAS 89 benchmark suite

circuit name	cycle No	Total Components	No of Testcases	No of Fault Candidates
s208	4	2360	1	219
			2	105
			3	105
			4	75
			5	75
	8	5720	1	219
			2	105
			3	105
			4	75
			5	75
s344	4	3748	1	59
			2	20
			3	20
			4	20
	8	7496	1	87
			2	75
			3	75
			4	75
			5	75
s349	4	3688	1	27
			2	27
			3	27
			4	27
			5	27
	8	7376	1	57
			2	40
			3	40
			4	40
			5	40
s382	4	3904	1	53
			2	20
			3	20
	8	7808	1	53
			2	53
			3	20
			4	20
s386	4	3568	1	234
			2	31
			3	31
	8	7136	1	142
			2	142
			3	131
			4	49
			5	49
s444	4	4488	1	70
			2	22
			3	22
	8	8976	1	75
			2	75
			3	22
			4	22
			5	22
s510	4	4928	1	189
			2	167
			3	166
			4	166
			5	166
	8	9856	1	431
			2	327
			3	299
			4	294
			5	283
s526	4	4844	1	40
			2	28
			3	28
	8	9688	1	31
			2	31
			3	28
			4	28
			5	28

Table 2. Maximum and Minimum no of Diagnosis candidates returned by a single testcase

circuit name	cycle No	Fault Candidates Max	Fault Candidates Min
s208	4	219	105
	8	219	137
s344	4	62	33
	8	202	84
s349	4	33	27
	8	139	45
s382	4	53	20
	8	58	20
s386	4	234	77
	8	252	78
s444	4	70	22
	8	79	22
s510	4	246	183
	8	491	374
s526	4	40	28
	8	40	28

For every introduced fault we verified that the introduced error is among the reported diagnosis candidates. Depending on the discrimination capability of the testcase on a specific circuit, the number of reported diagnosis candidates can be reduced significantly. In this respect, our simple setting can further be improved: Rather than performing a simple circuit equivalence check to obtain the error revealing input (and thus the testcase), one has to obtain a set of testcases that allows for optimal discrimination of the obtained diagnosis candidates. To achieve this, a number of testcases, where each individual testcase affects different variables, is highly desirable. The authors of [3] outline on how to obtain so called definitely discriminating testcases. However, even in our simple experimental setting, the results indicate that, given appropriate testcases, multiple testcases are capable of improving the fault localization process considerably.

6 Conclusion

Whereas there is lot of research work on verification and model checking in the Verilog domain, research work on fault localization is either done on the gate level or in the VHDL domain. This impedes the integration of verification and fault localization technologies. In summary

- The novel work we present herein briefly points out the notable semantic differences between Verilog and VHDL and reports on novel debugging models incorporating Verilog semantics.
- We discuss an advancement in terms of dealing with multiple testcases.
- First empirical results on the ISCAS 89 benchmarks indicate the practical applicability of our novel approach.

References

1. Brayton, R.K., et al.: VIS: A system for verification and synthesis. In: Alur, R., Henzinger, T.A. (eds.) CAV 1996. LNCS, vol. 1102, pp. 428–432. Springer, Heidelberg (1996)
2. de Kleer, J., Williams, B.C.: Diagnosing multiple faults. Artificial Intelligence 32(1), 97–130 (1987)
3. Esser, M., Struss, P.: Fault-model-based test generation for embedded software. In: IJCAI, pp. 342–347 (2007)
4. Gordon, M.J.C.: Relating event and trace semantics of hardware description languages. The Computer Journal 45(1), 27–36 (2002)
5. Greiner, R., Smith, B.A., Wilkerson, R.W.: A correction to the algorithm in Reiter's theory of diagnosis. Artificial Intelligence 41(1), 79–88 (1989)
6. Griesmayer, A., Staber, S., Bloem, R.: Automated fault localization in c programs. In: Proceedings of the First Workshop on Verification and Debugging (V&D 2006), pp. 95–111 (2006)
7. Hou, A.: A theory of measurement in diagnosis from first principles. Artificial Intelligence, pp. 281–328 (1991)
8. IEEE. IEEE Standard VHDL Language Reference Manual LRM Std 1076-1987, 1988. Institute of Electrical and Electronics Engineers, Inc. IEEE
9. IEEE. IEEE Standard Verilog Language Reference Manual LRM Std 11364-1995, 1995. Institute of Electrical and Electronics Engineers, Inc. IEEE
10. Meerwijk, A., Priest, C.: Using multiple tests for model-based diagnosis. In: Proceedings of the Third International Workshop on Principles of Diagnosis, Washington, pp. 30–39 (1992)
11. Peischl, B., Köb, D., Wotawa, F.: Debugging VHDL designs using temporal process instances. In: Chung, P.W.H., Hinde, C.J., Ali, M. (eds.) IEA/AIE 2003. LNCS, vol. 2718, pp. 402–415. Springer, Heidelberg (2003)
12. Peischl, B., Wotawa, F.: Automated source-level error localization in hardware designs. IEEE Des. Test 23(1), 8–19 (2006)
13. Reiter, R.: Towards a logical reconstruction of relational database theory. In: Brodie, M.L., Mylopoulos, J., Schmidt, J.W. (eds.) On Conceptual Modelling, Springer, Heidelberg (1984)
14. Reiter, R.: A theory of diagnosis from first principles. Artificial Intelligence 32(1), 57–95 (1987)
15. Wotawa, F.: On the relationship between model-based debugging and program mutation. In: Proceedings of the Twelfth International Workshop on Principles of Diagnosis, Sansicario, Italy (2001)

Spyware Prevention by Classifying End User License Agreements

Niklas Lavesson, Paul Davidsson, Martin Boldt, and Andreas Jacobsson

Blekinge Institute of Technology,
Dept. of Software and Systems Engineering,
Box 520, SE–372 25 Ronneby, Sweden
{niklas.lavesson,paul.davidsson,martin.boldt,andreas.jacobsson}@bth.se
http://www.bth.se/tek/disl

Abstract. We investigate the hypothesis that it is possible to detect from the End User License Agreement (EULA) if the associated software hosts spyware. We apply 15 learning algorithms on a data set consisting of 100 applications with classified EULAs. The results show that 13 algorithms are significantly more accurate than random guessing. Thus, we conclude that the hypothesis can be accepted. Based on the results, we present a novel tool that can be used to prevent spyware by automatically halting application installers and classifying the EULA, giving users the opportunity to make an informed choice about whether to continue with the installation. We discuss positive and negative aspects of this prevention approach and suggest a method for evaluating candidate algorithms for a future implementation.

Keywords: EULA spyware classification prevention.

1 Introduction

The amount of spyware has increased substantially over the last years, much due to the mainstream usage of Internet. Spyware is designed to collect user information for marketing campaigns without the informed consent of the users. This type of software is commonly secretly included with popular applications available for public download and it is typically difficult to remove once it has infected a computer system. Consequently, we recognize the need for an efficient prevention method. Spyware distributors usually disguise their software as legitimate to reach as many users as possible. However, to avoid legal repercussions they often mention in the End User License Agreement (EULA) that spyware will indeed be installed. However, this information is given in a way most users find hard to understand. Even legitimate EULAs can be hard to fully comprehend due to their length and the juridical terminology used.

We investigate whether it is possible to take advantage of the fact that the installation of spyware is mentioned in the EULAs and address the spyware problem by applying supervised learning algorithms to classify EULAs of both legitimate and spyware-hosting applications in order to determine if it is possible to detect from the EULA whether the associated software hosts spyware or not.

N.T. Nguyen, R. Katarzyniak (Eds.): New Chall. in Appl. Intel. Tech., SCI 134, pp. 373–382, 2008.
springerlink.com

1.1 Related Work

Adaware is an anti-spyware product [2] that can be used to scan computers for spyware and remove the found instances. However, whereas most anti-spyware products are reactive, thus trying to remove something which has already infected a system, we examine the possibility of a preventive approach that detects the presence of spyware by examining the EULA, even before the application is installed. There are a few applications available for this purpose today. We have identified one web-based tool called the EULA analyzer [3]. The user of the tool pastes the content of a EULA into a text field and the tool then analyzes the EULA and assigns a credit point that could indicate the likelihood of spyware inclusion in a particular application. It also shows some statistics about language complexity. However, it is important to notice that the credit point merely corresponds to the number of spyware related keywords or phrases that are found. There is no automatic learning of new patterns, there is no way to represent, e.g., non-trivial rules. Perhaps most notably, the tool makes no classification. Thus, this crucial task is left entirely to the user. A low credit point supposedly indicates legitimate software and a high point indicates spyware. The credit limit that separates bad EULAs from good EULAs is dependent on the particular set of submitted EULAs and has to be manually found by the user in order to be able to use the EULA analyzer for classifying software applications based on their corresponding EULAs.

Another product, with similar functionality, is EULAlyzer [4]. EULAlyzer is an installable application and the main difference in functionality, in comparison to EULA Analyzer, is only available in the professional version of the software. This functionality, called EULA watch, halts user-oriented application installers and analyzes the application EULA automatically. Unfortunately, both state-of-the-art tools are proprietary and thus it is impossible to know for sure which techniques or algorithms are used. However, the tool descriptions indicate that they rely on a static set of keywords. We have not been able to identify any studies that apply supervised learning algorithms to classify software on the basis of their EULAs. However, much work has been done in the related area of E-mail filtering, i.e., the classification of E-mail messages as legitimate or spam depending on the subject, body or other properties, using different learning algorithms, e.g.: rule learners and support vector machines [5, 6]. A more recent study investigates the performance of random forests for the same type of problem, claiming that this algorithm outperforms some of the earlier mentioned algorithms on several problems [7]. We have previously outlined a large number of related studies, which have been conducted within the security research field [1].

1.2 Outline

The remainder of our study is organized as follows. In Sect. 2 we describe the data gathering process and the experiment. This is concluded by a review of the experimental results. Section 3 contains a discussion about some consequences of these results. Finally, conclusions and pointers to future work are reviewed in the last section.

2   Experiments

Our hypothesis is that it is possible to detect from the EULA whether the associated software contains spyware or not, i.e., if it should be classified as good or bad. It is obvious that one could employ a nominal scale of more than two alternatives, for instance to indicate different threat levels. However, we argue that two alternatives are enough to investigate the stated hypothesis.

2.1   Data Collection

Our classification problem is quite analogous to that of classifying E-mail messages as either spam or legitimate. Thus, we choose to adopt a simple, yet successful way of representing the data from this area of research by representing each EULA with a word frequency vector. Thus, the data instances are represented by pairs of word frequency vectors and classes. The following search strategy was adopted to collect applications; they should be easily downloaded from the Internet and present the user with a EULA that could be copied and pasted as ASCII text. The good software instances were collected by downloading the 50 most popular Windows applications from Download.com [8] and the bad applications were collected from SpywareGuide.com [9]. After the installation of each application the operating system was scanned with Adaware to verify the classification [2]. This verification showed that all applications associated with bad EULAs were detected by Adware, while no hits were found for the legitimate applications.

2.2   Data Representation

We stored the data set using the Weka ARFF format in which each word frequency is represented by a numeric attribute and the class is represented by one nominal attribute with two possible values; good or bad [10]. The data set features 50 instances classified as good and 50 instances classified as bad, thus we did not have to deal with problems associated with a skewed class distribution. However, we believe that an equal class distribution will be difficult to achieve when creating larger data sets due to the simple fact that it is much harder to find bad applications and their corresponding EULAs. A future study could instead try to achieve a distribution that is close to the real-world distribution, which is generally perceived to have a much higher ratio of good software. The word frequency vector was generated using Wekas StringToWordVector filter with the settings adjusted as in the study by Frank and Bouckaert [11]. Thus, the TF IDF weight was applied, all characters were converted to lowercase, only alphabetic tokens were considered, and stop words as well as hapax legomena were removed. We furthermore employed a form of feature selection in that we used the Weka default setting of storing a maximum of 1000 words per class to generate the data set.

2.3   Algorithm Evaluation

The main objective of this paper is not to determine the most suitable algorithm for the studied problem, which would most certainly involve extensive parameter tuning of each featured algorithm, but rather we want to determine if the formulated hypothesis should be accepted or not. To maximize the probability of finding a useful pattern we included a diverse population of 15 algorithms from different learning categories (e.g. perceptron and kernel functions, lazy learners, Bayesian learners, trees, meta-learners, rules, etc.). We used Weka algorithm implementations and applied the default configuration for each algorithm. To test the hypothesis we needed to assess the accuracy of the algorithms.

Since we had a limited amount of data for training and testing (100 instances), we chose to perform repeated holdout tests to estimate prediction accuracy using two metrics; accuracy (correctly classified instances divided by total number of classified instances), and the area under the ROC curve (AUC). These metrics are by far the most widely used although one should keep in mind that there are issues both regarding accuracy and AUC as with most other metrics [12]. Many studies have shown the applicability of AUC for a wide range of learning problems, cf. Provost and Fawcett [13]. Intuitively, if our hypothesis holds, it should be possible to generate a classifier that performs better on average than randomly guessing the class. Hence, we formulate the hypothesis test as follows; if anyone of the featured algorithms is significantly better than a random guesser on the featured data set for both accuracy and AUC we accept the hypothesis, otherwise we reject it.

To investigate which algorithms performed significantly better than a random guesser we used repeated holdouts and the corrected paired t-test, which is a common setup used in similar applications [14]. We calculated the mean and standard deviation of 10 repeated holdouts with a 66% training set / 34% test set randomized split for each of the following metrics; accuracy (percent correct), AUC (including true positives rate and false positives rate), training time, and testing time. We used the corrected paired t-test (confidence 0.05, two-tailed) to compare each featured algorithm with a Weka baseline classifier called ZeroR, which classifies all instances as belonging to the same class. Thus, it shares the same results for both accuracy and AUC with a random guesser for a Boolean problem.

2.4   Comparison with the State-of-the-Art

We also compared the performance, in terms of accuracy, of the 15 featured algorithms with the state-of-the-art EULA analyzer tool according to the following procedure; we generated ten folds for testing by sampling, without replacement, 17 bad instances and 17 good instances for each fold (since the holdout procedure used to evaluate the 15 algorithms uses a 66/34 split) from the collection of EULAs. Obviously we did not generate any training folds since the EULA analyzer is a static model in the sense that it does not learn. Since the EULA analyzer works by looking for keywords in plain text we used text document instances instead of word vector representations, which have been subjected to, e.g., feature selection. Thus, it is important to keep this difference in mind when comparing the results later. More importantly, one should recognize that in a real-world scenario, the accuracy of EULA analyzer is dependent on the interpretation capabilities of the user concerning the resulting credit score for a particular EULA. For our experiment we used the optimal credit score cut-point which means that the published accuracy results of the EULA analyzer are more than likely to be higher than what could be achieved by the average user.

Since the testing folds for the EULA analyzer and the testing folds for the algorithms are not identical, we used a corrected non-paired t-test (confidence 0.05, two-tailed) for this part of the experiment. The objective was to find out for which algorithms there are significant improvements or degradations in performance compared to the state-of-the-art. However, as clearly mentioned earlier, we did not try to tune any of the learning algorithms to maximize performance (i.e., increasing the probability of finding significant improvements over the state-of-the-art). A secondary priority was to measure performance in terms of training and testing time. These priorities coincide with the objective of investigating if indeed a tool can be designed that builds upon the classification method presented in this paper to prevent the covert installation of spyware.

2.5 Experimental Results

The results are presented in Tab. 1. It is clear that our hypothesis should be accepted since at least one classifier achieves a significant improvement, with regard to both AUC and accuracy, in comparison to the baseline classifier. There are, in fact, significant improvements of both accuracy and AUC for 13 out of 15 featured algorithms (Ridor and DecisionStump are the only exceptions). Moreover, when comparing with the accuracy of the state-of-the-art method it is also shown that 10 algorithms outperform this method on the studied data set. However, only the improvements of NaiveBayesNominal (abbrievated NaiveBayesNom in Tab. 1) and SMO (Support Vector Machines) are statistically significant. For one algorithm, KStar, the accuracy is significantly degraded in comparison to the state-of-the-art. The high false positive rate of Kstar (0.77) might be alarming but it is important to recognize that our study merely features 100 instances. As more data is gathered for future work our guess is that the performance will increase even for the worst performing algorithms.

We further observe that NaiveBayesNominal is the best performing algorithm, achieving the best AUC and accuracy followed by SMO and VotedPerceptron. SMO has the highest training time of these three candidates; however, the testing time does not differ significantly between them. Only KStar stands out, with regards to the measured testing time, with a mean result of approximately 10 seconds, while the other algorithms achieve results close to zero seconds. Regarding training time, there is no algorithm needing more than 5 seconds and, in particular, HyperPipes, IBk, KStar, NaiveBayesNominal, and VotedPerceptron

Table 1. Experimental results in terms of: accuracy, true positives rate (TPR), false positives rate (FPR), area under the ROC curve (AUC), training time, testing time, and CEF. Significant improvements in accuracy, compared to the EULA analyzer, are shown with + while significant degradations are shown with −.

Algorithm	Accuracy	TPR	FPR	AUC	Training Time	Testing Time	CEF
AdaBoostM1	0.74(0.06)	0.72(0.08)	0.24(0.09)	0.78(0.04)	3.55(0.28)	0.00(0.01)	0.67
DecisionStump	0.69(0.11)	0.54(0.16)	0.16(0.19)	0.69(0.11)	0.33(0.08)	0.00(0.00)	0.00
HyperPipes	0.76(0.08)	0.91(0.14)	0.38(0.19)	0.90(0.07)	0.04(0.01)	0.07(0.09)	0.00
IBk	0.78(0.06)	0.71(0.07)	0.15(0.10)	0.78(0.06)	0.04(0.01)	0.13(0.02)	0.72
J48	0.73(0.10)	0.72(0.16)	0.26(0.18)	0.73(0.10)	1.29(0.23)	0.00(0.01)	0.81
JRip	0.71(0.05)	0.71(0.14)	0.29(0.12)	0.72(0.07)	2.02(0.23)	0.00(0.00)	0.79
KStar$^-$	0.60(0.04)	0.96(0.03)	0.77(0.09)	0.68(0.07)	0.00(0.00)	9.20(0.42)	0.00
NaiveBayes	0.79(0.10)	0.91(0.09)	0.32(0.17)	0.80(0.10)	0.31(0.02)	0.11(0.05)	0.00
NaiveBayesNom$^+$	0.88(0.06)	0.88(0.11)	0.12(0.11)	0.93(0.06)	0.03(0.01)	0.00(0.01)	0.77
PART	0.73(0.11)	0.72(0.15)	0.26(0.15)	0.72(0.11)	2.41(2.15)	0.00(0.01)	0.81
RandomForest	0.75(0.07)	0.79(0.09)	0.28(0.09)	0.83(0.08)	3.64(0.20)	0.00(0.00)	0.68
RBFNetwork	0.77(0.08)	0.75(0.12)	0.21(0.13)	0.78(0.09)	1.46(0.19)	0.17(0.02)	0.71
Ridor	0.68(0.11)	0.63(0.14)	0.28(0.13)	0.68(0.11)	0.87(0.11)	0.00(0.01)	0.00
SMO$^+$	0.84(0.04)	0.78(0.08)	0.11(0.08)	0.84(0.04)	0.25(0.08)	0.00(0.00)	0.75
VotedPerceptron	0.81(0.07)	0.85(0.13)	0.22(0.10)	0.87(0.07)	0.04(0.01)	0.02(0.01)	0.72
ZeroR (baseline)	0.50(0.00)	1.00(0.00)	1.00(0.00)	0.50(0.00)	0.00(0.01)	0.00(0.00)	0.00
EULA analyzer	0.73(0.04)	0.71(0.05)	0.22(0.12)	0.80(0.08)	N/A	N/A	N/A

need close to zero seconds. Exactly what words the best behaving algorithms used for distinguishing between good and bad software EULAs could not be easily grasped because of their implicit representation of the learned classifiers. However, in our limited data set of 100 instances, tree and rule based algorithms identified combinations of words such as 'search' and 'advertisements' for distinguishing between good and bad EULAs.

2.6   Multi-criteria Evaluation

One of the most fundamental properties of an intended prevention tool would arguably be to minimize the possibility of misclassifying spyware-hosting applications as legitimate, since those misclassifications could result in an increased security risk. Consequently, we prefer as candidates for implementing this tool, those algorithms that achieve a low FPR. Overall, algorithms included in the prevention tool should be as accurate as possible. In order to ensure high accuracy and a low FPR we consider the use of multi-criteria optimization and evaluation of learning algorithms.

For this purpose we suggest the generic multi-criteria metric, CEF [18], which can be used to trade-off multiple evaluation metrics when evaluating or selecting between different learning algorithms or classifiers. Each included metric can be associated with an explicit weight and an acceptable range. When analyzing the prevention tool requirements it is clear that we need to use evaluation metrics for accuracy (of classifying both good and bad applications), time (classification response time), and explainability (for visualization). Mapping the requirements to the available experiment data and making informed choices (see [18]) about bounds and explicit weighting, we can calculate the CEF score for all algorithms included in the experiment. This score is presented in the last column of Tab. 1. For the selected setting, 6 algorithms get a CEF score of 0 since at least one metric value for each algorithm was outside of the acceptable range. JRip and PART are ranked higher than NaiveBayesNominal since they scored higher in explainability.

3   Discussion

The experiment raises several interesting issues which will now be addressed. We first bring forth some technical aspects related to the featured algorithms and their performance on EULA classification, and continue discussing the importance of automatic EULA analysis. This is followed by a proposal of a novel software tool for spyware prevention. The results of the top three candidates seem to be well-aligned with results in related work; NaiveBayesNominal is known to perform very well on large vocabularies [15]. However, it is usually acknowledged that SMO outperforms NaiveBayesNominal on many problems [16]. Still, it should be considered that NaiveBayesNominal does not need to be tuned for a particular problem, while SMO includes a large number of configurable parameters. This would favor the former algorithm in this study since only default configurations are used.

It is typically hard for the average user to know if an application hosts spyware or not. It is evident that many users would benefit from using an automated tool, which could assist them in analyzing the EULA by predicting if the related software hosts spyware or not. In order to implement a successful tool, it is neccessary to collect data for a larger empirical experiment since a data set consisting of 100 instances is only acceptable for an initial study. It is also plausible to assume that legitimate applications greatly outnumber the applications that host spyware in a real-world setting. Thus, it would be beneficial to perform the larger experiment using a data set with a skewed class distribution in order to properly represent this assumed setting. In other words, if 1000 EULAs are collected, it may be sufficient if only 5% or 10% of them are spyware EULAs.

3.1   A Novel Tool for Spyware Prevention

The EULA classification method outlined in this paper can be implemented as a software tool for spyware prevention. We argue that this tool should be designed as a middleware that operates between the operating system and the application installers. The tool should be executed as a background process set to identify and analyze a EULA as soon as it appears on the screen during an installation. Based on the result from the EULA analysis, the tool will provide the user with recommendations about the classification of the application. This allows the tool to assist users in making informed decisions about the installation of software without forcing them to read (and understand) the lengthy and intricate EULAs. Should a EULA be classified as bad, a user can take appropriate actions against it, e.g., by disagreeing with the EULA and exiting the installation process. It should be noted that any tool based on our method should not be used in isolation, but in combination with other approaches, e.g., anti-spyware software.

One possible difference between the proposed tool and the state-of-the-art lie in the application of learning algorithms, which can be trained on large amounts of recent EULAs in order to make more accurate classifications than static keyword spotting services. Additionally, many of these learning algorithms generate classifiers which can be used for visualizing which parts of the EULA that significantly contributed to its classification as either good or bad software. This visualization could, for instance, be implemented by using extracted rules or generated trees. We outline a design of the prevention tool as follows. In order for it to work properly, there are certain requirements that need to be fulfilled. First, we need to make sure that the tool is accurate in its classifications since this is the main functionality. The tool should essentially be able to detect all, or a very high quantity of, bad software but it is also desirable that it manages to classify good software correctly. Furthermore, we want the tool to respond quickly when an application presents a EULA. However, the actual training phase could be done offline and the resulting classifier could then be downloaded by the tool periodically. Thus, there are no specific requirements related to training time. Finally, it is desirable that the tool can visualize what parts of the EULA text that prompted the actual classification. However, in relation to the earlier stated requirements, this is a secondary priority.

3.2   Arguments Against the Tool

It could be argued that, if the prevention tool was made available, the spyware authors could tailor their EULA around it. For instance, virus writers have difficulties writing viruses that avoid detection of scanners because viruses must contain executable code that cannot be arbitrarily changed. Since an EULA does not contain any executable code the spyware authors could be more creative in changing its content. This argument, however, does not hold, since the spyware authors are very aware of the fact that they need to mention in the EULA that spyware will be installed, in order to avoid legal repercussions. We simply exploit this fact and use it against the spyware distributors. Another argument could be that the intended tool would be no better than, for example, Adaware, in detecting spyware since the training set classifications are validated using this product. However, we argue that this validation is sufficient for the scope of our experiment and, more importantly, the real classification of the training set was carried out by downloading good software from a well-known source of non-spyware hosting software and bad software from a site which only features spyware-hosting applications.

Additionally, the intention is not for the tool to be a replacement for products like Adaware but rather it would be a complement that could be used to detect spyware from EULAs in software that has not yet been classified by such products. The high false positives rate is a serious concern since the user has to be able to trust the preventative tool to be able to make the correct choice of either continuing with the installation of a particular application or not. As mentioned before, we strongly believe that future experiments with larger data sets will result in lower false positives rates for most algorithms. Nevertheless, this issue should be further addressed by optimizing potential prevention tool algorithms to minimize the false positives rate. Additionally, one could more deeply investigate the performance of ensemble learners on this problem, since they have proven to be both accurate and robust in similar settings.

4   Conclusions and Future Work

The results of the conducted experiment reveal that 13 out of the 15 featured algorithms significantly outperformed a random guesser (i.e., a user that has not read or understood the EULA). Two algorithms also significantly outperformed the state-of-the-art EULA analysis method in terms of accuracy. Most notably, NaiveBayesNominal, SMO, and VotedPerceptron achieve the highest AUC and accuracy and also share a low false positives rate, which is crucial if EULA classification should indeed be used in a preventive tool. The results strongly support our hypothesis that EULAs can indeed be used to classify the corresponding software as good or bad. Based on this conclusion and the low training and testing times of most algorithms, we also conclude that it would be quite possible to use the EULA classification method in a spyware prevention tool that classifies the EULA when it is shown to a user during an application installation. The result from such an analysis gives the user a recommendation about the legitimacy of the application before the installation continues as well as some type of visualization of what information in the EULA that triggered this classification. We also suggest a multi-criteria approach for evaluating candidate algorithms to be included into the implementation of the prevention tool and calculate test scores for demonstration purposes.

For future work we intend to implement and evaluate the proposed tool using a refined multi-criteria evaluation approach. Additionally, we will collect data for a larger experiment to further validate our hypothesis and the results in general. It would also be interesting to tune the parameters of the algorithms to find the best candidate problem solvers, cf. Lavesson and Davidsson [17]. We would also involve more algorithms, computational linguistic methods, other EULA text document representations except for word frequency vectors, as well as an analysis and visualization of words that trigger classification into legitimate software and spyware, respectively.

References

1. Boldt, M., Jacobsson, A., Lavesson, N., Davidsson, P.: Automated Spyware Detection Using End User License Agreements. In: 2nd International Conference on Information Security and Assurance, IEEE Press, New York (2008)
2. Adaware, http://www.lavasoft.com
3. EULA Analyzer, http://www.spywareguide.com/analyze

4. EULAlyzer, http://www.javacoolsoftware.com/eulalyzerpro.html
5. Cohen, W.: Learning Rules that Classify E-Mail. In: Advances in Inductive Logic Programming, IOS Press, Amsterdam (1996)
6. Drucker, H., Wu, D., Vapnik, V.: Support Vector Machines for Spam Categorization. IEEE Transactions on Neural Networks 10(5), 1048–1054 (1999)
7. Koprinska, I., Poon, J., Clark, J., Chan, J.: Learning to Classify E-Mail. Information Sciences 177, 2167–2187 (2007)
8. CNET Download.com, http://www.download.com
9. Spyware Guide, http://www.spywareguide.com
10. Witten, I.H., Frank, E.: Data Mining: Practical Machine Learning Tools and Techniques, 2nd edn. Morgan Kaufmann, San Francisco (2005)
11. Frank, E., Bouckaert, R.R.: Naive Bayes for Text Classification with Unbalanced Classes. In: Fürnkranz, J., Scheffer, T., Spiliopoulou, M. (eds.) PKDD 2006. LNCS (LNAI), vol. 4213, pp. 503–510. Springer, Heidelberg (2006)
12. Provost, F., Fawcett, T., Kohavi, R.: The Case against Accuracy Estimation for Comparing Induction Algorithms. In: 15th International Conference on Machine Learning, pp. 445–453. Morgan Kaufmann, San Francisco (1998)
13. Provost, F., Fawcett, T.: Analysis and Visualization of Classifier Performance: Comparison under Imprecise Class and Cost Distributions. In: 3rd International Conference on Knowledge Discovery and Data Mining, pp. 43–48. AAAI Press, Menlo Park (1997)
14. Nadeau, C., Bengio, Y.: Inference for the Generalization Error. Machine Learning 52(3), 239–281 (2003)
15. McCallum, A., Nigam, K.: A Comparison of Event Models for Naive Bayes Text Classification. In: AAAI 1998 Workshop on Learning for Text Categorization, pp. 41–48. AAAI Press, Menlo Park (1998)
16. Kibriya, A.M., Frank, E., Pfahringer, B., Holmes, G.: Multinomial Naive Bayes for Text Categorization Revisited. In: 7th Australian Joint Conference on Artificial Intelligence, pp. 488–499. Springer, Berlin (2004)
17. Lavesson, N., Davidsson, P.: Quantifying the Impact of Learning Algorithm Parameter Tuning. In: 21st National Conference on Artificial Intelligence, pp. 395–400. AAAI Press, Menlo Park (2006)
18. Lavesson, N., Davidsson, P.: Generic Methods for Multi-criteria Evaluation. In: SIAM International Conference on Data Mining (2008)

An Application of LS-SVM Method for Clustering in Wireless Sensor Networks

Jerzy Martyna

Institute of Computer Science, Jagiellonian University, ul. Nawojki 11,
30-072 Cracow, Poland
martyna@softlab.ii.uj.edu.pl

Abstract. We consider the problem of estimating the clustering of nodes in wireless sensor networks (WSNs). A solution to this problem is proposed, which uses Least Squares Support Vector Machines (LS-SVM). Using mixtures of kernels and the image energy distribution of the sensor field surface, we have been solved the clustering problem in WSNs. Some computer experiments for the simulated sensor fields are carried out. Through comparing with classical clustering scheme we state that LS-SVM method has a better improvement in clustering accuracy in these networks.

1　Introduction

Wireless sensor networks (WSNs) [1] are consisted of inexpensive sensing devices called as sensors. These devices have limited energy resources and radio transmission range. They are responsible for collecting data from the environment and sending them to the sink(s). In most cases, they are organized into clusters which prevent large amounts of packet transmission and save energy power.

Clustering in WSNs come into being by introduce a hierarchy: some nodes as belonging to a backbone are dominating set. They are often referred to as clusterheads and are natural places to aggregate and compress traffic converging from many ordinary sensors. All of them form the lower layer of WSNs and they collect the data (such as temperature, humidity, etc.) from sensor field to the clusterhead nodes.

The clustering has been proposed by D.J. Baker et al. [2]. This technique was used to manage the cluster and relay the collected data. As an example for clusterhead rotation protocol is the Low-Energy Adaptive Clustering Hierarchy (LEACH) introduced by W.R. Heinzelman et al. [6], [7]. Its target scenario is a WSN with known number of nodes and known location of data sink.

A clustering algorithm that allows find the the clusters in WSN and to ensure sharing of the load between several nodes was proposed by Chatterjee et al. [4]. More recent example of clustering technique used in WSN are given by Tang and Li [17]. In the first of them clustering was used for QoS supporting and an optimal energy allocation. In paper by Krishnan [9] some algorithms which involve the so called local "growth budgets" to neighbours were proposed.

N.T. Nguyen, R. Katarzyniak (Eds.): New Chall. in Appl. Intel. Tech., SCI 134, pp. 383–392, 2008.
springerlink.com

Nevertheless, none of these algorithms aim at minimizing the energy during the formation of clusters.

The support vector machines (SVM), based on statistical learning theory, as a new tool for data classification and function estimation has been developed by Cortes and Vapnik [5]. This method maps input data into a high dimensional feature space where it is linearly separable. The SVM method has been used to solving many problems, such as pattern recognition [3], function estimation [18], face recognition [12], etc. The SVM method has been modified into Least Squares Support Vector Machines (LS-SVM) by Suykens and Vandewolle [14] and Suykens et al. [16]. In this technique a set of linear equations instead of a quadratic programming (QP) problem is used. The LS-SVM method was used in a number of industrial applications, such as optimal control [15], damage detection [19], etc.

Main goal of this paper is to introduce a new technique of clustering in WSNs which is based on LS-SVM method. By use the mapping function of sensor field into the sensor energy intensity surface and application of a mixture of kernels, we obtain with the help of LS-SVM method good estimate of clusters in a WSN. A number of numerical experiments confirme the usefulness of proposed technique.

The rest of this paper is organized as follows. In the next section, we describe the clustering problem in WSN. Section 3 presents a LS-SVM technique for solving this problem. In section 4, we provide some numerical experiments. Section 5 summarizes the paper.

2   Clustering Problems in WSNs

In this section we formulate the clustering problem in WSNs.

Clustering in WSNs come into being by introduce a hierarchy: some nodes as belonging to a backbone are dominating set. They are often referred to as clusterheads and are natural places to aggregate and compress traffic converging from many ordinary sensor nodes. All of them form the lower layer of WSNs and they collect the data (such as values of temperature, humidity, etc.) from the sensor field to the clusterhead nodes. An exemplary two-tier WSN with the single sink node is shown in Fig. 1.

By use clustering in WSNs we obtain higher scalability of higher-layer protocols (since the size and complexity of the network as seen by higher layers is reduced by clustering) and energy consumed by whole WSNs is reduced [8]. However, the clustering problems in WSNs requires the answer for following questions:

1) *How many clusters should be created?* The pertitioning of WSN into several clusters does not mandate anything about the number of a cluster.
2) *May clusters overlap?* This question arises when a cluster is adjacent to two clusterheads. One option is to assign these nodes to both clusters, what is equivalent to overlap of clusters. Because it is not desirable, some decisions to avoidance of overlapping are required.

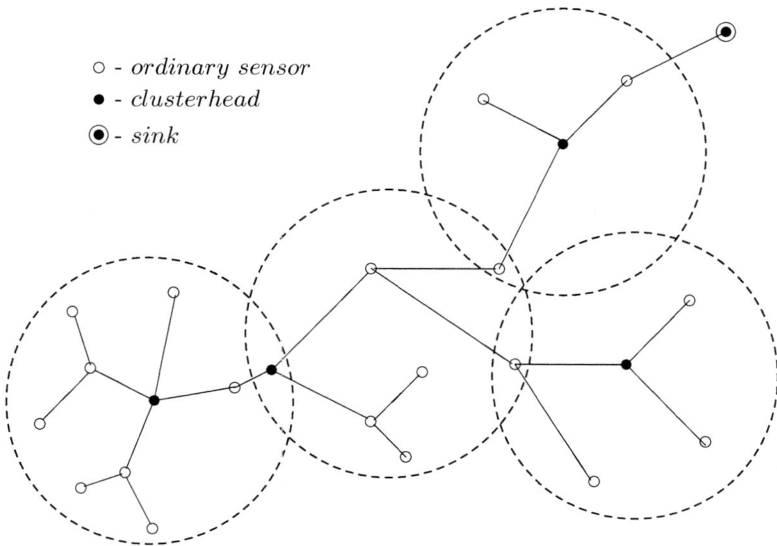

Fig. 1. An example of clustered layers in sensor network

3) *How do clusters save the energy?* All clusters should be energetically balanced. It means that the energy stored in the batteries must be consumed in uniformly way or the clusters must be form again and again.
4) *Is there a hierarchy of clusters?* Usually, a two-level hierarchy is sufficient. Nevertheless, it is possible to consider more level of hierarchy in the WSN.

Our approach into clustering in WSNs supposes that at the initial energy stored in the batteries of all sensor nodes is given. The consumption of energy in each sensor proceeds in three phases, namely:

a) *the sensing phase*, in which each sensor node collects the data from the environment. The energy needed to the data sensing is equal to

$$E_1 = c_1 \cdot b \tag{1}$$

where c_1 is the speed of the data sensing, b is the number of the sensing bits from the environment.

b) *the computation phase*, in which the sensors deplete their energy for coding, decoding, data aggregation, etc. The coding/decoding energy depends on the block length m and the number of bits b as follows

$$E_2 = (2m \cdot b + 2b^2) \cdot (E_{add} + E_{mult}) \tag{2}$$

where E_{add} and E_{mult} are the energy needed to carry out the addition and the multiplication in the Galois field $GF(2^k)$ with $k = \lfloor \log_2 m + 1 \rfloor$, m is the length of the channel block.

c) *the communication phase.* The energy consumption in the communication
phase is given by

$$E_3(n_1, n_2) = (c_2 \cdot d(n_1, n_2))^{\xi} + c_3 \cdot E(n_1) \cdot E(n_2) + c_4 \cdot \left(\frac{E_b}{N_o}\right)^{(required)} \quad (3)$$

where c_2, c_3, c_4 are positive constants, ξ is the coefficient which depends on
the environment. For the air ξ is equal to 2. $d(n_1, n_2)$ is the distance between
the two nodes n_1 and n_2. $E(n_1)$ and $E(n_2)$ are the energy of the sensor node
sending the data and sensor node receiving it, respectively. $\frac{E_b}{N_o}$ in db is the
ratio of energy per bit and the noise energy. This ratio has a close relationship
to the SNR (Signal-to-Noise Ratio) or SINR (Signal-to-Inference and Noise
Ratio).

Looking at the energy consumption by a single sensor, we can summarize all
three components, namely

$$E_{cons}(j) = E_1(j) + E_2(j) + \sum_{k \in N^{(1)}} E_3(j, k) \quad (4)$$

where $N^{(1)}$ is the set of neighbouring sensors of the first order of the neighbourhood for sensor j. We assume that the set of sensors of the first order of
the neighbourhood contains all neighbours which can communicate in directly
communication. On the other hand, all the sensors of the neighbouring set of
the first order of the neighbourhood for a given sensor are direct neighbours for
him, because, for example, of the limits on the transmission power.

3   Least Squares SVM Method for Clustering in WSNs

In this section, we present our LS-SVM method used to clustering in WSNs.

Given a training set $\{(x_i, y_i)\}_{i=1}^{N}$ with the input data $x_i \in \Re^n$ and output
data $y_i \in \Re$ with class labels $y_i \in \{-1, 1\}$. The linear classifier can be written as

$$y(x) = sign[w^T x + b] \quad (5)$$

where vector w determines hyperplane with the minimal norm and b is an offset
vector.

By use the LS-SVM method the minimalization problem is as follows

$$\min_{w,b,e} J(w, b, e) = \frac{1}{2} w^T w + \gamma \frac{1}{2} \sum_{k=1}^{W} e_k^2$$

$$\text{subject to} \quad y_k[w^T \phi(x_k) + b] = 1 - e_k, \quad k = 1, 2, \dots, N \quad (6)$$

The corresponding Lagrangian for Eq. (6) is given by

$$L(w, b, e; \alpha) = J(w, b, e) - \sum_{k=1}^{N} \alpha_k \{y_k[w^T \phi(x_k) = b] - 1 + e_k\} \quad (7)$$

where the α_k are the Lagrange multipliers. The optimality condition leads to the following $(N+1) \times (N+1)$ linear system:

$$\begin{bmatrix} 0 & Y^T \\ Y & \Omega^\star + \gamma^{-1}I \end{bmatrix} \begin{bmatrix} b \\ \alpha \end{bmatrix} = \begin{bmatrix} 0 \\ \vec{1} \end{bmatrix} \tag{8}$$

where $Z = [\phi(x_1)^T y_1, \ldots, \phi(x_N)^T y_N]$, $Y = [y_1, \ldots, y_N]$, $\vec{1} = [1, \ldots, 1]$, $\alpha = [\alpha_1, \ldots, \alpha_N]$ and $\Omega^\star = ZZ^T$. Due to the Mercer's condition [16] there exists a mapping and the expansion

$$\Omega^\star_{kl} = y_k y_l \phi(x_k)^T \phi(x_l) = y_k y_l K(x_k, x_l) \tag{9}$$

The LS-SVM model for the function estimation is given by

$$y(x) = \sum_{k=1}^{N} \alpha_k y_k \cdot K(x, x_k) + b \tag{10}$$

where parameters α_k and b are based on the solution to Eqs. (7) and (8).

To obtain a better performance for the function estimation we used a mixture of kernel functions. After Smits et al. [13], we applied a mixture of the RBF and polynomial kernels given in the form, namely

$$K_{mix} = \rho \cdot K_{poly} + (1 - \rho)K_{RBF} \tag{11}$$

where ρ is the mixing coefficient treated as a constant scalar.

In our approach the LS-SVM method was transformed into a clustering problem. All input data are split up into blocks of 2×2, 4×4, 8×8, etc. units.

Thus, the Eq. (8) can be transformed and solved. The solution gives the values

$$\begin{bmatrix} 0 & 1^T \\ 1 & \Omega \end{bmatrix} \begin{bmatrix} b \\ \alpha \end{bmatrix} = \begin{bmatrix} 0 \\ Y \end{bmatrix} \tag{12}$$

where $\Omega = K(x_i, y_i) + \gamma^{-1}I$, $Y = [y_1, \ldots, y_N]$, $1^T = [1_1, \ldots, 1_N]$ $\alpha = [\alpha_1, \ldots, \alpha_N]$. Thus, the Ω is given by

$$\Omega = K(x_i, x_j) + \gamma^{-1}I \tag{13}$$

The solution of Eq. (12) gives the values

$$\begin{cases} b = \frac{1^T \Omega^{-1} Y}{1^T \Omega^{-1} 1} \\ \alpha = \Omega^{-1}(Y - b1) \end{cases} \tag{14}$$

By setting $A = \Omega^{-1}$ and $B = \frac{1^T \Omega^{-1}}{1^T \Omega^{-1} 1}$ we obtain

$$\begin{cases} b = BY \\ \alpha = A(Y - b1) \end{cases} \tag{15}$$

where A and B are precalculated matrixes that depend only on the input vector (x_k) but not on the vector y_k.

The sensors are usually correlated due to the high probability that the adjacent pixels will contain the sensors in the cluster. We assume that the sensor field is two-dimensional and the image energy distribution of the sensor field on the surface, known as the point spread function (PSF), can be approximated by the Gausian PSF. On the other hand, the center point of the PSF corresponds to the measured sensor position.

The LS-SVM with the RBF and polynomial kernels transformed into a sensor clustering problem has the fitted energy intensity surface function over the constant vector space as follows

$$g(r,z) = \sum_{k=1}^{N} \alpha_k \{ (\rho[(rr_k + zz_k) + 1]^q + (1-\rho) \cdot e^{-\frac{|r-r_k|^2 + |z-z_k|^2}{\sigma^2}} \} + b \qquad (16)$$

where (r, z) are the coordinates of the pixels. The function $g(r, z)$ gives the corresponding energy intensity value, b and α are obtained as a solution of Eq. (15).

4   Numerical Experiments

We have simulated the clustering process with use the LS-SVMLab [10]. included into MATLAB system. In each simulation experiment, we have found the location of each sensor in the sensor field and its energy power using the random number generators. Two of them was applied to the generation of sensor localization with uniformly distribution in $[0, 2\alpha]$, where α is the length of a side of the square area. The third random number generator was defined the current value of energy power uniformly in $[0, E_{max}]$. In all of the experiments, the communication range of each sensor was assumed to be 1 unit.

Our goal is to build clusters for WSNs. Each of the sensors must belong to one of them. We propose the following algorithm for this purpose (see Fig. 2). The proposed algorithm can provide the energy balance between all the obtained clusters. As result, we obtain well energy balanced cluters with all of the sensor field. In order to find clusters in WSNs we used one of the most popular clustering technique, namely the Low-Energy Adaptive Clustering Hierarchy given in the papers [6], [7]. In this algorithm all information about the sensor nodes in WSN is

```
procedure clustering_of_WSN;
  begin
    while sensor_state(i) = ACTIVE do
      begin
        if energy_sensor(i) > minimal_energy_level;
        then add_sensor(i)_to_cluster(j);
      end;
    energy_balance_for_all_clusters;
  end;
```

Fig. 2. Pseudo-code of the clustering procedure for WSN

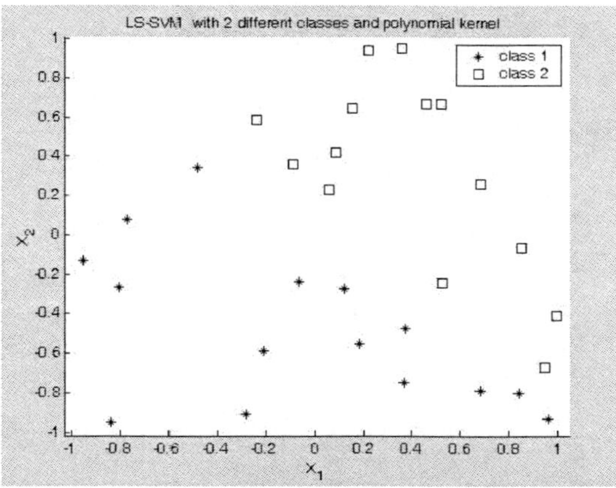

Fig. 3. The clustering process with the polynomial kernel

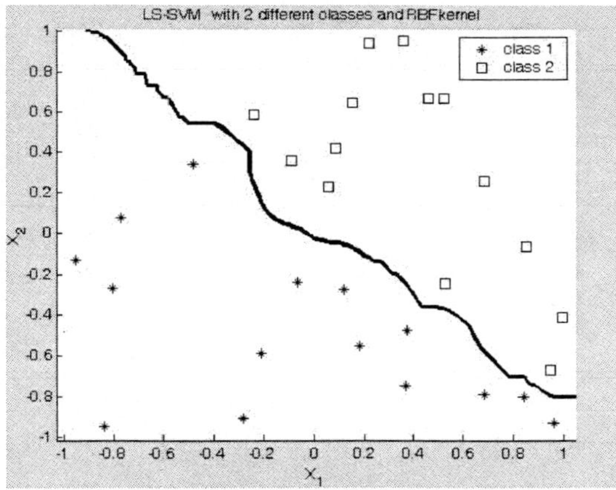

Fig. 4. The clustering process with the RBF kernel

given. This information can be used to derive the information about clusters. The obtained data used for the clusterization consisted of 845 sessions, of which 425 were used for training and the remaining 420 for the evaluation of the LS-SVM performance. Using the method reffered to as an incremental learning with the SVM [11], we employed among others, two measures of our LS-SVM classifier, namely: accuracy and precision. Both of them are standard metrics commonly used in the context of multiclass problems. Each metric with the respect of test examples is defined as follows:

$$precision = \frac{TP}{TP + FN} \times 100\% \tag{17}$$

$$accuracy = \frac{TP + TN}{TP + FN + TN + FP} \times 100\% \qquad (18)$$

where TP stands for true positive, TN true negative, FP for false positive and FN for false negative clustering in WSN. For example, when the random given clusters are concerned, TP represents the obtained clusters as good, TN represents the obtained clusters in a false way. Otherwise, FP represents misclassified clustering and FN represents non-clusters being misclassified as non-clusters.

Figures 3 and 4 show the exemplary output of the clusterization process with the polynomial and the RBF kernel, respectively, for a sensor network of 30

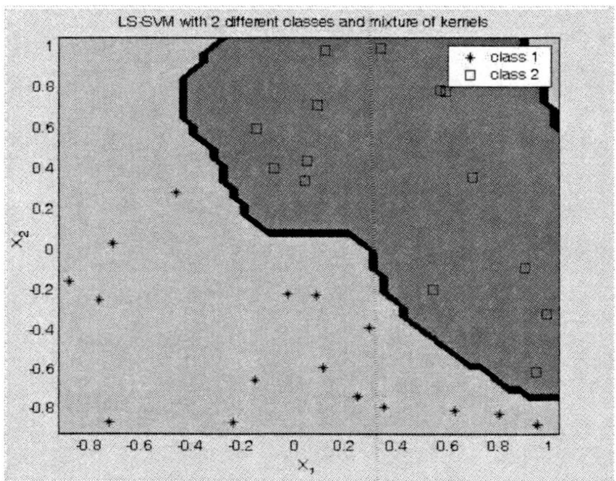

Fig. 5. The clustering process with the mixture of kernels

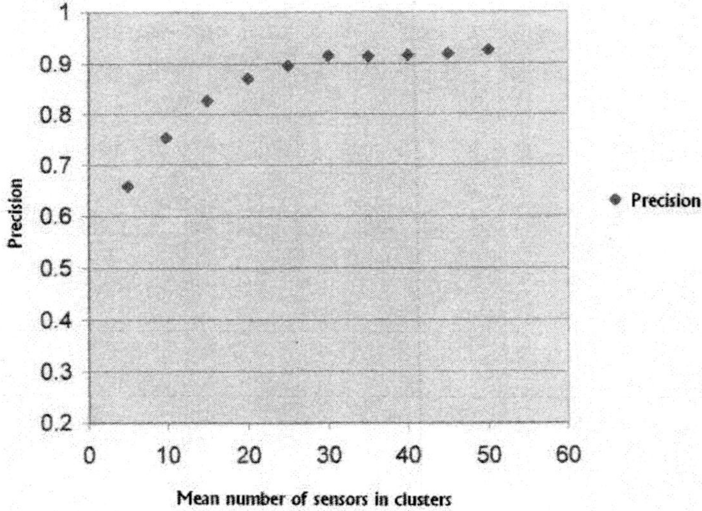

Fig. 6. The precision of clustering process in WSN based on LS-SVM method with the mixture of kernels

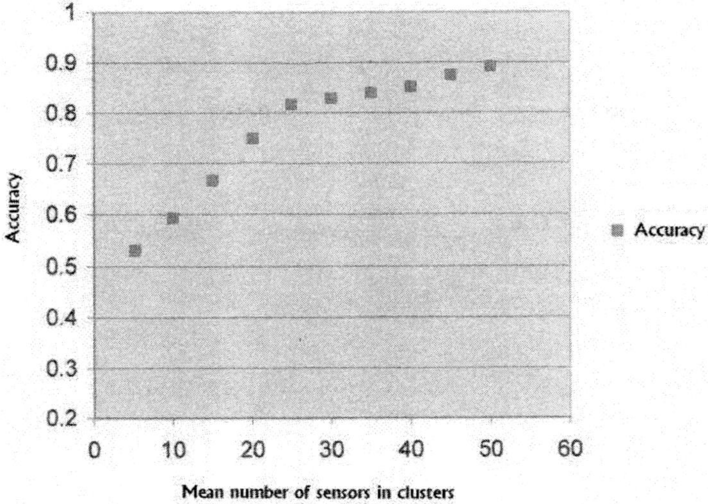

Fig. 7. The accuracy of clustering process in WSN based on LS-SVM method with the mixture of kernels

sensors that are distributed uniformly in a square of 100 square units. Figure 5 gives the output of our algorithm with the mixture of both kernels. It was observed that the clustering process with a mixture of kernels gives the better results.

The graph in Fig. 6 shows the obtained precision for clustering in WSNs with the LS-SVM classifier with a mixture of kernels in dependence of the given number of sensors in the cluster. The graph in Fig. 7 shows the accuracy of clustering process with the same classifier in the number of sensor nodes belonging to the clusters. We have observed that the improve of all measures is visible for a middle number of sensors in the clusters.

5   Conclusion

We have proposed a new method for clustering process in WSNs that is performed by LS-SVM method. With the use of the image energy distribution of the sensor field surface and the mixture of kernels, we have been able to solve the clustering problem in WSNs with acceptable effectiveness. Moreover, it is possible to maximize the accuracy and precision measures of clustering.

The proposed method has a large significant in the design and implementation of WSNs. It is very sparse and can be implemented in the low-cost chips. In the future works, this method can be extended to sensor nodes localization in WSN.

References

[1] Akyildiz, I.F., Su, W., Sankarasubramaniam, Y., Cayirci, E.: Wireless Sensor Networks. Computer Networks 38, 393–422 (2002)

[2] Baker, D.J., Ephremides, A.: The Architectural Organization of a Mobile Radio Network via a Distributed Algorithm. IEEE Trans. on Communications 29(11), 1694–1701 (1981)

[3] Burges, C.J.C.: A Tutorial on Support Vector Machines for Pattern Recognition. Knowledge Discovery and Data Mining 2(2), 121–167 (1998)

[4] Chatterjee, M., Das, S., Turgut, D.: WCA: A Weighted Clustering Algorithm for Mobile Ad Hoc Networks. Cluster Computing Journal 5, 193–204 (2002)

[5] Cortes, C., Vapnik, V.N.: Support Vector Networks. Machine Learning 20, 273–297 (1995)

[6] Heinzelman, W.B., Chandrakasan, A., Balakrishnan, H.: Energy-Efficient Communication Protocol for Wireless Microsensor Networks. In: Proc. of the 33rd Hawaii International Conference on System Science, January 2000, pp. 174–185 (2000)

[7] Heinzelman, W.B., Chandrakasan, A.P., Balakrishnan, H.: An Application-Specific Protocol Architecture for Wireless Microsensor Networks. IEEE Trans. on Wireless Networking 1(4), 660–670 (2002)

[8] Karl, H., Willig, A.: Protocols and Architectures for Wireless Sensor Networks. John Wiley and Sons, Hoboken (2005)

[9] Krishnan, R., Starobinski, D.: Efficient Clustering Algorithms for Self-organizing Wireless Sensor Networks. Ad Hoc Networks 4, 36–50 (2006)

[10] Pelckmans, K., Suykens, J.A.K., van Gestel, T., de Brabanter, J., Lukas, L., Hamer, B., de Moor, B., Vandewalle, J.: LS-SVMLab Toolbox User's Guide, ver. 1.5, http://www.esat.kuleuven.ac.be/be/sista/lssvmlab

[11] Ruping, S.: Incremental Learning with Support Vector Machines. In: Proc. IEEE Int. Conf. on Data Mining, San Jose, CA, USA, November 2001, pp. 641–642 (2001)

[12] Safari, M., Harandi, M.T., Araabi, B.N.: A SVM-based Method for Face Recognition Using a Wavelet PCA Representation of Faces. In: Int. Conf. on Image Processing (ICIP), pp. 853–856. IEEE Press, Los Alamitos (2004)

[13] Smits, G.F., Jprdan, E.M.: Improved SVM Regression Using Mixtures of Kernels. In: Proc. of IJCNN 2002 on Neural Networks, vol. 3, pp. 2785–2790 (2002)

[14] Suykens, J.A.K., Vandewalle, J.: Least Squares Support Vector Machine Classifiers. Neural Processing Letters 9(3), 293–300 (1999)

[15] Suykens, J.A.K., Vandewalle, J., De Moor, B.: Optimal Control by Least Squares Support Vector Machines. Neural Networks 14(1), 23–35 (2001)

[16] Suykens, J.A.K., Van Gestel, T., De Brabanter, J., De Moor, B., Vandewolle, J.: Least Squares Support Vector Machines. World Scientific, New Jersey (2002)

[17] Tang, Sh., Li, W.: QoS Supporting and Optimal Energy Allocation for a Cluster Based Wireless Sensor Network. Computer Communications 29, 2569–2577 (2006)

[18] Vapnik, V., Golowich, S., Smola, A.: Support Vector Method for Function Approximation, Regression Estimation, and Signal Processing. In: Mozer, M., Jordan, M., Petsche, T. (eds.) Advances in Neural Information Processing Systems, vol. 9, pp. 281–287. MIT Press, Cambridge (1997)

[19] Xie-Jian-hong: LS-SVM Method Applied to Detect Damage for Piezoelectric Smart Structures. Chinese Journal of Sensors and Actuators 20(1), 164–167 (2007)

Author Index